Methods in Molecular Biology

Series Editor
John M. Walker
School of Life and Medical Sciences,
University of Hertfordshire, Hatfield,
Hertfordshire AL10 9AB, UK

For further volumes:
http://www.springer.com/series/7651

Microbial Proteomics

Methods and Protocols

Edited by

Dörte Becher

Department of Microbial Proteomics, Institute for Microbiology, University Greifswald, Greifswald, Germany

Humana Press

Editor
Dörte Becher
Department of Microbial Proteomics
Institute for Microbiology
University Greifswald
Greifswald, Germany

ISSN 1064-3745 ISSN 1940-6029 (electronic)
Methods in Molecular Biology
ISBN 978-1-4939-8693-4 ISBN 978-1-4939-8695-8 (eBook)
https://doi.org/10.1007/978-1-4939-8695-8

Library of Congress Control Number: 2018955428

This Humana Press imprint is published by the registered company Springer Science+Business Media, LLC, part of Springer Nature.
The registered company address is: 233 Spring Street, New York, NY 10013, U.S.A.

Dedication

This book is dedicated to Michael Hecker, a pioneer in the field of microbial proteomics, in recognition of his pivotal scientific contributions

Preface

Considering the fact that microbes are by far the most abundant organisms on earth, there is no doubt that microbes and their activities are important to almost all processes on our planet. They affect all aspects of human life, as they are ubiquitously present in the environment and human microbiota, and contribute to human health and disease. Therefore, it is not surprising that microorganisms and microbiota are the focus of research to understand their physiology. The understanding of metabolic processes, stress adaptation, and survival strategies will lead to optimal conditions to cultivate microorganisms or to the development of new strategies to combat pathogens. Furthermore, it will expand our knowledge about global environmental processes in which microbes are involved.

In this research, proteomics is of increasing interest, because technical capabilities have greatly expanded during recent years and since the proteome covers the proteins, the main players of life, and thus represents the current physiological state of microorganisms. Depending on the problem or the question which should be solved or answered, plenty of different methods are available in the field of proteomics. This edition of *Microbial Proteomics* in the Springer series *Methods in Molecular Biology* summarizes state-of-the-art methods for the identification, quantification, and characterization of microbial proteins.

Most of the global proteome analysis experiments aim to identify proteins and to quantify dynamic changes of protein amounts in response to environmental challenges or genetic manipulations. Therefore, the first part deals with global sample preparation and protein enrichment. In such experiments, the compatibility of different buffers necessary for efficient lysis of microbes and extraction of proteins with subsequent analysis techniques such as LC-MS might be challenging. A method considering this challenge is presented in Chapter 1. Another challenge might be the enrichment of extremely diluted proteins, e.g., naturally secreted proteins or proteins which are diluted during the sample preparation process as described in Chapter 2.

Despite the fact that microorganisms display a rather limited number of proteins compared to human samples, there are several reasons why a subcellular fractionation can be useful and is therefore considered in the second part on subcellular fractionation. Thus, it is known that proteins exposed to the cell surface or embedded in the cell membrane are involved in first cell-cell contact during infection processes and therefore target of specific analyses of the cell envelope. Therefore, methods for the specific analysis of cell surface or membrane proteins are presented in Chapters 3 and 4. Some microbes have very specific organelles, which are focused on by research as in case for *Magnetospirillum gryphiswaldense* (Chapter 5). Other microbes are able to induce the formation of organelles like vacuoles in host cells, which is intensively studied to understand the signal trafficking in this process (Chapter 6).

However, most of the proteome studies aim to quantify the proteins in a relative or absolute manner, and therefore, the third part of this book, on protein quantification, is the largest part of the book. As 2D PAGE is still one of the most powerful tools to analyze different isoforms in microbial proteomes, the relative quantification of low-abundance proteins (Chapter 7) as well as the multiplexing of samples via DIGE (Chapter 8) is discussed. How the synthesis and stability of proteins in the bacterial cell can be analyzed is described in Chapters 9 and 10. Besides the relative quantification, especially in systems biology, the absolute protein quantification is of interest. The absolute quantification of bacterial proteins described in Chapter 11 bridges gel-based and gel-free methods. Frequently protein quantification methods are based on metabolic labeling for which two examples are given in Chapters 12 and 13. Label-free quantification methods are described in Chapters 14 and 15. In Chapter 14, furthermore, similar to Chapter 10, protease targets are in the focus, whereas the method described in Chapter 15 has to be assigned to the field of ultraproteomics, as the analyses are carried out with rather small sample amounts after cell sorting.

Another challenging task in microbial proteomics is still the analysis of posttranslational protein modifications, which is therefore included in the fourth part. For many years, several research groups have tried to solve the problem of the analysis of bacterial protein phosphorylation, as phosphorylated protein are in rather low abundance in bacteria despite their importance in cell signaling and metabolism. Two examples for the analysis of phosphorylated proteins are explained in Chapters 16 and 17. As oxidative stress is one of the most important stresses bacteria face during the infection processes, the resulting protein modifications are of special interest. A method for the analysis of reversible thiol oxidations is given in Chapter 18.

A relatively new branch of microbial proteomics does not investigate bacteria in pure culture anymore but the proteins of all microbes comprising an environmental consortium. The fifth part of the book deals with this branch, metaproteomics. As metaproteomics is usually applied in combination with metagenomics, Chapter 19 describes a method which enables the analysis of DNA, RNA, protein, and metabolites from the same sample targeting the human microbiome. Another challenging environment, which is increasingly investigated, is the soil. Methods describing the extraction and analysis of proteins derived from microorganisms living in soil and leaf litter are provided in Chapters 20 and 21. Chapter 22 completes this part of the book with a method enabling the enrichment of bacteria from invertebrate hosts, which is progressively used to investigate bacteria-invertebrate symbiosis.

In summary, this book provides a collection of methods for qualitative and quantitative analysis of proteins beginning with synthesis via modification until degradation in pure microbial culture and to a certain extent in microbial communities. Unfortunately, there are still some important topics which certainly deserve to be included as well but had to be left out to focus on microbial proteomics.

I would like to thank all contributors for providing detailed protocols and for sharing their expertise. Finally, I hope that this book will become a valuable and stimulating source for all beginners and advanced researchers in the field of microbial proteomics and beyond.

Greifswald, Germany *Dörte Becher*

Contents

Preface .... *vii*
Contributors .... *xiii*

Part I Global Sample Preparation and Protein Enrichment

1 Filter-Aided Sample Preparation for Proteome Analysis 3
Jacek R. Wiśniewski

2 Protein Enrichment from Highly Dilute Samples with StrataClean 11
Florian Bonn and Andreas Otto

Part II Subcellular Fractionation

3 Membrane Proteomics in Gram-Positive Bacteria: Two Complementary Approaches to Target the Hydrophobic Species of Proteins 21
Susanne Sievers

4 Enrichment of Cell Surface-Associated Proteins in Gram-Positive Bacteria by Biotinylation or Trypsin Shaving for Mass Spectrometry Analysis 35
Florian Bonn, Sandra Maaß, and Jan Maarten van Dijl

5 Preparation of Bacterial Magnetosomes for Proteome Analysis 45
Oliver Raschdorf, Dirk Schüler, and René Uebe

6 Analysis of Legionella Metabolism by Pathogen Vacuole Proteomics 59
Christian Manske, Ivo Finsel, Christine Hoffmann, and Hubert Hilbi

Part III Protein Quantification

7 Detection and Identification of Low-Abundant Proteins Using HPE Gels, Fluorescent Stains, and MALDI-ToF-ToF-MS 79
Martin Moche, Dirk Albrecht, Reiner Westermeier, and Knut Büttner

8 Applications of Difference Gel Electrophoresis (DIGE) in the Study of Microorganisms 95
Kathleen Trautwein and Ralf Rabus

9 Proteomic Signatures in *Staphylococcus aureus* 113
Susanne Engelmann and Stephan Fuchs

10 How to Assess Protein Stability: Half-Life Determination of a Regulatory Protein in *Bacillus subtilis* 131
Alexander Reder, Stephan Michalik, and Ulf Gerth

11 Absolute Protein Quantification Using AQUA-Calibrated 2D-PAGE 141
Sandra Maaß

12 Sulfur-^{34}S and ^{36}S Stable Isotope Labeling of Amino Acids for Quantification (SULAQ34/36) of Proteome Analyses 163
Florian-Alexander Herbst, Nico Jehmlich, Martin von Bergen, and Frank Schmidt

13 Metabolic Labeling of Microorganisms with Stable Heavy Nitrogen Isotopes (^{15}N) 175
Andreas Otto

14 Next-Generation Trapping of Protease Substrates by Label-Free Proteomics 189
Claudia Lindemann, Nikolas Thomanek, Katja Kuhlmann, Helmut E. Meyer, Katrin Marcus, and Franz Narberhaus

15 In vivo Proteomics Approaches for the Analysis of Bacterial Adaptation Reactions in Host–Pathogen Settings 207
Henrike Pförtner, Maren Depke, Kristin Surmann, Frank Schmidt, and Uwe Völker

PART IV ANALYSIS OF POST TRANSLATIONAL PROTEIN MODIFICATIONS

16 Phosphopeptide Enrichment from Bacterial Samples Utilizing Titanium Oxide Affinity Chromatography 231
Boumediene Soufi, Christoph Täumer, Maja Semanjski, and Boris Macek

17 Phosphoproteomics in Microbiology: Protocols for Studying *Streptomyces coelicolor* Differentiation 249
Angel Manteca, Beatriz Rioseras, Adelina Rogowska-Wrzesinska, and Ole N. Jensen

18 Thiol-Redox Proteomics to Study Reversible Protein Thiol Oxidations in Bacteria 261
Martina Rossius, Falko Hochgräfe, and Haike Antelmann

PART V METAPROTEOMICS

19 Sequential Isolation of DNA, RNA, Protein, and Metabolite Fractions from Murine Organs and Intestinal Contents for Integrated Omics of Host–Microbiota Interactions 279
Pranjul Shah, Emilie E.L. Muller, Laura A. Lebrun, Linda Wampach, and Paul Wilmes

20 Utilization of a Detergent-Based Method for Direct Microbial Cellular Lysis/Proteome Extraction from Soil Samples for Metaproteomics Studies 293
Karuna Chourey and Robert L. Hettich

21 Sample Preparation for Metaproteome Analyses of Soil and Leaf Litter 303
Katharina M. Keiblinger and Katharina Riedel

22 Centrifugation-Based Enrichment of Bacterial Cell Populations for Metaproteomic Studies on Bacteria–Invertebrate Symbioses 319
Tjorven Hinzke, Manuel Kleiner, and Stephanie Markert

Correction to: Thiol-Redox Proteomics to Study Reversible Protein Thiol Oxidations in Bacteria E1

Index *335*

The original version of this book was revised. The correction is available at https://doi.org/10.1007/978-1-4939-8695-8_23

Contributors

DIRK ALBRECHT • *Department of Microbial Physiology and Molecular Biology, Institute for Microbiology, University Greifswald, Greifswald, Germany*

HAIKE ANTELMANN • *Institute for Biology-Microbiology, Freie Universität Berlin, Berlin, Germany*

FLORIAN BONN • *Institute of Biochemistry II, University Hospital, Goethe University School of Medicine, Frankfurt, Germany*

KNUT BÜTTNER • *Department of Microbial Physiology and Molecular Biology, Institute for Microbiology, University Greifswald, Greifswald, Germany*

KARUNA CHOUREY • *Chemical Sciences Division, Oak Ridge National Lab, Oak Ridge, TN, USA*

MAREN DEPKE • *Interfaculty Institute for Genetics and Functional Genomics, University Medicine Greifswald, Greifswald, Germany*

SUSANNE ENGELMANN • *Institute for Microbiology, Technical University Braunschweig, Braunschweig, Germany; Microbial Proteomics, Helmholtz Centre for Infection Research, Braunschweig, Germany*

IVO FINSEL • *Max von Pettenkofer Institute, Ludwig-Maximilians University, Munich, Germany*

STEPHAN FUCHS • *Robert Koch Institute, Wernigerode, Germany*

ULF GERTH • *Department of Microbial Physiology and Molecular Biology, Institute for Microbiology, University Greifswald, Greifswald, Germany*

FLORIAN-ALEXANDER HERBST • *Department of Chemistry and Bioscience, Aalborg University, Aalborg East, Denmark*

ROBERT L. HETTICH • *Chemical Sciences Division, Oak Ridge National Lab, Oak Ridge, TN, USA*

HUBERT HILBI • *Institute of Medical Microbiology, University of Zürich, Zürich, Switzerland*

TJORVEN HINZKE • *Institute of Marine Biotechnology, Greifswald, Germany*

FALKO HOCHGRäFE • *Junior Research Group Pathoproteomics, Institute for Microbiology, University Greifswald, Greifswald, Germany*

CHRISTINE HOFFMANN • *Max von Pettenkofer Institute, Ludwig-Maximilians University, Munich, Germany*

NICO JEHMLICH • *Department of Molecular Systems Biology, Helmholtz Centre for Environmental Research–UFZ, Leipzig, Germany*

OLE N. JENSEN • *Department of Biochemistry and Molecular Biology, University of Southern Denmark, Odense, Denmark*

KATHARINA M. KEIBLINGER • *Department of Forest and Soil Sciences, Institute of Soil Research, University of Natural Resources and Life Sciences Vienna (BOKU), Vienna, Austria*

MANUEL KLEINER • *Department of Plant and Microbial Biology, North Carolina State University, Raleigh, NC, USA*

KATJA KUHLMANN • *Medizinisches Proteom-Center, Ruhr-Universität Bochum, Bochum, Germany*

LAURA A. LEBRUN • *Luxembourg Centre for Systems Biomedicine, University of Luxembourg, Esch-sur-Alzette, Luxembourg*

CLAUDIA LINDEMANN • *Medizinisches Proteom-Center, Ruhr-Universität Bochum, Bochum, Germany*
SANDRA MAAß • *Department of Microbial Proteomics, Institute for Microbiology, University Greifswald, Greifswald, Germany*
BORIS MACEK • *Proteome Center Tübingen, University of Tübingen, Tübingen, Germany*
CHRISTIAN MANSKE • *Max von Pettenkofer Institute, Ludwig-Maximilians University, Munich, Germany*
ANGEL MANTECA • *Area de Microbiologia, Departamento de Biologia Funcional, Facultad de Medicina, IUOPA, Universidad de Oviedo, Oviedo, Spain*
KATRIN MARCUS • *Medizinisches Proteom-Center, Ruhr-Universität Bochum, Bochum, Germany*
STEPHANIE MARKERT • *Institute of Marine Biotechnology, Greifswald, Germany; Department of Pharmaceutical Biotechnology, Institute of Pharmacy, University Greifswald, Greifswald, Germany*
HELMUT E. MEYER • *Medizinisches Proteom-Center, Ruhr-Universität Bochum, Bochum, Germany*
STEPHAN MICHALIK • *Interfaculty Institute for Genetics and Functional Genomics, Greifswald University, Greifswald, Germany*
MARTIN MOCHE • *Waters GmbH, Eschborn, Germany*
EMILIE E. L. MULLER • *Luxembourg Centre for Systems Biomedicine, University of Luxembourg, Esch-sur-Alzette, Luxembourg*
FRANZ NARBERHAUS • *Lehrstuhl für Biologie der Mikroorganismen, Ruhr-Universität Bochum, Bochum, Germany*
ANDREAS OTTO • *Department of Microbial Proteomics, Institute for Microbiology, University Greifswald, Greifswald, Germany*
HENRIKE PFÖRTNER • *Interfaculty Institute for Genetics and Functional Genomics, University Medicine Greifswald, Greifswald, Germany*
RALF RABUS • *Institute for Chemistry and Biology of the Marine Environment (ICBM), Carl von Ossietzky University Oldenburg, Oldenburg, Germany*
OLIVER RASCHDORF • *Ludwig Maximilian University, Munich, Germany*
ALEXANDER REDER • *Interfaculty Institute for Genetics and Functional Genomics, Greifswald University, Greifswald, Germany*
KATHARINA RIEDEL • *Department of Microbial Physiology and Molecular Biology, Institute for Microbiology, University Greifswald, Greifswald, Germany*
BEATRIZ RIOSERAS • *Area de Microbiologia, Departamento de Biologia Funcional, Facultad de Medicina, IUOPA, Universidad de Oviedo, Oviedo, Spain*
ADELINA ROGOWSKA-WRZESINSKA • *Department of Biochemistry and Molecular Biology, University of Southern Denmark, Odense, Denmark*
MARTINA ROSSIUS • *Institute for Biology-Microbiology, Freie Universität Berlin, Berlin, Germany*
FRANK SCHMIDT • *Interfaculty Institute for Genetics and Functional Genomics, University Medicine Greifswald, Greifswald, Germany*
DIRK SCHÜLER • *Department of Microbiology, University of Bayreuth, Bayreuth, Germany*
MAJA SEMANJSKI • *Proteome Center Tübingen, University of Tübingen, Tübingen, Germany*
PRANJUL SHAH • *Luxembourg Centre for Systems Biomedicine, University of Luxembourg, Esch-sur-Alzette, Luxembourg*
SUSANNE SIEVERS • *Department of Microbial Physiology and Molecular Biology, Institute for Microbiology, University of Greifswald, Greifswald, Germany*

BOUMEDIENE SOUFI • *Proteome Center Tübingen, University of Tübingen, Tübingen, Germany*
KRISTIN SURMANN • *Interfaculty Institute for Genetics and Functional Genomics, University Medicine Greifswald, Greifswald, Germany*
CHRISTOPH TÄUMER • *Proteome Center Tübingen, University of Tübingen, Tübingen, Germany*
NIKOLAS THOMANEK • *Medizinisches Proteom-Center, Ruhr-Universität Bochum, Bochum, Germany*
KATHLEEN TRAUTWEIN • *Institute for Chemistry and Biology of the Marine Environment (ICBM), Carl von Ossietzky University Oldenburg, Oldenburg, Germany*
RENÉ UEBE • *Department of Microbiology, University of Bayreuth, Bayreuth, Germany*
JAN MAARTEN VAN DIJL • *Department of Medical Microbiology—Molecular Bacteriology, University Medical Center Groningen (UMCG), Groningen, The Netherlands*
UWE VÖLKER • *Interfaculty Institute for Genetics and Functional Genomics, University Medicine Greifswald, Greifswald, Germany*
MARTIN VON BERGEN • *Department of Molecular Systems Biology, Helmholtz Centre for Environmental Research-UFZ, Leipzig, Germany; Department of Metabolomics, Helmholtz Centre for Environmental Research-UFZ, Leipzig, Germany*
LINDA WAMPACH • *Luxembourg Centre for Systems Biomedicine, University of Luxembourg, Esch-sur-Alzette, Luxembourg*
REINER WESTERMEIER • *Freising, Germany*
PAUL WILMES • *Luxembourg Centre for Systems Biomedicine, University of Luxembourg, Esch-sur-Alzette, Luxembourg*
JACEK R. WIŚNIEWSKI • *Biochemical Proteomics Group, Department of Proteomics and Signal Transduction, Max-Planck Institute for Biochemistry, Martinsried, Germany*

Part I

Global Sample Preparation and Protein Enrichment

Chapter 1

Filter-Aided Sample Preparation for Proteome Analysis

Jacek R. Wiśniewski

Abstract

Quantitative protein extraction and high-yield generation of peptides from biological samples are the prerequisite for successful bottom-up type proteomic analysis. Filter aided sample preparation (FASP) is a method for processing of SDS-solubilized cells in a proteomic reactor format. In FASP, disposable centrifugal ultrafiltration units allow for detergent depletion, protein digestion, and isolation of peptides released by proteases from undigested material. Consecutive protein digestion with two or three proteases enables generation of peptide fractions with minimal overlap and considerably increases the number of identifications and protein sequence coverage. FASP is useful for analysis of samples varying in size from a few micrograms to several milligrams of total protein.

Key words Proteomic sample preparation, Sample preparation, Proteomic reactor, Protein digestion, Filter-aided sample preparation (FASP), Multienzyme digestion FASP, Bottom-up proteomics

1 Introduction

Liquid chromatography tandem mass spectrometry (LC-MS/MS) is the key method in proteomics. Irrespective of the type of instrumentation, the quality and the depth of proteomic analysis depend on the purity of the used protein digests which preferentially should contain only peptides. Preparation of samples satisfying this requirement appears trivial, but in fact, it has been a subject of endless number of publications, describing specific methods for processing of various organisms, tissues, and organelles. The major difficulty in the sample preparation process is the fitting together of quite different conditions needed for sample lysis and protein digestion. Whereas for tissue lysis strong denaturing conditions are required, typically, proteinases used for protein digestion do not tolerate high concentrations of denaturants. Typically, solutions compromising between these conditions result in a partial solubilization of the biological material and a digestion of proteins at low yield. Faced with these problems we have developed an almost universal sample preparation method applicable to any

Dörte Becher (ed.), *Microbial Proteomics: Methods and Protocols*, Methods in Molecular Biology, vol. 1841,
https://doi.org/10.1007/978-1-4939-8695-8_1,

type of biological sample with satisfying peptide yields. The method, called filter aided sample preparation (FASP) is well suited to process protein lysates prepared in the presence of high concentrations of sodium dodecyl sulfate (SDS) [1] or other detergents [2].

The FASP method uses commercially available centrifugal concentrators in a chemical reactor setup. In the reactor proteins are depleted from the detergent, modified, digested, and finally pure peptides filtered through the ultrafiltration membrane. In the first step of the procedure the lysate is depleted from the detergent using 8 M urea (Fig. 1a). High concentration of urea is not only necessary for maintaining proteins in solution but also for a shrinkage and dissociation of detergent micelles which is the prerequisite for fast and quantitative removal of the detergent. In the second step, the denatured proteins can be modified chemically (Fig. 1b). In most cases cysteine residues are carboamidomethylated with iodoacetamide. At a desired time of the reaction, the excess of the reagent can be removed quickly from the filtration unit before protein cleavage (Fig. 1c).

Digestion of protein can be carried out in the presence or absence of urea. Cleavage in urea is recommended when samples contain mainly water-insoluble proteins or endogenous proteases, such as lysates of mouse intestinal mucosa [3]. While using the denaturant attention should be paid to the fact that many proteases loose enzymatic activity and specificity under denaturing conditions. Digestion in the absence of urea abrogates this limitation and allows for using of many different proteases. However, among many tested proteases, cleavage with trypsin, and endoproteinases LysC and ArgC results in the highest peptide identification rates during the LC-MS/MS analysis [1, 4].

Following the digestionstep, peptides released from the proteins are collected by ultrafiltration. This allows for retaining of undigested protein fragments and other genuine macromolecules

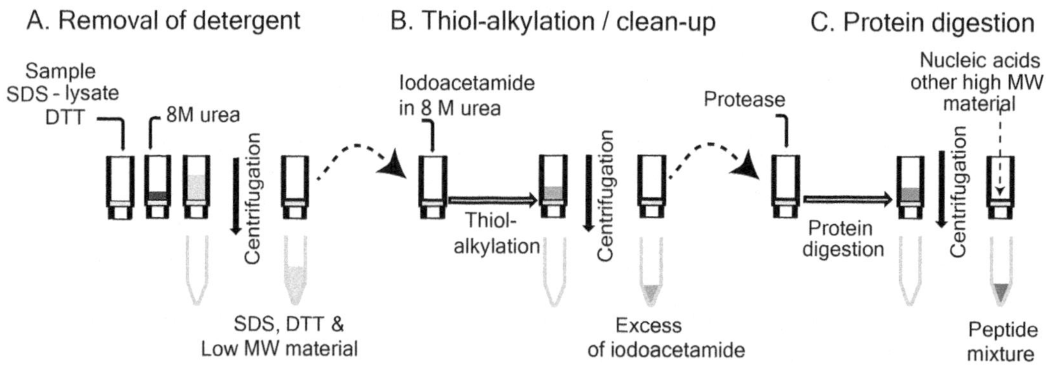

Fig. 1 Workflow of the filter aided sample preparation. (**a**) Removal of detergent. (**b**) Thiol alkylation and protein cleanup. (**c**) Protein digestion and isolation of peptides

present in the lysate, such as nucleic acids and oligosaccharides. The purity of the protein digest is the prerequisite for high-resolution liquid chromatography of the peptides and the run-to-run reproducibility, and slows down column deterioration.

FASP was initially developed using filtration units with molecular weights cut offs (MWCO) of 3000 and 10,000 molecular weight, but better results can be achieved using membranes with larger pores with MWCO of 30,000 without any loss of proteins with low molecular weights in the range of 10,000. This phenomenon reflects large Stokes' radii of denatured proteins compared to those in native form. The filtration units with MWCO of 10,000 are recommended for analysis where a complete separation of the generated peptides from a protease is required, for example, when the peptides are subjected to further enzymatic manipulations such as lectin-affinity enrichment or treatment with phosphatases.

FASP can be used over a wide range of the sample amount. For standard proteomic experiments requiring digestion of 20–200 μg protein, 0.5 mL filtration units are appropriate. The same type of filtration units is useful for processing of minute amounts of protein in range of few micrograms and below. In this application spontaneous adsorption of proteins and peptides to the filter surface can be minimized by supplementing of lysate with polyethylene glycol with MW of 20,000 [5]. For studies requiring digestion of large protein amounts, such as phosphoproteomic analyses, FASP can be carried out in 15-mL units, allowing for processing up to 20 mg protein [6].

In a standard FASP protocol, proteins are digested with a single protease. In the multi enzyme digestion FASP (MED-FASP) proteins are consecutively cleaved with various proteinases and the released peptides are collected after each digestion step. This approach generates peptide fractions with only a few percent of peptides present in more than one fraction. Compared to the standard FASP procedure, separate LC-MS/MS analyses of the fractions significantly improve the total number of protein identifications and result in higher sequence coverage [4]. Analysis of *E. coli* whole lysate using MED-FASP and two 4 h LC-MS/MS shots allowed for identification of 19,000 unique peptides and 2200 proteins per sample [7]. This protocol is also advantageous when only limited amount of biological material is available because more unique peptides can be generated from the given amount of sample [8].

In this chapter, I describe three FASP protocols—two basic protocols with single enzyme cleavage for small and large protein amounts and the MED-FASP protocol involving two-step digestion with endoproteinase LysC and trypsin.

2 Materials

2.1 Preparation of Lysates from Tissue and Cultured Cells

1. Branson-type sonifier.
2. Bath with boiling water.
3. Benchtop centrifuge.
4. Lysis buffer: 0.05 M Tris–HCl, pH 7.6, 2% (v/w) SDS, 0.05 M DTT.

2.2 Standard FASP/MED–FASP

1. Microcon-30 kDa centrifugal filter unit with Ultracel-30 membrane (Previously sold as Microcon YM-30 and Forensic 30 k) (Merck-Millipore, Cat. No.MRCF0R030).
2. Benchtop centrifuge, temperature 20 °C.
3. Wet chamber (box) with a rack for Eppendorf-type disposal tubes.
4. Incubator set to 37 °C.
5. UA: 8 M urea in 0.1 M Tris–HCl, pH 8.5. Prepare 1 mL per sample.
6. IAA solution: 0.05 M iodoacetamide in UA. Prepare 0.1 mL per sample (*see* **Note 1**).
7. Trypsin, stock 0.4 μg/μL.
8. Endoproteinase LysC. Stock 1 μg/μL.
9. Digestion buffer: 0.05 M Tris–HCl, pH 8.5. Prepare 0.5 mL per sample.

2.3 Large Protein Scale FASP

1. Amicon ultracentrifugal filter Ultracel-30 K (Millipore UCF903096).
2. Benchtop centrifuge with a swing out rotor for 50 mL tubes ("Falcon" type) with the temperature set to 18 °C.
3. Incubator set to 37 °C.
4. UA: 8 M urea in 0.1 M Tris–HCl, pH 8.5. Prepare 60 mL per sample.
5. IAA solution: 0.05 M iodoacetamide in UA. Prepare 5 mL per sample (*see* **Note 1**).
6. Trypsin, stock 0.4 μg/μL.
7. Digestion buffer (DB): 0.05 M Tris–HCl, pH 8.5. Prepare 3 mL per sample.

3 Methods

3.1 Preparation of Lysates in SDS

1. The major advantage of FASP over many other proteomic digestion protocols is its applicability to the proteomic sample processing of SDS-containing lysates.

2. Mix the tissue or cells with about fivefold excess (volume) of the lysis buffer (*see* **Note 2**).
3. Facilitate disintegration of the cell wall by sonication (*see* **Note 3**).
4. Place the tubes with the homogenate in a bath with boiling water and incubate them for 3 min. Cool the sample to the room temperature.
5. Clarify the lysate by centrifugation at 10,000 × *g* for 5 min (*see* **Note 4**).

3.2 FASP Standard Protocol (10–200 μg of Total Protein)

1. Mix up to 30 μL of a protein extract with 200 μL of UA in the filter unit and centrifuge at 10000 × *g* until less than 10 μL the sample remains in the filter. This usually requires centrifugation time of 10–15 min. This applies to all further centrifugation steps (*see* **Note 5**).
2. Add 200 μL of UA to the filter unit and repeat the centrifugation.
3. Discard the flow-through form the collection tube.
4. Add 100 μL IAA solution and mix at 600 rpm in a thermomixer for 1 min.
5. Centrifuge the filter units at 10000 × *g* for 10 min.
6. Add 100 μL of UA to the filter unit and centrifuge at 10000 × *g* for 15 min. Repeat this step twice.
7. Add 100 μL of DB to the filter unit and centrifuge at 10000 × *g* for 10 min. Repeat this step twice.
8. Add 40 μL DB with trypsin (enzyme-to-protein ratio 1:100) and mix at 600 rpm in thermomixer for 1 min.
9. Incubate the units in a humid chamber at 37 °C for 1.5–18 h. Transfer the filter units to new collection tubes.
10. Centrifuge the filter units at 10000 × *g* until the solution completely passed the filter membrane (about 5 min).
11. Add 100 μL of DB and centrifuge the filter units at 10000 × *g*.
12. Determine peptide concentration by A_{280} measurement in a spectrophotometer (*see* **Note 6**) or by the fluorometric assay as described in [9].

3.3 MED-FASP with LysC and Trypsin Digestion

1. Follows **steps 1**–7 as described in Subheading 3.2.
2. Add 40 μL DB with LysC (enzyme-to-protein ratio 1:100) and mix at 600 rpm in thermomixer for 1 min.
3. Incubate the units in a humid chamber at 37 °C for 1.5–18 h.
4. Transfer the filter units to new collection tubes.
5. Centrifuge the filter units at 10000 × *g* until the solution completely passed the filter membrane (about 5 min).

6. Add 100 μL of DB and centrifuge the filter units at 10000 × *g*.
7. Add 40 μL DB with trypsin (enzyme-to-protein ratio 1:100) and repeat **steps 2–6**.
8. Continue with **steps 9–13** as described in Subheading 3.2.

3.4 FASP Protocol for Large Amounts of Protein (up to 20 mg of Total Protein)

1. Mix up to 5 mL of a protein extract with 10 mL of UA in the filter unit and centrifuge at 4000 × *g* until the sample is concentrated to about 2 mL. Usually this requires centrifugation for 15–20 min. Discard the flow-through.
2. Add 10 mL of UA to the filter unit and centrifuge at 4000 × *g* until the sample is concentrated. Discard the flow-through form the collection tube.
3. Add 1 mL IAA solution and with sample concentrate.
4. Centrifuge the filter units at 4000 × *g* until the sample is concentrated.
5. Add 10 mL of UA to the filter unit and centrifuge at 4000 × *g* until the sample is concentrated. Repeat this step twice.
6. Add 10 mL of DB to the filter unit and centrifuge at 4000 × *g* until the sample is concentrated. Repeat this step twice.
7. Add 1 mL DB with trypsin (enzyme-to-protein ratio 1:100), mix, and incubate the unit at 37 °C for 1.5–18 h.
8. Transfer the filter units to new collection tubes.
9. Centrifuge the filter units at 4000 × *g* until the sample nearly completely passes the filter membrane.
10. Repeat this centrifugation after adding of 2 mL DB.
11. Determine peptide concentration by A_{280} measurement in a spectrophotometer (*see* **Note 7**) or by the fluorometric assay as described in [9].

4 Notes

1. UA and IAA solutions must be freshly prepared and used within a day.
2. To achieve the best results of lysis it is important to use a sufficient excess of the lysis buffer over the sample. Typically a fivefold excess is recommended. Protein concentrations of the lysates below 1 mg/mL should be avoided. Too large amounts of the detergent require additional detergent depletion steps in the FASP procedure. The lysis buffer should be stored at above 15 °C to avoid precipitation of SDS.
3. Condition for cell wall disintegration vary between organisms. Hence should be optimized experimentally. A Branson-type

sonicator with submersible tips fitting in Eppendorf type tubes is recommended.

4. Potassium and other metals form insoluble salts with SDS. They can appear as a film over the clarified lysate.
5. The centrifugation in each step has to be continued until less than 5% of the initial solution remains in the filter. The time needed to achieve it varies between the filter types, lysate properties and the total protein concentration.
6. Currently on the market are two basic types of ultrafiltration units differing in the direction of the filtration membrane. Units with the horizontally positioned membrane are useful for standard (10–100 μg total protein) [10] and microscale (1–10 μg) FASP [5]. The filters with vertically arranged membranes are suited to larger sample amounts: those with a volume of 15 mL for treating of sample containing up to 20 mg of protein.
7. Concentration of the peptides can be estimated by UV spectrometer assuming that 0.1% solution of vertebrate proteins has at 280 nm an extinction of 1.1 units. (1 mg/mL solution has an au of 1.1). Always record a spectrum from 240 to 340 nm. It should have a distinct peak with a maximum at 270–280 nm and the extinction at 320 nm has to be 0. This measurement results in reliable data only for peptide concentrations at A_{280} above 0.1 in a cuvette with a 10 mm path. Extinction values below 0.1 can be used only for rough protein estimation due to increasing contribution of light scattering to the spectrum.

Acknowledgments

This work was supported by the Max-Planck Society for the Advancement of Science.

References

1. Zielinska DF, Gnad F, Jedrusik-Bode M, Wisniewski JR, Mann M (2009) Caenorhabditis elegans has a phosphoproteome atypical for metazoans that is enriched in developmental and sex determination proteins. J Proteome Res 8(8):4039–4049
2. Wisniewski JR, Zielinska DF, Mann M (2011) Comparison of ultrafiltration units for proteomic and N-glycoproteomic analysis by the filter-aided sample preparation method. Anal Biochem 410(2):307–309
3. Wisniewski JR, Friedrich A, Keller T, Mann M, Koepsell H (2014) The impact of high-fat diet on metabolism and immune defense in small intestine mucosa. J Proteome Res 14 (1):353–365
4. Wisniewski JR, Ostasiewicz P, Mann M (2011) High recovery FASP applied to the proteomic analysis of microdissected formalin fixed paraffin embedded cancer tissues retrieves known colon cancer markers. J Proteome Res 10 (7):3040–3049

5. Wisniewski JR, Nagaraj N, Zougman A, Gnad F, Mann M (2010) Brain phosphoproteome obtained by a FASP-based method reveals plasma membrane protein topology. J Proteome Res 9(6):3280–3289
6. Wisniewski JR, Mann M (2012) Consecutive proteolytic digestion in an enzyme reactor increases depth of proteomic and phosphoproteomic analysis. Anal Chem 84(6):2631–2637
7. Wisniewski JR, Rakus D (2014) Multi-enzyme digestion FASP and the 'Total protein Approach'-based absolute quantification of the Escherichia coli proteome. J Proteome 109:322–331
8. Wisniewski JR, Dus K, Mann M (2013) Proteomic workflow for analysis of archival formalin-fixed and paraffin-embedded clinical samples to a depth of 10 000 proteins. Proteomics Clin Appl 7(3–4):225–233
9. Wisniewski JR, Gaugaz FZ (2015) Fast and sensitive total protein and peptide assays for proteomic analysis. Anal Chem 87 (8):4110–4116
10. Wisniewski JR (2016) Quantitative evaluation of filter aided sample preparation (FASP) and multienzyme digestion FASP protocols. Anal Chem 88(10):5438–5443

Chapter 2

Protein Enrichment from Highly Dilute Samples with StrataClean

Florian Bonn and Andreas Otto

Abstract

Efficient and reproducible protein enrichment is a prerequisite for the proteomic analysis of highly dilute solutions like culture supernatant or body fluids. Traditionally several precipitation strategies were used for this purpose, but they have certain drawbacks. Here, a protocol for StrataClean enrichment—a very efficient and easy-to-use method for the enrichment of proteins from highly dilute samples—which is compatible to gel-free proteomics, is presented.

Key words Protein enrichment, StrataClean, Solid-phase enrichment, Highly dilute samples

1 Introduction

In modern proteomic studies analysis of highly dilute samples has become an important task. Although mass spectrometry has become increasingly sensitive, protein enrichment is still necessary for the analysis of dilute samples. Besides traditional methods like trichloroacetic acid–acetone precipitation or methanol–chloroform extraction dilute proteins can be obtained by ultrafiltration or solid phase extraction (SPE). All established protocols have certain drawbacks, making them unsuitable as a universal method for proteomic analyses [1].

SPE provides one-pot enrichment with minimal use of hazardous chemicals and little increase in sample volume. StrataClean was introduced in the early 1990s as a tool for removal of proteins from PCR samples [2]. The molecular basis of the StrataClean system are hydroxylated groups which are immobilized on silica beads and interact with proteins similar to phenol. Although StrataClean was used in high impact proteomic studies [3, 4], other studies described it as unfeasible for efficient protein enrichment [1]. A new protocol for StrataClean enrichment was established, making StrataClean enrichment a highly effective and universal method for

Dörte Becher (ed.), *Microbial Proteomics: Methods and Protocols*, Methods in Molecular Biology, vol. 1841,
https://doi.org/10.1007/978-1-4939-8695-8_2,

the analysis of highly dilute proteins [5]. The optimized protocol is easy to use, very efficient for the enrichment of extremely dilute samples and highly reproducible.

With StrataClean highly dilute complex proteome samples (0.1 μg protein /mL) can be enriched efficiently [5]. Compared to detergent assisted TCA precipitation, 50% more proteins could be reproducibly identified. In addition, an *in-solution* digest protocol for unbiased, label-free protein quantification has been developed. Finally, it could be shown that after desiccation complex proteome samples can be stored at ambient temperature on the beads without significant loss of proteins. Therefore, StrataClean enrichment is a universal method for preparation of dilute samples in gel-free proteomics. Due to the strong denaturing conditions needed for protein elution from the beads, the optimized protocol for StrataClean enrichment cannot be used for protein enrichment in 2D gel based experiments.

2 Materials

Prepare all solutions with ultrapure water (LC-MS grade). Store all solutions at 4 °C, unless otherwise indicated. When working with hazardous chemicals always follow local bylaws and hazardous regulations.

2.1 Protein Isolation

1. StrataClean resin: Purchased from Agilent.
2. Concentrated hydrochloric acid (12 M).
3. TE: 50 mM Tris, 10 mM EDTA, pH 7.0. Dissolve 6 g Tris base and 2.92 g EDTA in 0.9 L of water, afterward adjust the pH to 7.0 (hydrochloric acid) and add water to a final volume of 1 L.

2.2 Gel-Based Workflow

1. 1 M Tris pH 6.8: Dissolve 12.11 g Tris in 80 mL water. Adjust pH to 6.8 (hydrochloric acid) and add water to a final volume of 100 mL.
2. Gel loading buffer: 125 mM Tris, 20% (w/v) glycerol, 4% (w/v) SDS, 3.75% (v/v) β-mercaptoethanol, 20 mM DTT, 0.04% (w/v) bromophenol blue. Dissolve 40 mg SDS, 0.4 mg bromophenol blue, 3.1 mg DTT and 125 μL 1 M Tris pH 6.8 in 645 μL water. Add 200 μL glyercol and 37.5 μL β-mercaptoethanol and mix carefully. Store at −20 °C. Before usage thaw at elevated temperature (30–50 °C) with gentle shaking until all SDS is dissolved again.
3. Laemmli gel, SDS gel puffer, fixing solution, coomassie staining solution: Selfcast gels as well as precast gels can be used without specific drawbacks. Use your laboratory standards.

2.3 Gel-Free Workflow

1. TEAB (50 mM) Triethylammonium bicarbonate buffer: Purchased from Sigma (T7408). The 1 M stock solution is diluted with water to 50 mM and can be stored for 2 weeks.
2. RapiGest stock solution: Purchased from Waters. Prepare a 0.5% (w/v) stock solution in 50 mM TEAB which can be stored in aliquots at −70 °C for several months.
3. TEAB with 0.2% RapiGest: Either dissolve RapiGest in 50 mM TEAB (for example 1 mg (1 small vial) in 500 μL) or mix three parts 50 mM TEAB with two parts of a 0.5% RapiGest stock solution.
4. 500 mM TCEP: Prepare a small amount of 500 mM Tris (2-carboxyethyl) phosphine (TCEP) in 1 M TEAB freshly (for example dissolve 10 mg TCEP in 70 μL TEAB).
5. 1 M IAA (Iodoacetamide): Prepare a small amount of 1 M IAA in 50 mM TEAB freshly (for example dissolve 10 mg IAA in 54 μL TEAB) (*see* **Note 1**).
6. Sequencing grade activated trypsin: Purchased from Promega. Activate the trypsin by incubation in resuspension buffer (is delivered with the enzyme, 100 μL for every 20 μg vial) for 15 min at 37 °C. You can prepare trypsin aliquots in resuspension buffer (before activation) and store them for several weeks at −20 °C. Thaw and activate directly before the digest.
7. Trifluoroacetic acid (LC-MS grade).
8. 30% acetonitrile: Mix 300 μL acetonitrile (LC-MS grade) with 700 μL water.

2.4 Desalting of In-Solution Digested Samples

1. Bulk C18 material, particle size smaller than 5 μm.
2. Methanol, LC-MS grade.
3. StageTips, C18 material, GELoader pipette tip: Purchased from Thermo Fisher (SP 101).
4. Buffer A: Add 10 μL of LC-MS grade acetic acid to 9.99 mL water.
5. Buffer B: Add 10 μL of LC-MS grade acetic acid to 9.99 mL acetonitrile (LC-MS grade).
6. Elution buffer: Mix 300 μL buffer A with 700 μL buffer B.

3 Methods

3.1 Protein Enrichment

1. Vortex the StrataClean resin thoroughly and transfer 20 μL slurry to a reaction tube (one per sample). When analyzing multiple samples, vortex after every removal or after two removals, as the resign is sedimenting quickly (*see* **Note 2**).

2. Sediment the beads by centrifugation (2 min, 6000 × *g*, ambient temperature) and wash twice with 500 μL TE.
3. For cleaning and hydrolyzing all impurities which may be loaded on the beads add 200 μL concentrated HCl to the beads and incubate for at least 6 h at 100 °C in an extractor hood. Close the tubes thoroughly and take care on opening (*see* **Note 3**).
4. Sediment the beads by centrifugation (2 min, 6000 × *g*). Wash twice with 1 mL TE.
5. Add the beads to the sample by resuspending them in a small amount of sample and then transfer the suspension back to the remaining sample.
6. Incubate overnight at 4 °C with overhead rotation; (alternatively by gentle shaking; the beads must not sediment).
7. Sediment the loaded beads by centrifugation (4 °C, 10,000 × *g*). For small samples up to 2 mL 5 min is sufficient. Larger volumes should be sedimented for at least 20 min. Extremely large samples (more than 100 mL) may require longer centrifugation times. Remove the supernatant carefully, especially in case of large samples (*see* **Note 4**).
8. Wash the beads twice with 500 μL TE. Before the first centrifugation step transfer the suspended beads to a small reaction tube, if necessary.
9. If you want to store your samples on the beads at ambient temperature evaporate to dryness under vacuum. Afterward the samples can be stored for several weeks at ambient temperature (*see* **Note 5**).
10. Otherwise use your samples immediately or freeze them at −20 °C or −70 °C.

3.2 Gel-Based Workflow

1. Remove excessive buffer by drying the beads 15 min in a vacuum centrifuge.
2. Add 20 μL gel loading buffer to every sample, mix carefully and incubate for 5 min at 95 °C.
3. Cool the samples on ice and centrifuge for 30 s (6000 × *g*) to remove condensate from the lid.
4. Carefully mix the slurry with a pipette tip and load onto a Laemmli gel. The slots need to have a minimal volume of 40 μL. You can also use precast gels (*see* **Note 6**).
5. Perform the electrophoresis with your standard protocol, for example at 150 V.
6. Stain the gel (for example with colloidal coomassie).
7. For further analysis, perform an in-gel digest followed by LC-MS analysis.

3.3 Gel-Free Workflow

1. Resuspend the loaded beads in 100 μL TEAB with 0.2% RapiGest.
2. Add 4 μL 500 mM TCEP (final concentration of 20 mM) and incubate for 30–45 min at 60 °C with gentle shaking to reduce disulfide bonds (*see* **Note 7**).
3. Cool the sample, add 4 μL 1 M IAA (final concentration of 40 mM) and incubate for 15 min at ambient temperature in the dark.
4. Add 2.5 μL activated trypsin (0.5 μg) and incubate for 3 h at 37 °C with shaking at 900 rpm (*see* **Note 8**).
5. Add 2.5 μL activated trypsin (0.5 μg) and incubate for another 3 h at 37 °C with shaking at 900 rpm.
6. Sediment the beads by centrifugation (1 min, 10,000 × *g*, 4 °C) and transfer the supernatant to another reaction tube (store on ice).
7. Wash the beads twice with 30 μL 30% acetonitrile and pool the supernatants with the supernatant from **step 6**.
8. Add 1 μL trifluoroacetic acid to the pooled supernatants ($pH < 2$) to hydrolyze the RapiGest. Mix thoroughly and check the pH. Incubate at 37 °C without shaking for 45 minutes; mix carefully every 5–10 min.
9. Remove RapiGest by centrifugation (15 min, 12,000 × *g*, 4 °C). Repeat at least once to be sure that no RapiGest can contaminate your LC-MS system.
10. Remove the acetonitrile by reducing the volume to 70–80 μL in a vacuum centrifuge.
11. Especially if you do not use a precolumn it is recommended to desalt your sample with StageTip purification (see next part of the chapter).
12. Perform the LC-MS analysis.

3.4 Desalting In-Solution *Digested Samples*

1. Resuspend approximately 10 μL of bulk C18 material per sample in 20 μL methanol.
2. Prepare a mount for each StageTip by puncturing of a hole of approximately 2 mm diameter in the middle of the lid of a 2 mL reaction tube. Cut the tip 2 mm below the C18 and place it in the holder (*see* Fig. 1, *see* **Note 9**).
3. Add C18 suspension to the tip and sediment the C18 material by centrifugation (2000–5000 × *g*). Repeat until the lower tampered part over the original C18 is filled with C18 material (*see* Fig. 1, *see* **Note 10**).
4. Wash the C18 material with 100 μL buffer A. Fill it into the tip with a GelLoader tip and squeeze it through the C18 by centrifugation (8000 × *g*, 1–2 min). Repeat once.

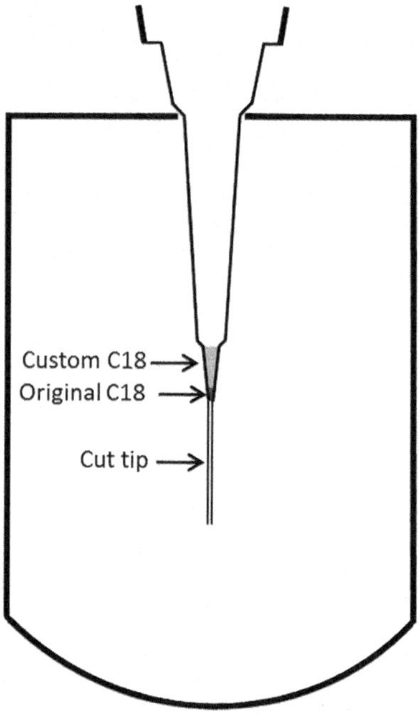

Fig. 1 StageTip in the mount. This figure shows a schema of the custom-loaded StageTip in the mount (not proportional)

5. Wash the C18 material two times with 100 μL buffer B, as described for buffer A. *See* **Note 11**.
6. Equilibrate the C18 material with 100 μL buffer A as described for the washing. Do not let the C18 material run dry until elution. The final liquid level should be ~ 1 mm above the custom C18.
7. Add the digested sample to the StageTip. Place it with a Gel-Loader tip directly above the C18, without an air bubble in between.
8. Squeeze it through the C18 material by centrifugation (8000 × *g*, 1–2 min).
9. Wash the loaded C18 material with 100 μL buffer A twice.
10. Squeeze the remaining buffer trough the C18 with a syringe (one syringe is with each box of tips) until the upper liquid level is directly above the C18.
11. Elute the peptides with 30 μL elution buffer. Squeeze it through the C18 material directly into a LC vial with a syringe.
12. Add 20 μL buffer A to the sample.
13. Reduce the sample volume in a vacuum centrifuge to 20 μL to remove the acetonitrile.
14. Perform the LC-MS analysis, in case the peptide concentration is too high, dilute the sample with water or buffer A.

4 Notes

1. IAA is light sensitive, store and incubate in the dark!
2. We recommend to aliquot the StrataClean beads in smaller samples of 200–300 μL in reaction tubes. The original 3 mL tube is both not easy to vortex and not fast to open, which is a prerequisite for drawing constant sample amounts as the beads sediment very fast.
3. Long-term boiling cause high pressure due to evaporation, which will open the lids of reaction tubes. The tubes can be sealed with plastic clips (for example: MCP LidLocks, Sorenson, 11870). As an alternative you can place a glass plate on the tubes in the heating block. Be careful when removing the plate and make sure that the tubes have been cooled down before, as otherwise the tubes can open spontaneously due to internal pressure.
4. If you use very large buckets (>500 mL) beads often get lost while pouring. It can be helpful to use 50 mL tubes and centrifuge stepwise. Just pour the supernatant of the first centrifugation cycle and add the next sample aliquot.
5. Complete dryness is absolutely necessary for storage at ambient temperature. For storage in the laboratory we recommend to freeze the beads. But for shipping protein samples the storage on desiccated beads at ambient temperature is an excellent choice.
6. Carefully resuspend the beads before loading on the gel. Gel-loader tips are not recommended, as the beads can block the capillary of the tip.
7. TCEP is an acid. Check the pH before the digest, as it has to be slightly basic for the digest. If you have problems with acidification increase the TEAB concentration.
8. The beads should not sediment during the digest. This may be a problem with mixers that have an orbit of less than 3 mm. In this case the shaking rate has to be raised.
9. The tip should fit tight in the holder. If the tip is cut directly below the C18, the C18 may get loose during the later centrifugation steps.
10. As the top of the tip is several millimeters above the tube, the inner lid of the centrifuge may not fit.
11. Acetonitrile has a lower backpressure on C18 than water, therefore centrifugation times for squeezing buffer B trough the C18 material will be shorter than for squeezing aqueous solutions.

Acknowledgments

The authors like to thank Jürgen Bartel and Knut Büttner for their support in developing of the protocol. This work was supported by grants from the Deutsche Forschungsgemeinschaft (Transregio 34 Pathophysiology of staphylococci in the post-genomic era).

References

1. Chevallet M, Diemer H, Van Dorssealer A et al (2007) Toward a better analysis of secreted proteins: the example of the myeloid cells secretome. Proteomics 7(11):1757–1770
2. McCormick RM (1989) A solid-phase extraction procedure for DNA purification. Anal Biochem 181(1):66–74
3. Zanivan S, Maione F, Hein MY et al (2013) SILAC-based proteomics of human primary endothelial cell morphogenesis unveils tumor Angiogenic markers. Mol Cell Proteomics 12 (12):3599–3611
4. Pasztor L, Ziebandt A, Nega M et al (2010) Staphylococcal major autolysin (Atl) is involved in excretion of cytoplasmic proteins. J Biol Chem 285(47):36794–36803
5. Bonn F, Bartel J, Büttner K et al (2014) Picking vanished proteins from the void: how to collect and ship/share extremely dilute proteins in a reproducible and highly efficient manner. Anal Chem 86(15):7421–7427

Part II

Subcellular Fractionation

Chapter 3

Membrane Proteomics in Gram-Positive Bacteria: Two Complementary Approaches to Target the Hydrophobic Species of Proteins

Susanne Sievers

Abstract

This protocol represents a detailed instruction how to prepare protein samples in order to raise mass spectrometry-based identification and quantification rates with respect to the challenging class of membrane proteins. This will increase comprehensiveness of global proteome studies on the one hand but could also be of interest for researchers targeting specific membrane proteins or membrane protein sequences on the other hand. The protocol is a composite of two parts, one focusing on the identification of protein sequences exterior to a cellular membrane (loops of integral membrane proteins, peripheral membrane proteins), and the other part targeting primarily protein domains spanning the lipid bilayer. The feasibility of the protocol, as it is described here, was originally shown for the gram-positive pathogenic bacterium *Staphylococcus aureus* but should be applicable to any kind of membrane protein.

Key words Integral membrane proteins, Membrane-spanning domains, Membrane protein loops, Membrane shaving, Proteinase K cleavage, Chymotryptic digestion

1 Introduction

Membrane proteins play a tremendous role in several cellular activities. They are involved in the perception of stimuli, in the transport of nutrients and other substances across the membrane, in energy conversion processes and mediate cell-to-cell interactions. 20–30% of all different protein species in a cell are partially located in the membrane with up to 20 integral membrane domains. This applies to bacterial cells but also to the ones of highly developed organisms as human [1]. The qualitative high proportion of one quarter of membrane proteins is not reflected in quantitative aspects. Membrane proteins are rather low-abundant and occulted by the high abundance of soluble proteins which is one of the main reasons for their underrepresentation in proteomic analyses. Another issue is their hydrophobic character resulting in a poor solubility in aqueous buffers. Additionally, the amino acid sequence in integral

Dörte Becher (ed.), *Microbial Proteomics: Methods and Protocols*, Methods in Molecular Biology, vol. 1841,
https://doi.org/10.1007/978-1-4939-8695-8_3,

membrane protein domains is characterized by a very low frequency of lysine and arginine residues. Hence, trypsin, the most common protease in mass spectrometry (MS) based studies, does not cut within these domains, and they consequently defy detection. Challenges in membrane proteomics are generally acknowledged [2, 3] and have been discussed in great detail [4, 5]. Protein preparation and digestion are thought to be the most crucial points for a successful analysis of membrane proteins [6], since it could be shown that ionization and fragmentation efficiencies of hydrophobic peptides are comparable to those of other peptides [7, 8]. Specifically, an enrichment of membrane proteins by depletion of the vast amount of soluble proteins is mandatory. This can be achieved by thorough washing steps of the membrane fraction using specific buffers [9, 10], or by an enzymatic shaving of the cellular membrane removing all material that is not protected by the lipid bilayer [11, 12]. Furthermore, hydrophobic proteins have to be relieved from interfering membrane lipids but kept in solution, and possibly unfolded to enhance their cleavage by proteases in order to obtain peptides for MS analysis. This has been put into practice by using for example organic solvents [3], MS-compatible detergents [12, 13] or bile acids [14, 15]. Alternative cleaving strategies, others than trypsin, have been proven to be much better suited with respect to the detection of membrane-spanning domains of proteins [3, 12, 16].

The recent progress made with respect to top-down approaches in MS brings great promise in terms of sequence coverage of membrane proteins. With a survey of entire protein molecules, the hydrophobic membrane-spanning domains will be completely covered as well [17]. However, at the present, a top-down approach works for selected proteins only. For global and most comprehensive proteome studies a bottom-up approach based on the detection of peptides still represents the method of choice. To this end, higher sequence coverage of proteins, including peptides from soluble and membrane-spanning domains, gives rise to most solid datasets. A bottom-up approach that exhibits especially great comprehensiveness and is also applicable to membrane proteins [18, 19] is to separate proteins via 1D-SDS-PAGE (one-dimensional sodium dodecyl sulfate polyacrylamide gel electrophoresis), fractionate according to protein size by cutting the entire gel lane into pieces and digest proteins in-gel before the extracted peptide fractions are analyzed by liquid chromatography-mass spectrometry (GeLC-MS approach).

This protocol describes two workflows both implementing several of the latest advances in membrane proteomics in order to get a most comprehensive picture on the membrane proteome. The first workflow makes use of several stringent washing steps of the membrane fraction, before membrane proteins are subjected to a GeLC-MS analysis. In the second method, peripheral membrane proteins and protein domains are shaved off the membrane, transmembrane protein domains solubilized in a MS-compatible detergent and digested with chymotrypsin (Fig. 1). Application of the two workflows on the membrane proteome of the human pathogen *S. aureus* revealed an almost perfect complementarity. Whereas the GeLC-MS approach is best suited for the detection of soluble domains of membrane proteins, the shaving approach gives rise to very hydrophobic peptides located integral to the membrane and is therefore customized for very hydrophobic proteins spanning the membrane multiple times (Fig. 2).

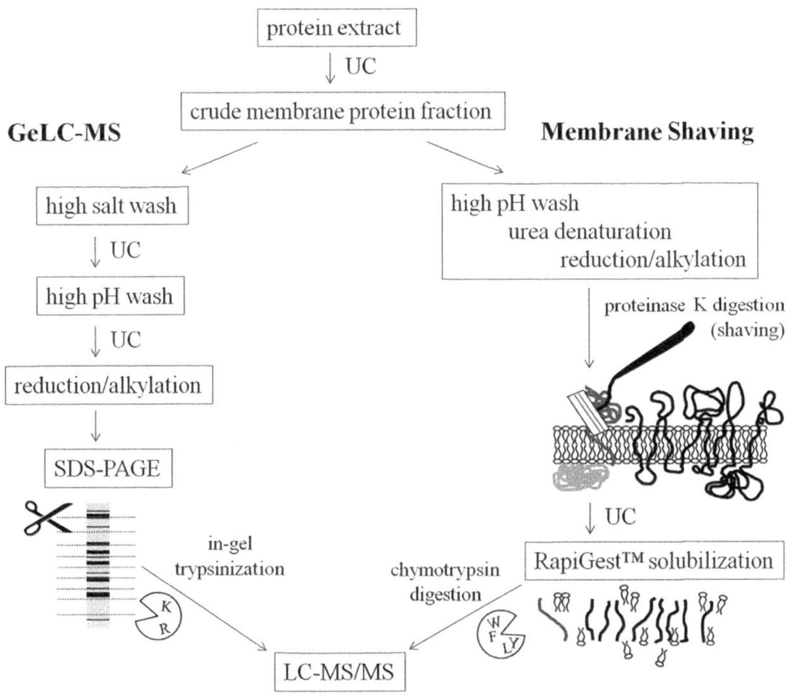

Fig. 1 Schematic overview of two complementary approaches for the analysis of membrane proteins. Both approaches start out with the ultracentrifugation (UC) of a raw protein extract. According the GeLC-MS approach (left hand branch), the crude membrane fraction is further purified from cytosolic contaminants by washing in high salt buffer and high pH buffer, respectively. The resulting membrane proteins are reduced and alkylated before they are separated via SDS-PAGE. The gel is cut down the entire lane and proteins digested in-gel with trypsin before the peptides are separated and detected by LC-MS/MS. The right hand branch describes the Membrane Shaving approach which starts with a wash in high pH buffer of a crude membrane fraction immediately followed by the denaturation of proteins in urea and their reduction and alkylation. The membranes are subsequently "shaved" by a proteinase K treatment and ultracentrifuged down. The remaining integral membrane protein domains are solubilized in RapiGest™ and digested with chymotrypsin before peptides are likewise analyzed via LC-MS/MS

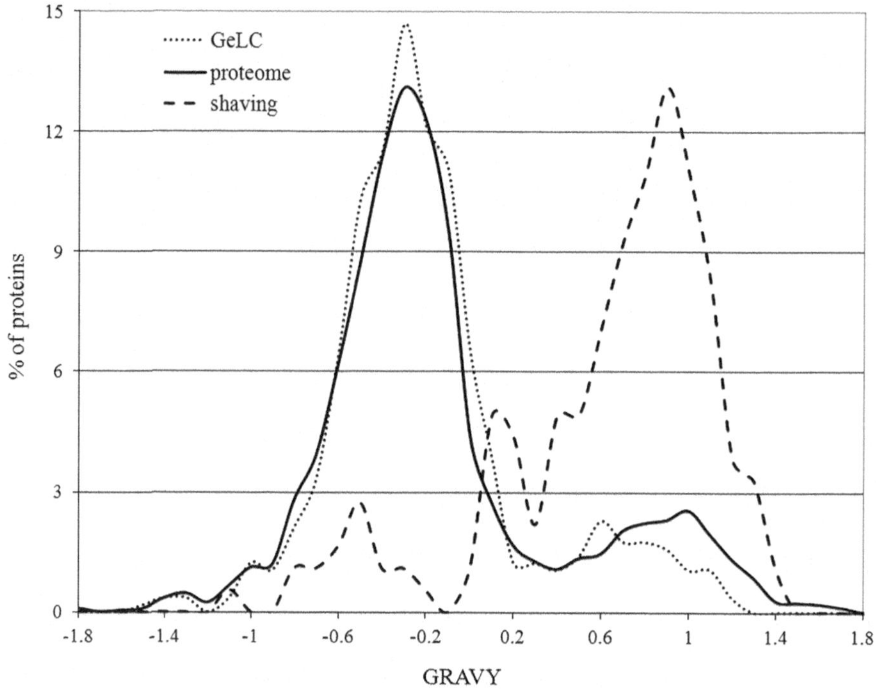

Fig. 2 Scope of the two different methods GeLC and shaving with respect to the hydrophobic nature of identified proteins. Proteins identified by the GeLC approach and by membrane shaving in a study of *S. aureus* [12], as well as all theoretical proteins of *S. aureus* COL (extracted from ncbi 2007) have been binned into 0.1 windows with regard to their GRAVY indices. Protein numbers in bins have been normalized by the number of protein identifications in the corresponding approach or by the total number of *S. aureus* proteins, respectively. Although the GeLC approach is capable of an identification of membrane proteins, hydrophilic proteins, characterized by a negative GRAVY index, are not depleted efficiently, since the distribution of identified proteins is comparable to the one of the entire proteome. In contrast, the shaving approach yields mainly hydrophobic proteins whereas proteins with negative GRAVY indices are rather underrepresented

2 Materials

The water used in all described materials and methods is purified water (i.e. distilled or deionized), but for simplicity it is termed "water".

2.1 GeLC-MS

1. Tris-buffered saline (TBS) buffer: 50 mM Tris(hydroxymethyl) aminomethane–HCl (Tris–HCl), 150 mM NaCl, pH 8.0.
2. Protease inhibitor: cOmplete ULTRA Tablets, Mini, EASY-pack (Roche).
3. Protease inhibitor without EDTA: cOmplete ULTRA Tablets, Mini, EDTA-free, EASYpack (Roche).
4. Cell lysis buffer: 20 mM Tris–HCl, 10 mM $MgCl_2$, 1 mM $CaCl_2$, pH 7.5.

5. Micro tubes: 2 mL.
6. Glass beads: diameter 0.1–0.11 mm.
7. DNAse I: grade II from bovine pancreas.
8. RNAse: crude mixture of RNAses from bovine pancreas (Roche).
9. Protein assay kit to determine protein concentration (*see* **Note 1**).
10. High salt buffer: 20 mM Tris–HCl, 1 M NaCl, protease inhibitor, pH 7.5.
11. Hand homogenizer or tissue grinder: 2 mL capacity.
12. Carbonate buffer: 100 mM Na_2CO_3, 100 mM NaCl, pH 11.
13. Long needle: Sterican® needle 0.8 × 120 mm.
14. Solubilization buffer: 50 mM Tris–HCl, 8 M urea, 1% (w/v) CHAPS (3-[(3-cholamidopropyl)dimethylammonio]-1-propanesulfonate), pH 7.5.
15. TCEP stock solution: 500 mM Tris(2-carboxyethyl)phosphine (TCEP) in 200 mM Tris–HCl, pH 8 (*see* **Note 2**).
16. Iodoacetamide stock solution: 500 mM iodoacetamide in water, freshly prepared.
17. 12% resolving gel: 3.35 mL water, 4.0 mL 30% (w/v) acrylamide–bisacrylamide, 2.6 mL of 1.5 M Tris–HCl pH 8.8, 106 μL 10% (w/v) sodium dodecyl sulfate (SDS), 100 μL 10% (w/v) ammonium persulfate (APS), and 10 μL tetramethylethylendiamine (TEMED).
18. Stacking gel: 6.1 mL water, 1.3 mL 30% (w/v) acrylamide–bisacrylamide, 2,6 mL 0.4% (w/v) SDS in 0.5 M Tris–HCl, pH 6.8, 100 μL 10% (w/v) APS, 20 μL TEMED.
19. SDS-PAGE running buffer: 25 mM Tris, 195 mM glycine, 0.1% (w/v) SDS, pH 8.3.
20. SDS-PAGE sample buffer for hydrophobic proteins: 125 mM Tris–HCl, pH 6.8, 20% (v/v) glycerol, 20% (w/v) SDS, a trace of bromphenol blue (*see* **Note 3**).
21. Gel fixing solution: 40% (v/v) ethanol /10% (v/v) acetic acid.
22. Blue silver stain: dissolve 100 g ammonium sulfate in 200 mL 50% (v/v) phosphoric acid. Fill up to 800 mL with water and add 1.2 g Coomassie G-250. This solution can be stored over a long period of time. Shortly before use add methanol to a final concentration of 20% (v/v).
23. Prelubricated tubes: 1.7 mL.
24. Gel washing solution: 50% (v/v) acetonitrile LC/MS grade (e.g. Optima™ LC/MS, Fisher Chemical), 50 mM NH_4HCO_3 in water.

25. Trypsin solution (5 ng/μL): Dissolve 1 μg trypsin (sequencing grade modified trypsin, Promega) in 5 μL trypsin resuspension buffer (Promega) and incubate 15 min at 30 °C. Mix trypsin with 200 μL freshly prepared 50 mM NH_4HCO_3.
26. Acetic acid for peptide extraction: Glacial acetic acid, ultrapure.
27. ZipTip: ZipTip with 0.2 μL or 0.6 μL C18 resin (Merck Millipore).

2.2 Membrane Shaving

Components that have already been mentioned under Subheading 2.1 are not listed again

1. Urea: >99.5% pure.
2. Proteinase K: proteinase K from the fungus *Tritirachium album* limber, reconstituted in 50 mM Tris–HCl, 10 mM $CaCl_2$, pH 8.0.
3. TEAB: 50 mM triethylammonium bicarbonate buffer (for HPLC grade, pH 8.4–8.6), pH adjusted to 7.8 with HCl.
4. Chymotryptic digestion buffer: 50 mM TEAB, 10 mM $CaCl_2$, and 0.5% (w/v) RapiGest™, pH 7.8.
5. RapiGest™: RapiGest SF Surfactant (Waters).
6. Chymotrypsin: bovine chymotrypsin, sequencing grade, reconstituted in 1 mM HCl.

3 Methods

The origin of proteins as well as the type of instrumentation for LC-MS/MS analysis can differ a lot among users. Therefore, this protocol concentrates on the processing from a raw protein extract to a sample that can be analyzed by LC-MS/MS.

If not indicated otherwise all procedures handling protein/peptide samples take place on ice.

All ultracentrifugation steps are carried out for 1 h at 100,000 × *g* at 4 °C.

3.1 GeLC-MS

1. Harvest cells of interest by centrifugation (8000 × *g*) for 10 min at 4 °C. Wash cell pellets with ice-cold TBS buffer with protease inhibitor (*see* **Note 4**) and a second time with TBS with protease inhibitor without EDTA (*see* **Note 5**).
2. Resuspend cells in lysis buffer containing protease inhibitor without EDTA and disrupt them in a bead homogenizer. For this purpose, a maximum of 1 mL cell suspension is filled into a tightly to close micro tube containing 500 μL glass beads (*see* **Note 6**). In the case of very tough bacteria as the gram-positive *S. aureus* three cycles of 30 s at maximum speed (6.5 m/s) should be run (cool samples on ice in between the cycles).

3. Add DNAse I (200 μ/mL sample) and RNAse (10 μ/mL sample) to the lysate and incubate 20 min at 37 °C to allow for nucleolytic digestion (*see* **Note 7**). Remove cell debris and glass beads by centrifugation (8000 × *g*, 4 °C, 10 min).
4. The supernatant containing the crude protein extract can be frozen at −20 °C or colder, or directly subjected to a determination of protein content by a method of choice (e.g. Bradford assay, BCA assay), (*see* **Note 1**).
5. Ultracentrifuge the crude protein extract. Cytosolic proteins in the supernatant can be discarded or analyzed separately.
6. The pellet containing the membrane fraction is homogenized in 2 mL of ice-cold high salt buffer using a hand homogenizer, transferred to a 15 mL centrifuge tube and filled up to 6 mL with high salt buffer. Incubate the sample for 1 h at 4 °C on a rotary shaker or overhead shaker (*see* **Note 8**) before the membrane fraction is pelleted again by ultracentrifugation.
7. Solubilize the pellet in 2 mL of ice-cold carbonate buffer using the homogenizer, transfer to 15 mL tube and fill up to 6 mL with carbonate buffer. The extract is incubated 1 h at 4 °C on an overhead shaker. Every 15 min the sample is further homogenized by drawing it up and down a syringe five times [11] using a long needle to reach the bottom of the tube (*see* **Note 9**). Ultracentrifuge the sample after the carbonate wash.
8. The purified membrane pellet is resuspended in 500 μL solubilization buffer. TCEP stock solution is added to the sample to give a final concentration of 5 mM. Incubate the sample 1 h at 30 °C in order to reduce the proteins (*see* **Note 10**). Supply the reduced protein sample with iodoacetamide to get a final concentration of 10 mM (*see* **Note 11**). The carbamidomethylation of protein thiol groups is carried out for 30 min in the dark at room temperature (*see* **Note 12**). Determine protein concentration.
9. Prepare a 1D-gel for the separation of membrane proteins via SDS-PAGE. Cast a 1 mm thick resolving gel, leave space for a stacking gel and overlay gel with isopropanol.
10. Remove isopropanol from polymerized resolving gel, cast stacking gel on top and immediately insert 1 mm comb.
11. Set up electrophoresis chamber, insert gel and fill up with SDS-PAGE running buffer.
12. Mix 20 μg of proteins with one volume of SDS-PAGE sample buffer for hydrophobic proteins and load sample into gel well. Run electrophoresis at 150 V until running front reaches bottom of gel (about 1 h).
13. Rinse the gel with water and fix it in gel fixing solution for 1 h.

14. Wash the gel in water for 30 min, change water and wash another 30 min.
15. The separated proteins are visualized by an overnight staining with Blue Silver stain [20].
16. Destain background of the gel by several washing steps in water.
17. Cut a full gel lane of separated membrane proteins into 10 pieces. Expediently, gel pieces are not cut equidistantly, but according to intensity in protein stain. Each gel piece is cut into smaller pieces (approximately 1 mm^3) and transferred into a tube (*see* **Note 13**). Gel pieces are washed in gel washing solution until completely destained (change washing solution several times). Remove the washing solution and completely dehydrate gel pieces in pure acetonitrile. Remove acetonitrile. Each gel piece is supplied with 20 μL trypsin solution (trypsin–protein ratio 1:20). When the solution is completely absorbed by the dry gel pieces, add 50 mM NH_4HCO_3 in small portions. Eventually all gel pieces should be completely rehydrated, but not much additional liquid should be present in the tubes. In-gel tryptic digestion is carried out overnight at 37 °C.
18. Extract peptides from the gel in several steps similar to a protocol recently described [21]. Gel pieces are dehydrated in acetonitrile, rehydrated in 1% (v/v) acetic acid, dried in acetonitrile, 10 min ultrasonicated in 5% (v/v) acetic acid, 10 min ultrasonicated in acetonitrile and eventually completely dried in acetonitrile. A volume of 200 μL is used in each extraction step. The six supernatants are eventually pooled and reduced to dryness in a vacuum centrifuge. The dried peptides can be stored at −20 °C and should be dissolved in 10 μL of 1% (v/v) acetic acid and 2% (v/v) acetonitrile supported by 10 min ultrasonication shortly before LC-MSMS analysis (*see* **Note 14**).

3.2 Membrane Shaving

1. Start the protocol with a crude membrane protein pellet obtained as described in Subheading 3.1, **steps 1–5**.
2. The pellet is homogenized in 500 μL carbonate buffer. Determine the protein concentration (*see* **Note 1**) and adjust it to be 1 mg/mL by adding carbonate buffer (*see* **Note 15**). The extract is incubated 1 h at 4 °C on an overhead shaker. Every 15 min the sample is further homogenized by drawing it up and down in a syringe.
3. Bring the sample to room temperature and add solid urea to give a final concentration of 8 M (*see* **Note 16**). Add TCEP to the sample to give a final concentration of 5 mM and incubate 60 min at 30 °C. Iodoacetamide is added in order to

carbamidomethylate free thiol groups (final IAM concentration 10 mM) for 30 min in the dark at room temperature (*see* **Note 17**).

4. Add proteinase K in an enzyme:protein ratio of 1:50 (*see* **Note 18**). The sample is incubated for 15 h in a thermomixer (900 rpm) at 37 °C [16].
5. After addition of 1 volume of 10% (v/v) acetonitrile in water cool the sample on ice and ultracentrifuge again. The supernatant containing short amino acid sequences originating from soluble parts of membrane proteins is discarded.
6. Wash the pellet in TEAB to remove residual urea and high pH buffer, and ultracentrifuge once more (*see* **Note 19**).
7. Resuspend the pellet in 200 μL chymotryptic digestion buffer. Add chymotrypsin (enzyme:protein ratio estimated between 1:50 and 1:100) and digest proteins for 6 h at 30 °C and 900 rpm (*see* **Note 20**).
8. In order to stop digestion and to remove RapiGest™ add concentrated HCl to reduce pH below 2 (approximately 250 mM HCl final). Incubate the sample for 45 min at 37 °C to hydrolyze RapiGest™ into insoluble products. Cool on ice and centrifuge for 15 min, 4 °C at 20,000 × *g*. Keep the clear supernatant and repeat the centrifugation twice to have surely removed any remaining RapiGest™ products (*see* **Note 21**).
9. The chymotryptic digest should be aliquoted and frozen, and taken out shortly before mass spectrometric analysis. In our hands the peptide mixture proved to be not very stable at 4 °C.
10. The resulting peptide amount is very limited and can be as low as only 1/100 of the original protein starting material. In order to find a suitable amount to be injected in LC-MS/MS analyses, a test injection of a very small volume of the digest is advisable (*see* **Notes 22** and **23**).

4 Notes

1. The two most frequently used methods to determine the protein content of a solution are the Bradford and BCA assay. Both assays are easily and quickly performed. However, a Bradford is based on the binding of Coomassie G-250 dye to primarily alkaline amino acids as arginine, lysine and histidine. A protein sample enriched in membrane proteins can differ a lot from a cytosolic protein extract with respect to the content of specific amino acids. Thus, an amino acid sequence independent assay as the BCA test could be more expedient, also because of its higher tolerance toward soaps and detergents.

2. TCEP is extremely acidic when solved in water and needs to be buffered.
3. SDS-PAGE sample buffer for hydrophobic proteins contains more SDS than regular buffer and no mercaptoethanol or other reducing agent since proteins had already been reduced and alkylated. Load the sample without heat denaturation. Cyanates originating from urea would massively modify amino acids.
4. Parameters for cell harvest and cell lysis possibly need to be adapted with respect to the origin of biological material. With respect to bacteria, ultrasonication with constant cooling is preferred. However, some bacteria as *S. aureus* are not to lyse by ultrasonication, but can be cracked via bead homogenization. If there is sufficient material available one should start the protocol with a comparatively high protein amount of 50 mg, since most of the material will be washed away as cytosolic fraction. If material is limited the protocol can be downscaled.
5. One tablet of protease inhibitor is dissolved in 10 mL buffer. EDTA is omitted in the second pellet wash, because the enzymes in the following nucleic acids digestion step require metal ions.
6. Filling too much volume or too thick cell suspensions into the micro tube might lead to incomplete cell lysis. Use at least 1 mL lysis buffer for cells originating from 50 mL culture of OD = 1.
7. Degradation of nucleic acids can increase purity of the later membrane protein fraction, as DNA with bound proteins will otherwise be pelleted during ultracentrifugation steps. Furthermore, a cleavage of rRNA results in a disintegration of ribosomes and a more efficient removal of ribosomal proteins.
8. A high-salt wash will destroy ionic interactions between integral and peripheral membrane proteins thereby depleting the latter.
9. A very alkaline pH leads to conversion of membrane vesicles to membrane sheets [9, 23] and thereby to a release of enclosed proteins.
10. Usually the reduction of proteins is carried out at increased temperature (50–60 °C), but urea in the buffer could form cyanates that could modify amino acids. Compensatory, the incubation time is extended to 1 h instead of the usual 30 min.
11. Iodoacetamide should always be prepared shortly before use and kept in the dark due to its light sensitivity.
12. A reduction and alkylation of proteins can also be carried out in-gel after the proteins have been separated via SDS-PAGE. However, reduction and alkylation in solution in a denaturing

buffer proved to be more efficient and is not as labor-intensive as handling every gel piece separately.

13. The use of prelubricated tubes can limit the loss of protein/peptide material.
14. Purification of peptide samples using ZipTips is expedient if a lot of impurities or broad peak widths are experienced in LC-MS/MS analysis.
15. From experience, approximately only 10% of the starting protein material will end up in the pellet of the first ultracentrifugation step, i.e., 90% of the proteins can be found in the cytosolic fraction of the supernatant.
16. Urea can break hydrophobic interactions between proteins and the membrane and further cleanse the membrane fraction from peripheral membrane proteins. A urea extraction is usually carried out with a concentration of 2 M urea [22, 23]. The much higher concentration used in this protocol intensifies the wash effect and at the same time results in an extensive denaturation of proteins making their cysteine residues accessible for the following reduction and alkylation step.
17. An alkaline pH is mandatory for the alkylation of thiols, because thiol groups need to be deprotonated for an electrophilic compound to attack. However, the pH of the carbonate buffer is very high and not perfect for a reduction and alkylation of thiols [24]. Increased incubation times compensate for this shortcoming.
18. Proteinase K can act under highly denaturing conditions as 8 M urea and at a very high pH as in carbonate buffer of pH 11.
19. The membrane pellet after proteinase K digestion can be very small or even invisible. Continue with protocol even if there is no obvious pellet.
20. RapiGest™ is a detergent that supports solubilization and accessibility for enzymatic attack of the remaining very hydrophobic protein parts. The detergent can be removed by acidic hydrolysis and hence does not hamper LC-MS/MS analysis.
21. The RapiGest™ precipitate can be a white pellet at the bottom of the tube, but can also form a layer on the surface of the chymotryptic digest.
22. If a column oven is available, a heating of the analytical column up to 60 °C during reversed phase chromatography can result in an increased peptide recovery [11].
23. A database search of the MS/MS data resulting from the measurement of the proteinase K/chymotryptic digestion is not as straight forward as it is known from a typical tryptic protein digestion. These samples should be searched with three

different parameter files. Firstly, defining no enzyme specificity in order to identify peptides that have not been cut by chymotrypsin but only by sequence unspecific proteinase K. Secondly, with a partial chymotrypsin digestion (at tryptophane (W), tyrosine (Y), phenylalanine (F) and leucine (L)) to identify peptides with one chymotryptic terminus, and thirdly, with a full chymotryptic cleavage for peptides that hold both termini chymotryptically. It is reasonable to allow four missed cleavage sites due to transmembrane intrinsic leucine clusters that are poorly cut by chymotrypsin.

Acknowledgments

This work was supported by the German Research Foundation Grant SFB/TR34.

References

1. Fagerberg L, Jonasson K, Heijne v G et al (2010) Prediction of the human membrane proteome. Proteomics 10:1141–1149
2. Blonder J, Goshe MB, Moore RJ et al (2002) Enrichment of integral membrane proteins for proteomic analysis using liquid chromatography-tandem mass spectrometry. J Proteome Res 1:351–360
3. Fischer F, Wolters D, Rögner M, Poetsch A (2006) Toward the complete membrane proteome: high coverage of integral membrane proteins through transmembrane peptide detection. Mol Cell Proteomics 5:444–453
4. Gilmore JM, Washburn MP (2010) Advances in shotgun proteomics and the analysis of membrane proteomes. J Proteome 73:2078–2091
5. Helbig AO, Heck AJ, Slijper M (2010) Exploring the membrane proteome--challenges and analytical strategies. J Proteome 73:868–878
6. Speers AE, Wu CC (2007) Proteomics of integral membrane proteins--theory and application. Chem Rev 107:3687–3714
7. Cech NB, Enke CG (2000) Relating electrospray ionization response to nonpolar character of small peptides. Anal Chem 72:2717–2723
8. Tabb DL, Huang Y, Wysocki VH, Yates JR 3rd (2004) Influence of basic residue content on fragment ion peak intensities in low-energy collision-induced dissociation spectra of peptides. Anal Chem 76:1243–1248
9. Fujiki Y, Hubbard AL, Fowler S, Lazarow PB (1982) Isolation of intracellular membranes by means of sodium carbonate treatment: application to endoplasmic reticulum. J Cell Biol 93:97–102
10. Eymann C, Dreisbach A, Albrecht D et al (2004) A comprehensive proteome map of growing *Bacillus subtilis* cells. Proteomics 4:2849–2876
11. Speers AE, Blackler AR, Wu CC (2007) Shotgun analysis of integral membrane proteins facilitated by elevated temperature. Anal Chem 79:4613–4620
12. Wolff S, Hahne H, Hecker M, Becher D (2008) Complementary analysis of the vegetative membrane proteome of the human pathogen *Staphylococcus aureus*. Mol Cell Proteomics 7:1460–1468
13. Yu Y, Gilar M, Lee PJ et al (2003) Enzyme-friendly, mass spectrometry-compatible surfactant for in-solution enzymatic digestion of proteins. Anal Chem 75:6023–6028
14. Masuda T, Tomita M, Ishihama Y (2008) Phase transfer surfactant-aided trypsin digestion for membrane proteome analysis. J Proteome Res 7:731–740
15. Masuda T, Saito N, Tomita M, Ishihama Y (2009) Unbiased quantitation of *Escherichia coli* membrane proteome using phase transfer surfactants. Mol Cell Proteomics 8:2770–2777
16. Blackler AR, Speers AE, Ladinsky MS, Wu CC (2008) A shotgun proteomic method for the identification of membrane-embedded proteins and peptides. J Proteome Res 7:3028–3034

17. Whitelegge JP (2013) Integral membrane proteins and bilayer proteomics. Anal Chem 85:2558–2568
18. Taylor SW, Fahy E, Zhang B et al (2003) Characterization of the human heart mitochondrial proteome. Nat Biotechnol 21:281–286
19. Graham RL, Pollock CE, O'Loughlin SN et al (2007) Multidimensional analysis of the insoluble sub-proteome of *Oceanobacillus iheyensis* HTE831, an alkaliphilic and halotolerant deep-sea bacterium isolated from the Iheya ridge. Proteomics 7:82–91
20. Candiano G, Bruschi M, Musante L et al (2004) Blue silver: a very sensitive colloidal Coomassie G-250 staining for proteome analysis. Electrophoresis 25:1327–1333
21. Lassek C, Burghartz M, Chaves-Moreno D et al (2015) A metaproteomics approach to elucidate host and pathogen protein expression during catheter-associated urinary tract infections. Mol Cell Proteomics 14:989–1008
22. Schook W, Puszkin S, Bloom W et al (1979) Mechanochemical properties of brain clathrin: interactions with actin and alpha-actinin and polymerization into basketlike structures or filaments. Proc Natl Acad Sci U S A 76:116–120
23. Okamoto T, Schwab RB, Scherer PE, Lisanti MP (2001) Analysis of the association of proteins with membranes. In: Current Protocols in Cell Biology. John Wiley & Sons, Inc., Hoboken., Chapter 5, unit 5.4, pp 1–17
24. Rombouts I, Lagrain B, Brunnbauer M et al (2013) Improved identification of wheat gluten proteins through alkylation of cysteine residues and peptide-based mass spectrometry. Sci Rep 3:2279

Chapter 4

Enrichment of Cell Surface-Associated Proteins in Gram-Positive Bacteria by Biotinylation or Trypsin Shaving for Mass Spectrometry Analysis

Florian Bonn, Sandra Maaß, and Jan Maarten van Dijl

Abstract

In microbial cells surface-exposed proteins represent a physiologically important class of molecules as they enable these cells to interact with their environment both as free-living organisms and during interactions with a host. However, the characteristics of these proteins are quite divergent, which makes attempts to enrich, analyze, and quantify these molecules a challenging task. In this chapter two complementary methods for the enrichment and identification of cell surface-associated proteins, namely the biotinylation and the shaving approaches, are presented. Both protocols have been optimized for Gram-positive bacteria, and we provide a step-by-step guide for sample generation. Possible pitfalls during protein preparation are discussed.

Key words Proteins, Cell surface, Cell wall, Membrane, Lipid anchor, Shaving, Biotinylation, Virulence factor

1 Introduction

Analysis of bacterial surface proteins is of special interest, because these proteins are key players in the interaction of the bacterium with its environment as they detect and respond to diverse stimuli like temperature, nutrient availability, or antibiotics [1]. Moreover, in pathogens many cell surface-exposed proteins function as virulence factors allowing for tissue adhesion, immune evasion and invasion of the host [2, 3]. Hence, this protein class offers a rich pool of candidates for antibiotic development as well as for directed vaccinology.

Although only some tens of proteins belong to the surface proteome, the characteristics of these proteins are quite divergent (*see* Fig. 1). The surface proteome of Gram-positive bacteria consists of proteins with multiple transmembrane helices and large extracellular loops and/or C-terminal domains that span the peptidoglycan layer; proteins that are anchored with a transmembrane

Dörte Becher (ed.), *Microbial Proteomics: Methods and Protocols*, Methods in Molecular Biology, vol. 1841,
https://doi.org/10.1007/978-1-4939-8695-8_4, © Springer Science+Business Media, LLC, part of Springer Nature 2018

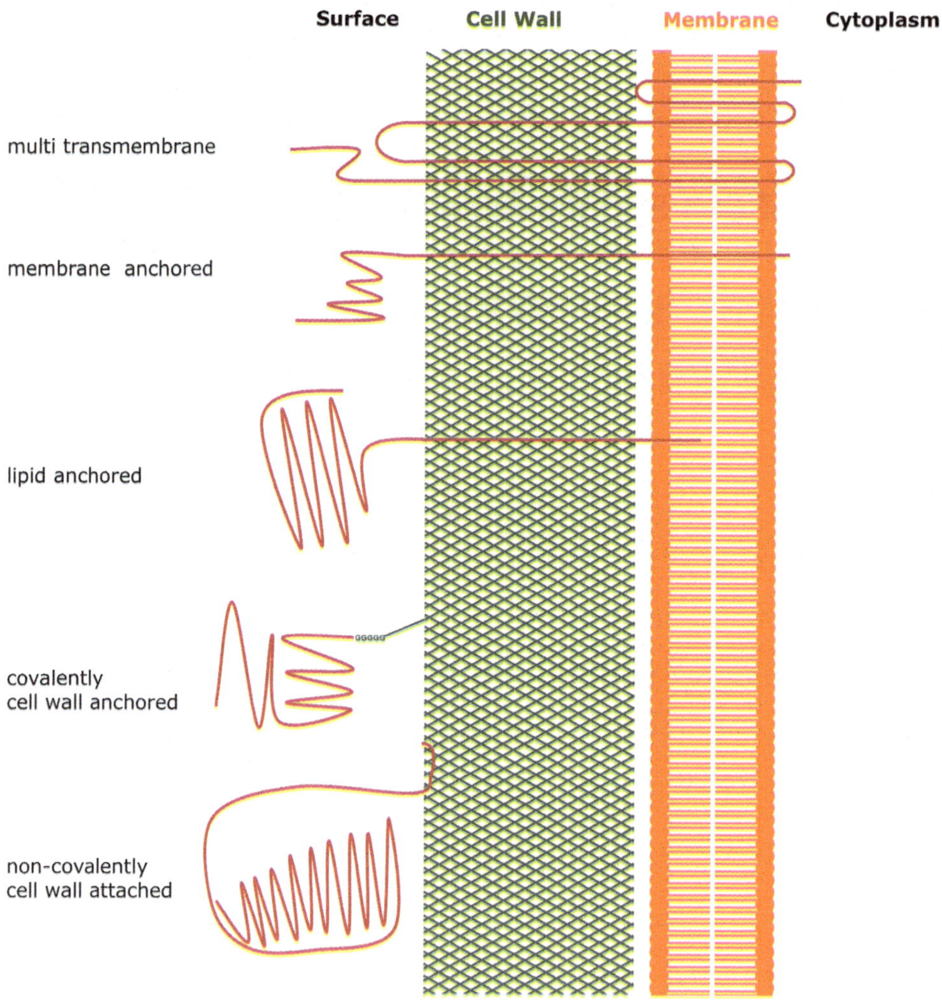

Fig. 1 Different classes of cell surface proteins in Gram-positive bacteria

domain; lipid-anchored proteins; sortase substrate proteins that are covalently anchored to the peptidoglycan in the cell wall and, finally, cell wall-attached proteins that are noncovalently bound to the peptidoglycan or teichoic acids [4]. Because of the different characteristics of these proteins, complementary approaches are needed for a complete coverage of the surface proteome. While lipoproteins are preferentially enriched in a biotinylation approach, the covalently and noncovalently cell wall-attached proteins can be analyzed best via surface shaving with trypsin [5].

1.1 Biotinylation

The biotinylation approach makes use of the high affinity between biotin and avidin (Kd = 10–15 M) [6] to enrich surface proteins previously labeled with a biotin moiety. Possible reagents for the binding of biotin to reactive residues in proteins usually comprise three components: a biotin moiety, a sufficiently sized spacer (to overcome steric hindrance), and a reactive group to interact with the targeted proteins. Careful selection of appropriate labeling reagents is made on the basis of a number of factors, including aqueous versus membrane solubility, spacer length, and the reactive properties of the functional group. In particular, due to the negative charge of N-hydroxysulfosuccinimide (NHS), the reagent Sulfo-NHS-SS-Biotin cannot be transported through the cell membrane and will therefore only react with surface-exposed protein residues. After covalent modification of the surface proteins of intact cells at primary amine residues with Sulfo-NHS-SS-biotin, cells are disrupted and biotin-labeled proteins are purified with an avidin-coated resin. In the following steps, contaminant proteins can be removed by stringent washing procedures. Biotinylated proteins are eluted by reduction of the disulfide bridge that is provided in the cleavable linker [7]. Subsequent sample fractionation or protein digestion can be performed by the here presented in-gel digestion protocol, or by following other established protocols.

1.2 Shaving Approach

The surface shaving approach was introduced in 2006 by Rodríguez-Ortega and coworkers [8] and has been recently reviewed in great detail [9]. Surface shaving can be performed with soluble and immobilized trypsin or other proteases and has been applied to several Gram-positive and Gram-negative bacteria as well as eukaryotes. As the cell wall of bacteria is highly divergent, shaving conditions need to be optimized for the organism of interest. The here presented protocol has been optimized for *Staphylococcus aureus* COL. It is based on a protocol by Dreisbach and coworkers [10] and has been further optimized by Hempel and coworkers [5]. Critical points are mentioned in Subheading 3.

2 Materials

Prepare all solutions and buffers with ultrapure water. Store all solutions at 4 °C unless indicated otherwise. When working with hazardous chemicals always follow local bylaws and regulations to protect laboratory staff and the environment. Use low-binding reaction tubes whenever working with proteins or peptides.

2.1 Biotinylation of Surface Proteins

1. 0.15 g bacterial cells (wet weight) from a living culture (*see* **Notes 1** and **2**).
2. Phosphate buffered saline (PBS): Dissolve 8.76 g/L NaCl, 2.6 g/L NaH_2PO_4, and 14.4 g/L Na_2HPO_4 in aqua bidest. Adjust pH to 8.0 using NaOH or HCl. The solution can be stored at room temperature (*see* **Note 3**).

3. 100 mM PMSF: Dissolve 0.1742 g phenylmethanesulfonyl-fluoride (PMSF) in 10 mL isopropanol (*see* **Note 4**).
4. PBS/1 mM PMSF: Dilute 0.1 mL 100 mM PMSF with 10 mL PBS (*see* **Note 5**).
5. Sulfo-NHS-SS-biotin: Use 1 mg per sample (*see* **Note 6**).
6. PBS/500 mM glycine: Dissolve 1.88 g glycine in 50 mL PBS. You need 3 mL of this solution per sample.
7. Iodoacetamide (IAA).
8. PBS/5% IAA: For every sample dissolve 0.15 g IAA in 3 mL PBS. This solution has to be prepared directly before use as IAA is unstable in solution.

2.2 Purification of Biotinylated Proteins

1. PBS/CHAPS: 20% (w/v) CHAPS, 20% (w/v) ASB-14 in PBS. Dissolve 0.2 g 3-(3-cholamidopropyl) dimethyl-ammonio)-1 propanesulfonate (CHAPS) and 0.2 g amidosulfobetaine-14 (ASB-14) in 1 mL PBS. This should be sufficient for ten samples.
2. NeutrAvidin-Agarose-Beads (e.g., by Pierce).
3. PBS/NP-40 1% (v/v): Dilute 5 mL NP-40 with 45 mL PBS. Dilute 5 mL of this solution again with 45 mL PBS. This solution can be stored.
4. PBS/NP-40/CHAPS: PBS/NP-40 1% (v/v), 6% (w/v) CHAPS. Solve 0.6 g CHAPS in 9 mL PBS/NP-40 1% (v/v). You need 4 mL of this solution per sample.
5. PBS/NP-40/SDS: PBS/NP-40 1% (v/v), 2% (w/v) SDS. Prepare a 10% (w/v) stock solution of ultrapure SDS (in *aqua bidest*). Add 2 mL of the SDS stock solution to 8 mL PBS/NP-40 1% (v/v). This solution can be stored (*see* **Note 7**).
6. Elution buffer: Dissolve 5% (v/v) 2-mercaptoethanol in aqua bidest. You need 2 mL per sample.
7. Acetone: Cool acetone down to −20 °C. You need 8 mL per sample.

2.3 Shaving Approach

1. About 10^9 bacterial cells from a living culture (*see* **Note 8**).
2. TEAB: 50 mM triethyl ammonium bicarbonate buffer (TEAB): Prepare from 1 M TEAB solution (e.g., Sigma-Aldrich) by diluting with 19 parts water. The diluted buffer can be stored for 2 weeks.
3. Trypsin solution: Dissolve 20 μg sequencing grade modified trypsin (Promega) in 1 mL water. Aliquots can be stored at −20 °C for several weeks. Thaw on ice and use directly after thawing (*see* **Note 9**).
4. Centrifuge filter: 0.22 μm cellulose acetate centrifuge tube filters, for example from Corning Inc.

3 Methods

3.1 Biotinylation and Purification of Surface Proteins

1. Harvest your bacterial cells in preweighted falcon tubes by centrifugation at 10,000 × *g* and 4 °C for 5 min. Store cells on ice during all steps of biotinylation (*see* **Note 3**).
2. Discard the supernatant. Wash the cell pellets by gentle resuspension in freshly prepared PBS–1 mM PMSF. Thereby pooling cells harvested under the same conditions. Centrifuge at 10,000 × *g* and 4 °C for 5 min.
3. Discard the supernatant. Weigh the wet cell pellets (*see* **Note 10**).
4. Resuspend the cells in PBS–1 mM PMSF (0.15 g cells per mL) (*see* **Note 2**) and transfer 1 mL of sample to a 2 mL reaction tube (*see* **Note 11**).
5. Dissolve 1 mg Sulfo-NHS-SS-Biotin in 100 μL PBS per sample. Add 100 μL of this biotin solution to 1 mL cell suspension (see previous step). Incubate for 1 h by gently shaking on ice (*see* **Note 12**).
6. Centrifuge at 20,000 × *g* and 4 °C for 1 min. Discard the supernatant.
7. Stop the biotinylation reaction by washing the cells in PBS/500 mM glycine three times. For each iteration resuspend the cells in 1 mL PBS/500 mM glycine and centrifuge at 20,000 × *g* and 4 °C for 1 min. Discard the supernatant.
8. Use 400 μL of freshly prepared PBS/5% IAA for resuspending the cells.
9. The samples can now be stored at −70 °C (*see* **Note 13**).
10. Start purification of biotinylated proteins by disrupting the bacterial cells using your preferred protocol (*see* **Note 14**).
11. Solubilize cell membranes by adding 100 μL PBS/CHAPS and homogenize (*see* **Note 14**).
12. Incubate on ice for 30 min.
13. Homogenize again.
14. Centrifuge at 15,800 × *g* and 4 °C for 15 min to remove cell debris. Transfer the supernatant to a new reaction tube.
15. Equilibrate 150 μL NeutrAvidin-Agarose-Beads per sample (*see* **Note 15**) by washing the beads two times with 300 μL PBS/NP-40. Centrifuge at 80 × *g* and 4 °C for 1 min and discard the supernatant.
16. Incubate the beads for binding of the biotin-labeled proteins for 1 h on ice with gentle shaking (*see* **Note 12**).
17. Centrifuge at 80 × *g* and 4 °C for 1 min and discard the supernatant.

18. In order to remove unbound unlabeled proteins, wash the beads with the bound biotinylated proteins four times with 1 mL PBS/NP-40/CHAPS. Vortex for about 10 s to properly mix beads and solution. Centrifuge at 80 × *g* and 4 °C for 1 min and discard the supernatant after every iteration.
19. Wash the beads with the bound biotinylated proteins two times with 1 mL PBS/NP-40/SDS. Vortex for about 10 s to properly mix beads and solution. Centrifuge at 80 × *g* and room temperature for 1 min and discard the supernatant after every iteration.
20. Add 1 mL elution buffer and incubate for 1 h at room temperature with gentle shaking (*see* **Note 12**).
21. Centrifuge at 80 × *g* and room temperature for 1 min. Transfer the supernatant to 8 mL precooled acetone (−20 °C).
22. Add 1 mL elution buffer to the beads again and mix thoroughly.
23. Centrifuge at 80 × *g* and room temperature for 1 min. Transfer the supernatant to the acetone supernatant mixture (*see* **step 21** in Subheading 3.1).
24. Incubate the acetone–supernatant mixture at −20 °C over night to enable protein precipitation.
25. Collect the precipitated proteins by centrifugation at 10,000 × *g* and 4 °C for 30 min. Discard the supernatant according to local bylaws and regulations for treatment of hazardous agents.
26. Wash the protein pellet in 1 mL precooled (4 °C) undenaturated ethanol and transfer the sample to 1.5 mL reaction tubes.
27. Centrifuge at 15,000 × *g* and 4 °C for 5 min. Discard the supernatant. Dry the pellet in a vacuum concentrator for 3–5 min (*see* **Note 16**) and store the proteins till further GeLC-MS analysis.

3.2 Shaving Approach

1. Harvest the cells by centrifugation (4000 × *g*, 5 min, 4 °C) (*see* **Notes 17** and **18**).
2. Resuspend the cells in 1.5 mL 50 mM TEAB and transfer them to a 2 mL reaction tube. Pellet them by centrifugation (4000 × *g*, 5 min, 4 °C) and wash two times with 1.5 mL 50 mM TEAB (*see* **Note 19**).
3. Wash a cellulose acetate centrifuge tube filter two times with 200 μL TEAB (8000 × *g*, 10 min).
4. Resuspend the cell pellet in 45 μL TEAB and transfer to the filter. Add 5 μL trypsin solution and incubate for 15 min at 37 °C while gentle shaking (the cells should not sediment, *see* **Note 20**).

5. Separate the supernatant by centrifugation (4000 × *g*, 10 min, 4 °C).
6. For complete digestion of extracellular domains incubate the isolated supernatant overnight in a low binding reaction tube with shaking (37 °C, 900 rpm, *see* **Note 21**).
7. Reduce the sample volume to 10 μL in a vacuum centrifuge and perform an LC-MS analysis. If you observe extensive low-molecular weight contaminations desalt the samples with for example ZipTip or StageTip purification.

4 Notes

1. This method is not suitable for analyzing surface proteins of bacteria that contain porins and other channels that allow for internalization of biotin reagents.
2. The amount of cells applied to the protocol should be adapted dependent on the specific amount of surface proteins. We found that 0.15 g (wet weight) of *Staphylococcus aureus* or *Bacillus subtilis* cells are sufficient to isolate 20–40 μg biotinylated proteins.
3. In order to avoid biotin labeling of unwanted intracellular proteins, cell lysis has to be prevented as much as possible. Therefore, cell harvesting conditions and the buffer used during biotinylation have to be adapted to the organisms analyzed. The protocol using PBS (presented here) was already applied to *Bacillus*, *Staphylococcus* and *Streptococcus spec* [5, 11, 12]. If other buffers shall be used those should not contain any primary amino groups as this will reduce the biotinylation efficiency of the targeted surface proteins.
4. Although the solution should be stored at 4 °C, it is beneficial to warm it to room temperature before use as PMSF may precipitate at 4 °C.
5. As PMSF will be rapidly inactivated in aqueous solutions, it is recommended to prepare the solution just before use.
6. Sulfo-NHS-esters of biotin are soluble in water up to 10 mM. They are prone to hydrolysis in an aqueous milieu and should be dissolved just before their use.
7. SDS will precipitate at 4 °C. It can be resolved at room temperature.
8. The protocol is scalable. If you adjust the cell number (up to 10^{10}) adjust also the buffer volumes.
9. Never freeze and reuse trypsin solution after thawing!

10. Make sure that there is no residual liquid on the cell pellet before weighing it. With a solid cell pellet this can be achieved by turning the falcon upside down on a tissue and let the fluid drip out gently.
11. If this protocol is coupled to a metabolic labeling approach, cells with different labels can be pooled at this step based on the cell wet weight.
12. Gently shaking of the sample works best if reaction tubes are positioned horizontally and the sample container is placed on a gel shaker. Select the speed that allows careful but sufficient mixing of the sample during incubation.
13. Biotin-labeled cells can be stored at −70 °C for up to 12 months.
14. For most Gram-positive bacteria mechanical treatment in a ribolyzer will give good results for cell disruption and homogenization during membrane solubilization. To do so, use 500 μL glass beads (diameter 0.1 mm) per sample. Run four iterations for 30 s at maximum speed for cell disruption and one iteration for 30 s at maximum speed for homogenization.
15. The volume of beads is calculated neglecting the sodium azide supernatant. Sodium azide can be removed by centrifuging the beads at 80 × *g* and 4 °C for 1 min. Residual sodium azide supernatant can be filled back to the stored NeutrAvidin-Agarose-Resin.
16. Dry protein pellets will have a pale white to yellow color, and appear shinier than the wet proteins. Drying the pellets for too long will hinder resolving of the proteins.
17. Harvesting conditions are organism-dependent. Choose as gentle conditions as possible for effective cell harvesting. Cell cultures with very low cell density will require higher centrifugation speed or longer duration of the centrifugation. If the peptide concentration in the final sample is not sufficient you can try to use more cells for the protocol.
18. If the results are inadequate, for example due to a high amount of cytosolic contaminants, you can perform a control experiment. For this, use the protocol as described above, but add trypsin after the isolation of the supernatant. With this you can verify whether you have a problem with cell lysis or unspecific extraction of proteins from the cell.
19. To minimize cell lysis resuspend the cells carefully with a pipette and avoid extensive vortexing. If the organism you work with is extremely sensitive to osmotic stress, you can try to reduce cell lysis by adding sucrose (up to 10% [w/v]) to the washing and digestion buffer. If you add sucrose or other protective agents to the digestion buffer you will need to

clean your samples by ZipTip or StageTip purification before analysis.

20. Some organisms may require longer incubation times, for example, if the cells are protected by a polysaccharide capsule. If you use more cells you should increase the sample volume and also the trypsin amount (use always nine volumes digestion buffer and one volume trypsin solution).
21. If you observe a high number of peptides with missed cleavage sites, you should increase the trypsin amount at least for the second digestion step.

Acknowledgments

This work was supported by grants from the Deutsche Forschungsgemeinschaft (SFB TRR 34 and GRK 1870) and Top Institute Pharma projects T4-213 and T4-502.

References

1. Lin J, Huang S, Zhang Q (2002) Outer membrane proteins: key players for bacterial adaptation in host niches. Microbes Infect 4 (3):325–331
2. Foster TJ, McDevitt D (1994) Surface-associated proteins of *Staphylococcus aureus*: their possible roles in virulence. FEMS Microbiol Lett 118(3):199–205
3. Gordon RJ, Lowy FD (2008) Pathogenesis of methicillin-resistant *Staphylococcus aureus* infection. Clin Infect Dis Off Publ Infect Dis Soc Am 46(Suppl 5):S350–S359
4. Dreisbach A, van Dijl JM, Buist G (2011) The cell surface proteome of *Staphylococcus aureus*. Proteomics 11(15):3154–3168
5. Hempel K, Herbst F-A, Moche M, Hecker M, Becher D (2011) Quantitative proteomic view on secreted, cell surface-associated, and cytoplasmic proteins of the methicillin-resistant human pathogen *Staphylococcus aureus* under iron-limited conditions. J Proteome Res 10 (4):1657–1666
6. Green NM (1975) Avidin. Adv Protein Chem 29:85–133
7. Scheurer SB, Rybak J-N, Roesli C, Brunisholz RA, Potthast F, Schlapbach R, Neri D, Elia G (2005) Identification and relative quantification of membrane proteins by surface biotinylation and two-dimensional peptide mapping. Proteomics 5(11):2718–2728
8. Rodríguez-Ortega MJ, Norais N, Bensi G, Liberatori S, Capo S, Mora M, Scarselli M, Doro F, Ferrari G, Garaguso I, Maggi T, Neumann A, Covre A, Telford JL, Grandi G (2006) Characterization and identification of vaccine candidate proteins through analysis of the group a *Streptococcus* surface proteome. Nat Biotechnol 24(2):191–197
9. Olaya-Abril A, Jiménez-Munguía I, Gómez-Gascón L, Rodríguez-Ortega MJ (2014) Surfomics: shaving live organisms for a fast proteomic identification of surface proteins. J Proteome 97:164–176
10. Dreisbach A, Hempel K, Buist G, Hecker M, Becher D, van Dijl JM (2010) Profiling the surfacome of *Staphylococcus aureus*. Proteomics 10(17):3082–3096
11. Otto A, Bernhardt J, Meyer H, Schaffer M, Herbst F-A, Siebourg J, Mäder U, Lalk M, Hecker M, Becher D (2010) Systems-wide temporal proteomic profiling in glucose-starved *Bacillus subtilis*. Nat Commun 1:137
12. Pribyl T, Moche M, Dreisbach A, Bijlsma JJE, Saleh M, Abdullah MR, Hecker M, van Dijl JM, Becher D, Hammerschmidt S (2014) Influence of impaired lipoprotein biogenesis on surface and exoproteome of *Streptococcus pneumoniae*. J Proteome Res 13(2):650–667

Chapter 5

Preparation of Bacterial Magnetosomes for Proteome Analysis

Oliver Raschdorf, Dirk Schüler, and René Uebe

Abstract

Magnetotactic bacteria form unique prokaryotic organelles, termed magnetosomes, which consist of membrane-enclosed magnetite nanoparticles. Analysis of magnetosome biogenesis has been greatly facilitated by proteomic methods. These, however, require pure, highly enriched magnetosomes. Here, we describe the purification of magnetosomes from *Magnetospirillum gryphiswaldense* using high pressure cell disruption, and sequential purification by magnetic enrichment and sucrose density ultracentrifugation. The resulting enriched magnetosomes can be subsequently subjected to proteomic analyses or biotechnological applications.

Key words Prokaryotic organelles, Magnetosomes, Magnetite nanoparticles, Magnetic separation, Ultracentrifugation, Biotechnological application

1 Introduction

For many years prokaryotes were thought to be very simple organisms that lack any intracellular compartmentalization. However, the discovery of protein-shell microcompartments (e.g., carboxysomes), thylakoid membranes or chlorosomes of photosynthetic bacteria showed that many bacteria are able to form intracellular compartments or even organelles [1]. One of the most intriguing examples of prokaryotic organelles can be found within magnetotactic bacteria (MTB). This diverse group of prokaryotes shares the ability to form linear chains of nanometer-sized, membrane-enclosed magnetic iron minerals (magnetite or greigite) termed magnetosomes. These organelles endow MTBs the ability to orient themselves along geomagnetic field lines within their natural habitat, the oxic–anoxic transition zone of aquatic sediments [2].

Based on work carried out with the α-proteobacterial model MTB *Magnetospirillum gryphiswaldense* (Fig. 1) it is known that magnetite biomineralization within magnetosome vesicles requires microaerobic or anaerobic conditions [3]. Biomineralization of

Dörte Becher (ed.), *Microbial Proteomics: Methods and Protocols*, Methods in Molecular Biology, vol. 1841, https://doi.org/10.1007/978-1-4939-8695-8_5,

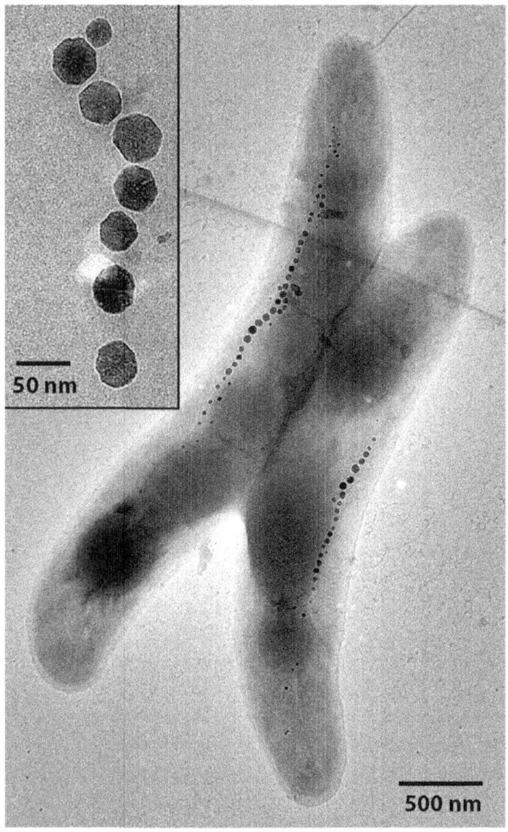

Fig. 1 Transmission electron microscopy (TEM) micrograph of dehydrated *M. gryphiswaldense* cells containing chains of nanometer-sized magnetosomes. Inlet shows magnetosome crystals in higher magnification. On average approximately 35 magnetite particles with a diameter of 35 nm are formed per cell

magnetite (Fe_3O_4) proceeds within phospholipid membrane vesicles by control of a set of >30 specific proteins in or at the magnetosome membrane [4, 5]. These magnetosome membrane proteins have been named Mam (*ma*gnetosome *m*embrane), Mms (*m*agnetosome *m*embrane *s*pecific), Mmx (*m*agnetosome *m*embrane unknown function), Mme (*m*agnetosome *me*mbrane), or Mtx (*m*agneto*tax*is) and share homology to tetratricopeptide repeat (TPR) proteins (MamA), transporters of the ubiquitous cation diffusion facilitator (CDF) (MamB and MamM) or major facilitator (MFS) (MamH and MamZ) superfamilies, HtrA-like serine proteases (MamE, MamO), cytochrome *c*-like proteins (magnetochromes MamE, MamP, MamT, MamX), and actin-like proteins (MamK). Additionally, several magnetosome specific proteins have no detectable sequence similarity to any other known protein family (e.g., MamC, Mms6). Recent reports showed that the assembly of magnetosomes occurs in a step-wise manner involving cytoplasmic membrane invagination, magnetosome vesicle formation, targeting of various specific proteins, transport of large amounts of iron into the vesicles, and crystallization of magnetite [6–9]. The controlled biomineralization within magnetosomes results in chemically and structurally perfect magnetite nanoparticles in the single-magnetic-domain size range [10]. Consequently, the exceptional magnetic properties of magnetosomes attracted many interests for their putative use in biomedical and biotechnological applications. For example, magnetosomes were already shown to be potent MRI contrast agents [11] and have been successfully tested for applications such as alternative magnetic field cancer therapy [12]. Additionally, the use of magnetosomes for biotechnological application might be extended since they can be genetically or chemically functionalized to display foreign proteins or other active components on their surface [13–16].

A key technique for the use of magnetosomes in biotechnological applications and to study magnetosome biogenesis or to analyze the lipid/protein composition of the magnetosome membrane is the isolation of intact magnetosome organelles. Initial attempts to isolate bacterial magnetite nanoparticles were performed by the use of hypochlorite solutions and dissolution of all organic material within the MTB sample [17], which did not allow analyses of the biological components of magnetosomes. First successful analysis of the magnetosome membrane composition was achieved after magnetosome purification by use of their magnetic attraction by a permanent magnet [18]. However, despite extensive washing magnetically enriched magnetosomes still contained numerous contaminating proteins from other cellular fractions as determined by sodium dodecyl sulfate–polyacrylamide gel electrophoresis (SDS-PAGE) [18]. Therefore, Grünberg et al. made use of the high density of the magnetite cores to further enrich magnetosomes by sucrose density ultracentrifugation [5, 19].

Here we describe a refined method for the isolation of highly purified magnetosomes from *M. gryphiswaldense* and present additional steps to further purify or solubilize magnetosome membrane proteins. Best results in terms of purity and yield were obtained using high pressure cell disruption methods, subsequent magnetic enrichment, and sucrose density ultracentrifugation. The purified magnetosomes can be directly applied for 1D and 2D SDS-PAGE and subsequently for proteomic analysis [5].

2 Materials

Several different MTB have been used to isolate membrane-enclosed magnetic nanoparticles [20, 21]. Here we describe a protocol that is optimized for the cultivation and isolation of magnetosomes from *M. gryphiswaldense*. For purification of magnetite nanoparticles from other MTB, growth media and conditions need to be adapted accordingly.

2.1 Growth Medium and Cultivation

1. HEPES (4-(2-Hydroxyethyl)piperazine-1-ethanesulfonic acid, 1 M, pH 7.0): Add about 50 mL ddH_2O to a 1-L graduated cylinder. Weigh 23.8 g HEPES and transfer to the cylinder. Add water to a volume of 90 mL. Mix and adjust pH with 10 M NaOH. Make up to 100 mL with ddH_2O. Autoclave 121 °C for 20 min. Store at 4 °C.
2. Sodium nitrate: 1 M $NaNO_3$. Weigh 8.5 g $NaNO_3$ and prepare a 100 mL solution as described in **item 1**. Autoclave 121 °C for 20 min. Store at 4 °C.
3. Potassium dihydrogen phosphate: 0.74 M KH_2PO_4. Weigh 10.07 g KH_2PO_4 and prepare a 100 mL solution as described in **item 1**. Autoclave 121 °C for 20 min. Store at 4 °C.
4. Magnesium sulfate heptahydrate: 0.1 M $MgSO_4{\cdot}7H_2O$. Weigh 3.47 g $MgSO_4{\cdot}7H_2O$ and prepare a 100 mL solution as described in **item 1**. Autoclave 121 °C for 20 min. Store at 4 °C.
5. Ferric citrate monohydrate: 10 mM ferric citrate monohydrate. Weigh 263 mg ferric citrate and add 100 mL ddH_2O. Autoclave 121 °C for 20 min (*see* **Note 1**). Store at 4 °C.
6. Soy bean peptone (e.g., Applichem GmbH, Darmstadt, Germany; *see* **Note 2**).
7. Yeast extract.
8. Potassium lactate (50% (w/v)).
9. Flask standard medium (FSM): Add 3 L ddH_2O to a beaker. Add 60 mL of 1 M HEPES, 18 mL 50% (w/v) potassium lactate, 24 mL 1 M sodium nitrate, 6 mL 0.74 M potassium dihydrogen phosphate, 30 mL 0.1 M magnesium sulfate heptahydrate ($MgSO_4{\cdot}7H_2O$), 30 mL ferric citrate monohydrate, 18 g soy bean peptone, and 0.6 g yeast extract. Mix and fill up with ddH_2O to 5.5 L and adjust pH to 7 with 10 M NaOH. Make up to 6 L with ddH_2O. Distribute medium: 1 × 5 L to a 10 L flask, 1 × 500 mL to a 1 L, 1 × 300 mL to a 500 mL flask and 1 × 100 mL to a 250 mL flask. Fill residual medium à 10 mL into 20 mL Hungate tubes, purge 3 min with 2% O_2/98% N_2 gas mixture and close tubes with butyl rubber stoppers. Add 4.5 g agar-agar to the flask containing 300 mL FSM. Autoclave 121 °C for 20 min. Pour agar into sterile petri dishes. Store at 4 °C.
10. 96- and 24-well plates.
11. Anaerobic jar.
12. Incubation shaker (e.g., New Brunswick, Innova 4330, Edison, USA).

2.2 Cell Harvest and Magnetosome Isolation

1. HEPES (500 mM, pH 7.4): Add about 500 mL ddH_2O to a 1-L graduated cylinder. Weigh 119.15 g HEPES and transfer to the cylinder. Add water to a volume of 800 mL. Mix and adjust pH with 10 M NaOH. Fill up to 1 L with water. Autoclave 121 °C for 20 min. Store at 4 °C.
2. EDTA (Ethylenediaminetetraacetic acid disodium salt, 500 mM): Weigh 18.61 g and dissolve in 60 mL ddH_2O at pH 4 to 7. Adjust to pH 7.4 by slowly adding NaOH-pellets. Fill to 100 mL with ddH_2O.
3. PMSF: 100 mM phenylmethanesulfonylfluoride. Weigh 17.4 mg and dissolve in 1 mL of 99.9% Isopropanol, store at −20 °C. To be added always directly before use (*see* **Note 3**)!
4. Sodium chloride: 1 M NaCl. Weigh 58.44 g NaCl and dissolve in 1 L ddH_2O.
5. Wash buffer (WB): 20 mM HEPES pH 7.4, 5 mM EDTA. Dilute HEPES stock solution (1:25) and EDTA stock solution (1:100) with ddH_2O.
6. Cell resuspension buffer (RB): 50 mM HEPES, 1 mM EDTA, 0.1 mM PMSF. Dilute HEPES (1:10) and EDTA (1:500) stock solutions. Add 1 μL/mL of PMSF stock immediately before use.
7. Extraction buffer (EB): 10 mM HEPES, 1 mM EDTA, 0.1 mM PMSF. Dilute HEPES stock solution (1:50) and EDTA stock solution (1:500). Add 1 μL/mL of PMSF stock immediately before use.
8. High salt buffer (HSB): 10 mM HEPES, 1 mM EDTA, 200 mM NaCl, 0.1 mM PMSF. Dilute stock solutions accordingly. Add 1 μL/mL of PMSF stock immediately before use.
9. Sucrose solution: 100 mL of 60% (w/w) sucrose solution in EB. Weigh 77.7 g of sucrose in a glass beaker and add ~50 mL of EB. Cover beaker with aluminum foil and place on a heated magnetic stirrer for approximately half an hour at 70 °C. After the sucrose is completely dissolved add EB to a final volume of 100 mL. Do not autoclave. Store at 4 °C.
10. High capacity/performance centrifuge (e.g., Beckman Avanti J-26XP with J-LITE JLA-8.1000 fixed angle rotor).
11. 1000 mL polypropylene centrifugation bottles (e.g., Beckman) and 50 mL Falcon tubes.
12. Laboratory benchtop centrifuge (e.g., Beckman Allegra X-15R with SX4750 swinging bucket rotor and 50 mL-Falcon tube adapters).
13. Cell disruption system (e.g., French press).

14. MACS cell separation column type CS (e.g., Art. No. 130-041-201, Miltenyi Biotec GmbH, Bergisch Gladbach, Germany).
15. Two neodymium iron boron cube permanent magnets.
16. Tripod/clamp stand.
17. Beaker.
18. Ultracentrifuge (e.g., Thermo Scientific Sorvall WX80 with T-1250 rotor and tubes).
19. Water-jet pump.
20. Syringe (60 mL).
21. Stopcock.

2.3 Magnetosome Membrane Enrichment

1. Carbonate buffer: 200 mM Na_2CO_3, 10 mM EDTA, 1 mM PMSF, pH 11.0. Weigh 2.12 g of Na_2CO_3 and add to approx. 80 mL of ddH_2O. Add 0.29 g of EDTA to the solution. Adjust pH to 11.0 using HCl, if necessary and fill up to 100 mL with ddH_2O. Solution can be stored at 4 °C for 4 weeks. Add 1 mL PMSF from 100 mM stock solution immediately before use.
2. TRIS-high-salt buffer (THB): 20 mM Tris, 1 M NaCl, 10 mM EDTA, 1 mM PMSF, pH 7.5. Weigh in 0.24 g of Tris Base and dissolve in approx. 80 mL of H_2O. Add 5.84 g of NaCl and 0.29 g of EDTA to solution. Adjust pH to 7.5 with HCl and fill up to 100 mL with ddH_2O. Solution can be stored at 4 °C for 4 weeks. Add 1 mL PMSF from 100 mM stock solution immediately before use.
3. TEAB buffer: 50 mM (triethylammonium bicarbonate), pH 7.8. Dilute 5 mL 1 M stock solution (e.g., Sigma-Aldrich) with 80 mL ddH_2O. Adjust pH to 7.8 with HCl and fill up to 100 mL with ddH_2O. Solution can be stored at 4 °C for 4 weeks.

2.4 Solubilization of Magnetosome Membrane Proteins

1. *n*-Dodecyl-ß-D-maltoside (DDM) stock solution (20%): Weigh in 100 mg of DDM and dissolve in 1 mL of ddH_2O, store at −20 °C for max. 1 year.

3 Methods

3.1 Cell Cultivation and Cell Harvest

1. Streak out *M. gryphiswaldense* from cryostocks on FSM-agar plates. Incubate plate under microaerobic atmosphere (2% O_2) in jars or incubators at 28 °C for 5 days.
2. Transfer a dark brownish looking single colony (indicates the presence of magnetosomes) to a 96-well plate filled with 100 μL FSM using a toothpick. Incubate at 28 °C for 24 h in anaerobic jar in microaerobic atmosphere (2% O_2).

3. Transfer complete culture volume of the 96-well culture to 900 μL fresh FSM in a 24-well plate. Incubate at 28 °C for 24 h in anaerobic jar in microaerobic atmosphere (2% O_2).
4. Transfer complete volume of the 24-well plate culture to a microaerobic Hungate tube filled with 10 mL FSM (*see* **Note 4**). Incubate at 28 °C for 24 h with mild shaking (120 rpm).
5. Transfer complete volume of the Hungate tube to a 250 mL flask filled with 100 mL fresh FSM. Incubate at 28 °C for 24 h with mild shaking (120 rpm).
6. Transfer 50 mL of the 100 mL culture to a 1 L flask with 500 mL fresh FSM. Incubate at 28 °C for 24 h with mild shaking (120 rpm).
7. Transfer complete 550 mL culture to a 10 L flask filled with 5 L FSM. Incubate at 28 °C for 17 h with moderate shaking (120 rpm).
8. Harvest cells by centrifugation at exponential growth phase (OD_{565} approx. 0.250) in a precooled (4 °C) high performance/capacity centrifuge. Therefore, transfer culture into 1000 mL centrifuge bottles and centrifuge at 10,000 × *g* for 15 min, 4 °C.
9. Resuspend cell pellets in ice-cold WB and transfer cell suspensions to a single centrifuge bottle. Centrifuge at 10,000 × *g* for 15 min, 4 °C.
10. Repeat this washing step once.
11. Weigh pellet to estimate cell yield, resuspend cells in 7 mL ice-cold RB per gram (fresh weight) of cells and proceed with Subheading 3.2. Alternatively, cell pellets can be frozen and stored at −20 °C until use.

3.2 Cell Disruption and Magnetosome Isolation

Carry out all steps at 4 °C and keep samples on ice!

1. To release magnetosomes disrupt cells by 3–5 passages through high pressure cell disruption systems (e.g., French press at 1.26×10^8 Pa; *see* **Note 5**). Check disruption efficiency by microscopy (*see* **Note 6**).
2. Transfer cell lysate to 50 mL falcon tubes and centrifuge for 10 min at 210 × *g* in a precooled laboratory benchtop centrifuge to remove cell debris and PHB granules (*see* **Note** 7).
3. Gently transfer supernatant to a new precooled tube or bottle.
4. Place a CS type MACS cell separation column between two permanent magnets at a clamp stand (*see* **Note 8**). Remove the plunger of a 60 mL syringe and connect a stopcock to the needle adapter of the barrel. Mount the barrel directly over the MACS column at the clamp stand. Place a beaker underneath the column (Fig. 2).

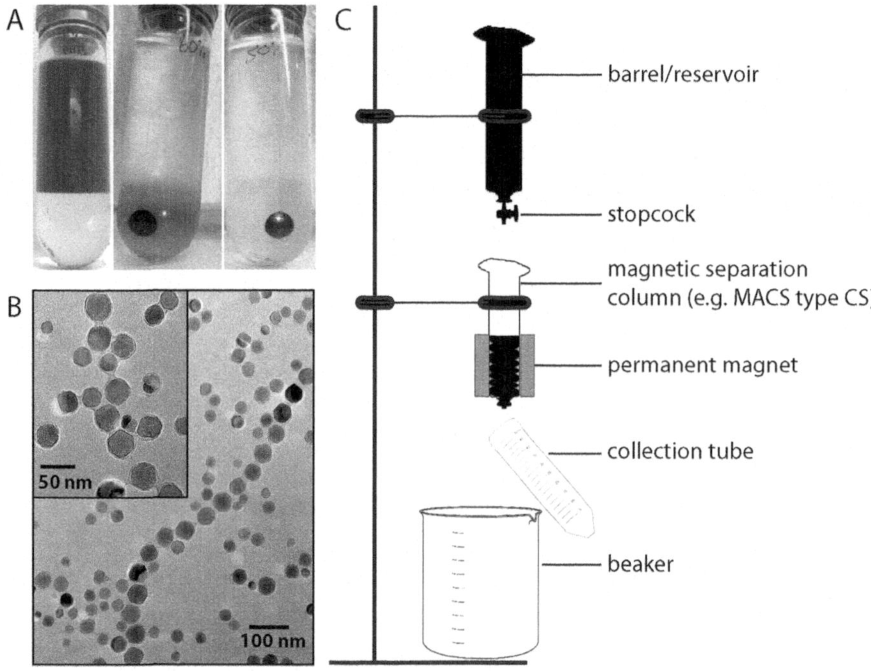

Fig. 2 Magnetosome isolation. (**a**) crude magnetosome suspension, obtained from magnetic separation, overlaying 60% sucrose cushion before ultracentrifugation (left image). Magnetosome pellets after centrifugation through 60% (center image) and 50% (right image) sucrose cushion. Lower sucrose concentrations may increase magnetosome yield (note brownish color of sucrose phase in center image, indicating nonpelleted magnetosomes) on the expense of purity of the sample. (**b**) TEM micrographs of isolated magnetosomes. Note that magnetosomes are often associated with organic material matrix (inlet) and are sometimes organized in chains. (**c**) Experimental setup for magnetosome separation. A MACS column is placed between two neodymium iron boron permanent magnets. A buffer reservoir is placed over the column and closed with a stopcock. Gravity flow is applied to load and wash the column

5. Close stopcock and fill barrel with 50 mL ice-cold EB. Slowly open stopcock to rinse and equilibrate column with EB.
6. Close stopcock and fill barrel with cleared cell lysate. Slowly open stopcock to rinse column with cell lysate. Magnetosomes are retained on the column. The nonmagnetic fraction flows through the column and might be collected with a falcon tube for further analyses (*see* **Note 9**).
7. Close stopcock and fill barrel with 50 mL ice-cold EB. Slowly open stopcock to wash the column.
8. Close stopcock and fill barrel with 50 mL ice-cold HSB. Slowly open stopcock to wash the column and remove contaminating proteins bound by electrostatic interactions.
9. Close stopcock and fill barrel with 50 mL ice-cold EB. Slowly open stopcock to remove salt from the high salt buffer from the column.

10. Remove magnets from the column (*see* **Note 10**).
11. Close stopcock and fill barrel with 30 mL ice-cold ddH_2O. Slowly open stopcock to elute magnetosomes from the column. Collect magnetosomes with Falcon tube.
12. Add HEPES, EDTA, and PMSF to enriched magnetosome suspension to final concentrations of 10, 1, and 0.1 mM, respectively.
13. Fill 7 mL sucrose (60% w/w) into ultracentrifuge tubes and carefully overlay with magnetosome suspension without destroying the sucrose phase. Carefully balance tubes with RB, if required (Fig. 2).
14. Centrifuge tubes at 4 °C with 230,000 × *g* for 2 h (*see* **Note 11**).
15. Remove supernatant and sucrose phase using a 1 mL pipette tip and a water jet pump.
16. Resuspend pelleted magnetosomes in EB. Purified magnetosome can now be subjected to downstream proteomic analyses or frozen at −20 °C until use (*see* **Note 12**).

3.3 Magnetosome Membrane Enrichment

All procedures are carried out at 4 °C.

1. Thaw/use 0.5 mL magnetosome suspension from standard magnetosome isolation protocol, spin down by ultracentrifugation at 100,000 × *g* for 30 min and discard supernatant. Resuspend magnetosomes in 2 mL carbonate buffer and incubate for 30 min with mild shaking (*see* **Note 13**).
2. Spin magnetosomes down with ultracentrifuge at 100,000 × *g* for 30 min and discard supernatant. Resuspend in 2 mL THB and incubate for 30 min with mild shaking.
3. Spin magnetosomes down with ultracentrifuge at 100,000 × *g* for 30 min and discard supernatant. Resuspend in 2 mL TEAB and immediately spin down using ultracentrifuge at 100,000 × *g* for 30 min. After discarding the supernatant, the pellet might be stored at 20 °C or directly used for further analysis (*see* **Note 14**).

3.4 Solubilization of Magnetosome Membrane Proteins

All procedures are carried at 4 °C.

1. Add 50 μL of 20% DDM solution to 1 mL of isolated magnetosomes (1% final concentration) and incubate at 4 °C for 10 min with mild shaking.
2. Spin down magnetosomes with ultracentrifuge at 100,000 × *g* for 30 min. Supernatant contains solubilized magnetosome proteins (Fig. 3) (*see* **Note 15**).

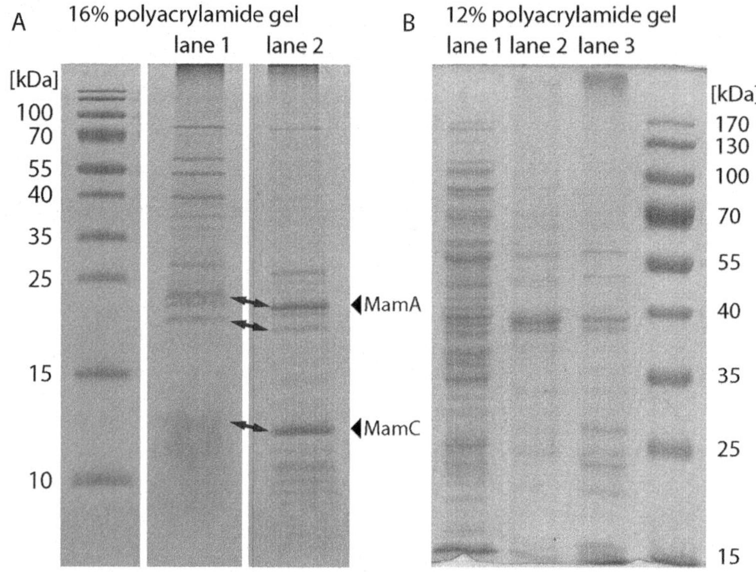

Fig. 3 SDS-PAGE of magnetosomes and other cellular fractions of *M. gryphiswaldense*. Approximate molecular weights of marker bands are given in kDa. (**a**) DDM-solubilized magnetosome proteins [lane 1] and residual (nonsolubilized and pelleted) magnetosomes [lane 2] on 16% polyacrylamide gel. Highly abundant proteins, like MamC and MamA can be directly detected in the gels. Note the size shift of comparable proteins bands in both samples, caused by the DDM solubilization. Many magnetosome proteins are of small size and therefore run close to the running front in SDS-PAGE. The solubilization efficiency of proteins varies greatly. (**b**) Different cellular fractions of *M. gryphiswaldense* on 12% polyacrylamide gel. From left to right: total soluble protein fraction [lane 1], total membrane fraction [lane 2], and magnetosome membrane fraction from isolated particles (MM) [lane 3]. Note the differential resolution of low molecular weight MM proteins, compared to 16% polyacrylamide gels (**a**). The most abundant specific MM proteins are found in between approx. 28 kDa and the running front at 15 kDa, abundant higher molecular weight bands are often shared between the MM and total membrane fractions and, therefore, most likely represent contaminations

4 Notes

1. Ferric citrate is only poorly soluble in water at room temperature. Therefore, autoclaving the solution is not only required for sterilization, but also for solubilization of ferric citrate.
2. Soy bean peptone is a complex medium component whose composition strongly varies between different suppliers or even different batches of the same supplier. Depending on the used peptone we observed huge differences in growth yield. Thus, different peptones sources should be tested prior to the magnetosome isolation.

3. PMSF is cytotoxic and should be handled with care. PMSF is highly unstable in aqueous solutions with reported half-lives as short as 35 min [22]. Therefore, PMSF should be dissolved in isopropanol, stored at 4 °C or −20 °C and added immediately before use.
4. Magnetosome formation in *M. gryphiswaldense* requires anoxic or microoxic conditions. Thus Hungate tubes and flasks should be purged with a 2% O_2/98% N_2 gas mixture. Alternatively, anoxic or microoxic conditions in the flasks are created by oxygen consumption of the cells at higher cell densities. Make sure that cultivation flasks are shaken only at moderate frequencies to minimize oxygen diffusion into the medium.
5. It has been found that the device used for cell lysis is critical to obtain reproducible results, e.g., with respect to the protein profile of the magnetosome membrane. Best and most reproducible results are obtained with a French press. However, depending on the downstream application, alternative methods such as constant cell disruption systems (CCDS), lysozyme treatment, and ultrasonication can be used alternatively. To guarantee highest magnetosome yield no clumps must be present in the cell suspension. Therefore, the cell suspension should be passed through a 20 G (yellow) needle using a 50-mL-syringe or resuspended using a Potter-Elvehjem homogenizer.
6. Cell lysis should be controlled by light microscopy and should be as complete as possible. Remaining, intact cells will copurify in the magnetic column and will also contaminate the magnetosome pellet after sucrose-ultracentrifugation and therefore could introduce protein contaminants in later mass-spectrometry analysis.
7. *M. gryphiswaldense* produces large amounts of polyhydroxybutyrate (PHB) granules that interfere with magnetosome preparation. To remove PHB granules from cell lysates centrifugation at very low *g*-forces is required as otherwise cosedimentation of magnetosomes, PHB granula, and cell debris would take place. Be careful with decanting the cleared cell lysate as the PHB pellet is very soft and is easily resuspended.
8. Be careful when placing the column between magnets. Neodymium iron boron magnets have a very strong attraction, which in case of unsafe handling could lead to bruises of fingers or skin between the magnets. Large magnets can even cause bone fractures due to their strength. Wear protective goggles as neodymium magnets are brittle and might shatter when they slam together.
9. Use gravity flow to load and wash magnetized column during the isolation process. Applied pressure might wash of

magnetosomes from column to decrease yield and will further contaminate nonmagnetic fraction. The nonmagnetic fraction will, however, always contain contaminating magnetosome membranes from empty magnetosome vesicles.

10. Do not forget to remove magnets before eluting magnetosomes! This dramatically reduces the yield of magnetosomes even when magnets are removed and water is applied for a second elution step.
11. Occasionally, magnetosomes become stuck in the sucrose cushion during ultracentrifugation. To improve yield, lower concentrated sucrose might be used at the expense of lower purity of the sample (Fig. 2a).
12. Even in highly enriched magnetosome membrane fractions contaminations with abundant cellular components like outer membrane proteins, ATPase subunits or ribosomal proteins are frequently observed.
13. 2% Triton X-100 [5] and 0.1 M CAPS-NaOH buffer (pH 11.0) [23] have also been used for selective solubilization/enrichment of magnetosome membrane proteins and might be used as an alternative for DDM and Na_2CO_3.
14. Several magnetosome proteins are very small and have a predicted or experimentally determined molecular weight of 7–15 kDa. If small magnetosome proteins are of special interest, higher concentrations (>15%) of polyacrylamide in SDS-PAGE gels might be advisable for protein separation.
15. Boiling magnetosomes in SDS-PAGE sample buffer might destabilize and change the magnetosome profile in comparison to nonboiled samples. It is therefore advisable to heat samples only to 60 °C for 20 min before use with SDS-PAGE.

Acknowledgment

We thank Florian Bonn and Dörte Becher (University of Greifswald) for help with developing membrane enrichment protocol. This work was supported by the Human Frontier Science Foundation (grant RGP0052/2012) and the Deutsche Forschungsgemeinschaft (grant Schu1080/15-1).

References

1. Murat D, Byrne M, Komeili A (2010) Cell biology of prokaryotic organelles. Cold Spring Harb Perspect Biol 2(10):a000422
2. Uebe R, Schüler D (2016) Magnetosome biogenesis in magnetotactic bacteria. Nat Rev Microbiol 14(10):621–637
3. Heyen U, Schüler D (2003) Growth and magnetosome formation by microaerophilic *Magnetospirillum* strains in an oxygen-controlled fermentor. Appl Microbiol Biotechnol 61 (5-6):536–544

4. Komeili A, Vali H, Beveridge TJ, Newman DK (2004) Magnetosome vesicles are present before magnetite formation, and MamA is required for their activation. Proc Natl Acad Sci U S A 101(11):3839–3844
5. Grünberg K, Müller EC, Otto A, Reszka R, Linder D, Kube M, Reinhardt R, Schüler D (2004) Biochemical and proteomic analysis of the magnetosome membrane in *Magnetospirillum gryphiswaldense*. Appl Environ Microbiol 70(2):1040–1050
6. Murat D, Quinlan A, Vali H, Komeili A (2010) Comprehensive genetic dissection of the magnetosome gene island reveals the step-wise assembly of a prokaryotic organelle. Proc Natl Acad Sci U S A 107(12):5593–5598
7. Lohße A, Borg S, Raschdorf O, Kolinko I, Tompa E, Posfai M, Faivre D, Baumgartner J, Schüler D (2014) Genetic dissection of the mamAB and mms6 operons reveals a gene set essential for magnetosome biogenesis in *Magnetospirillum gryphiswaldense*. J Bacteriol 196 (14):2658–2669
8. Lohße A, Ullrich S, Katzmann E, Borg S, Wanner G, Richter M, Voigt B, Schweder T, Schüler D (2011) Functional analysis of the magnetosome island in *Magnetospirillum gryphiswaldense:* the mamAB operon is sufficient for magnetite biomineralization. PLoS One 6 (10):e25561
9. Quinlan A, Murat D, Vali H, Komeili A (2011) The HtrA/DegP family protease MamE is a bifunctional protein with roles in magnetosome protein localization and magnetite biomineralization. Mol Microbiol 80 (4):1075–1087
10. Amor M, Busigny V, Durand-Dubief M, Tharaud M, Ona-Nguema G, Gelabert A, Alphandery E, Menguy N, Benedetti MF, Chebbi I, Guyot F (2015) Chemical signature of magnetotactic bacteria. Proc Natl Acad Sci U S A 112(6):1699–1703
11. Meriaux S, Boucher M, Marty B, Lalatonne Y, Preveral S, Motte L, Lefevre CT, Geffroy F, Lethimonnier F, Pean M, Garcia D, Adryanczyk-Perrier G, Pignol D, Ginet N (2015) Magnetosomes, biogenic magnetic nanomaterials for brain molecular imaging with 17.2 T MRI scanner. Adv Healthc Mater 4(7):1076–1083
12. Alphandery E, Faure S, Seksek O, Guyot F, Chebbi I (2011) Chains of magnetosomes extracted from AMB-1 magnetotactic bacteria for application in alternative magnetic field cancer therapy. ACS Nano 5(8):6279–6296
13. Alphandery E (2014) Applications of magnetosomes synthesized by magnetotactic bacteria in medicine. Front Bioeng Biotechnol 2:5
14. Pollithy A, Romer T, Lang C, Müller FD, Helma J, Leonhardt H, Rothbauer U, Schüler D (2011) Magnetosome expression of functional camelid antibody fragments (nanobodies) in *Magnetospirillum gryphiswaldense*. Appl Environ Microbiol 77(17):6165–6171
15. Lang C, Schüler D (2006) Biogenic nanoparticles: production, characterization, and application of bacterial magnetosomes. J Phys Condens Mat 18(38):S2815–S2828
16. Lang C, Schüler D, Faivre D (2007) Synthesis of magnetite nanoparticles for bio- and nanotechnology: genetic engineering and biomimetics of bacterial magnetosomes. Macromol Biosci 7(2):144–151
17. Mann S, Moench TT, Williams RJP (1984) A high-resolution electron-microscopic investigation of bacterial magnetite – implications for crystal-growth. Proc R Soc Ser B Biol 221 (1225):385–393
18. Gorby YA, Beveridge TJ, Blakemore RP (1988) Characterization of the bacterial magnetosome membrane. J Bacteriol 170 (2):834–841
19. Grünberg K, Wawer C, Tebo BM, Schüler D (2001) A large gene cluster encoding several magnetosome proteins is conserved in different species of magnetotactic bacteria. Appl Environ Microbiol 67(10):4573–4582
20. Matsunaga T, Nemoto M, Arakari A, Tanaka M (2009) Proteomic analysis of irregular, bullet-shaped magnetosomes in the sulphate-reducing magnetotactic bacterium *Desulfovibrio magneticus* RS-1. Proteomics 9 (12):3341–3352
21. Tanaka M, Okamura Y, Arakaki A, Tanaka T, Takeyama H, Matsunaga T (2006) Origin of magnetosome membrane: proteomic analysis of magnetosome membrane and comparison with cytoplasmic membrane. Proteomics 6 (19):5234–5247
22. James GT (1978) Inactivation of the protease inhibitor phenylmethylsulfonyl fluoride in buffers. Anal Biochem 86(2):574–579
23. Taoka A, Asada R, Sasaki H, Anzawa K, Wu LF, Fukumori Y (2006) Spatial localizations of Mam22 and Mam12 in the magnetosomes of *Magnetospirillum magnetotacticum*. J Bacteriol 188(11):3805–3812

Chapter 6

Analysis of Legionella Metabolism by Pathogen Vacuole Proteomics

Christian Manske, Ivo Finsel, Christine Hoffmann, and Hubert Hilbi

Abstract

The causative agent of Legionnaires' disease, *Legionella pneumophila*, replicates in free-living amoebae as well as in macrophages of the innate immune system within a distinct membrane-bound compartment, the "*Legionella*-containing-vacuole" (LCV). LCV formation is a complex process and requires the bacterial Icm/Dot type IV secretion system, which translocates approximately 300 different "effector" proteins. Intact LCVs from infected *Dictyostelium discoideum* amoebae or RAW 264.7 murine macrophages can be purified using a straightforward protocol. In the first step, the LCVs in cell homogenates are tagged with an antibody directed against an *L. pneumophila* effector protein specifically localizing to the pathogen vacuole membrane and isolated by immunomagnetic separation using a secondary antibody coupled to magnetic beads. In the second step, the LCVs are further enriched by density gradient centrifugation through a Histodenz cushion. LCVs thus purified are analyzed by mass spectrometry-based proteomics and characterized by biochemical and cell biological approaches.

Key words Amoeba, Density gradient centrifugation, *Dictyostelium discoideum*, Effector protein, Immunomagnetic separation, *Legionella pneumophila*, Macrophage, Metabolism, Pathogen vacuole, Type IV secretion

Abbreviations

ACES	*N*-(2-acetamido)-2-aminoethanesulfonic acid
GFP	Green fluorescent protein
HEPES	*N*-2-hydroxy-ethylpiperazine-*N′*-2-ethanesulfonic acid
Icm/Dot	Intracellular multiplication/defective organelle trafficking
LCV	*Legionella*-containing vacuole
T2SS	Type II secretion system
T4SS	Type IV secretion system

Dörte Becher (ed.), *Microbial Proteomics: Methods and Protocols*, Methods in Molecular Biology, vol. 1841, https://doi.org/10.1007/978-1-4939-8695-8_6, © Springer Science+Business Media, LLC, part of Springer Nature 2018

1 Introduction

1.1 Legionella pneumophila *Forms a Distinct Pathogen Vacuole*

Legionella pneumophila is a natural parasite of free-living protozoa. Upon inhalation of bacteria-containing aerosols, the opportunistic pathogen can infect and replicate in macrophages within the human lung, causing a severe pneumonia termed "Legionnaires' disease" [1]. Intracellular replication is accomplished using an apparently conserved mechanism that allows the bacteria to infect many different eukaryotic cells [1, 2]. In these cells, *L. pneumophila* replicates within a unique compartment, called the "*Legionella*-containing vacuole" (LCV). This membrane-bound pathogen compartment does not fuse with lysosomes, but communicates extensively with endosomal, secretory, and retrograde vesicle trafficking pathways, and thus constitutes the intracellular bacterial replication niche [3, 4]. Uptake of *L. pneumophila* and formation of the LCV are promoted by the Icm/Dot type IV secretion system (T4SS) that delivers approximately 300 different "effector" proteins into the host cell [5, 6]. Many of these effectors target central eukaryotic pathways by subverting small GTPases, phosphoinositide lipids, or a number of other host factors [7–10]. Given the plethora of effectors, *L. pneumophila* is a prime example of an intracellular pathogen that extensively manipulates and exploits host cell processes to ensure its own survival and growth.

1.2 Enrichment of Intracellular Pathogen Compartments

Intracellular pathogens in membrane-bound compartments avoid lysosomal degradation and gain access to a rich blend of potential carbon and energy sources [11]. In addition to *L. pneumophila*, intracellular pathogens such as *Salmonella*, *Chlamydia*, or *Mycobacterium* spp. form distinct vacuoles and thrive within eukaryotic cells. A major challenge for the analysis of the interactions between pathogen and host is that bacterial proteins are present in much lower concentrations compared to host cell proteins. This is of particular concern for proteomics or "omics" studies in general. The purification of pathogen vacuoles represents a powerful method to enrich bacterial proteins produced in the context of pathogen–host interplay. Pathogen compartments containing *Salmonella* [12, 13], *Chlamydia* [14], or *Mycobacterium* spp. [15] have been isolated and characterized to some extent. Recent proteomic analyses of distinct purified pathogen-modified vacuoles or membranes and a comparison of shared components focused on the role of host factors implicated in the formation of the pathogen compartments [16].

Here we describe the isolation and purification of intact LCVs, which are amenable to proteomics, biochemical, and cell biological investigations. We established a straightforward protocol to purify intact LCVs from different infected phagocytes [17–19]. To this end, we target the *L. pneumophila* Icm/Dot substrate SidC, which preferentially anchors to the cytoplasmic side of LCVs [20] by binding the host cell lipid phosphatidylinositol 4-phosphate (PtdIns(4)*P*) [21–23]. Using a primary antibody against SidC and a secondary antibody coupled to magnetic beads, intact LCVs are retained in a magnetic field and can be purified by immunomagnetic enrichment [17]. This step is followed by density gradient centrifugation to further purify LCVs. Quantity and integrity of purified LCVs can be monitored by fluorescence microscopy, as fluorescently labeled *L. pneumophila* and phagocytes are used (e.g., *D. discoideum* producing calnexin-GFP). This approach has been successfully applied to analyze the proteome of LCVs purified from *L. pneumophila*-infected *D. discoideum* amoebae [24] or RAW 264.7 macrophages [10].

1.3 Metabolism of L. pneumophila

L. pneumophila survives within various environmental niches, including biofilms and protozoa [25]. Free-living (planktonic or sessile) *L. pneumophila* encounter a plethora of possible carbon and energy sources; likewise, intracellular bacteria in infected host cells can also possibly use many different nutrients [11, 26–28]. Amino acids represent the preferred source of carbon and energy for *L. pneumophila*, and the pathogen is auxotrophic for cysteine, arginine, isoleucine, leucine, threonine, valine, methionine, and serine [29–31]. To compensate for these auxotrophies and to take up amino acids, *L. pneumophila* employs a variety of transporters, permeases, and proteases [32–35]. Moreover, during intracellular growth, genes for amino acid synthesis and transport are highly upregulated [36]. Finally, *L. pneumophila* exploits host cell amino acid transporters and the host proteolytic machinery [37, 38], resulting in a substantial uptake and the utilization of host cell-derived amino acids as carbon and energy sources [39].

In addition to amino acids, *L. pneumophila* is also able to metabolize carbohydrates and polysaccharides. The genomes of different *Legionella* species encode the components of the Embden–Meyerhof–Parnas (EMP) pathway, the Entner–Doudoroff (ED) pathway, and an incomplete pentose phosphate (PP) pathway [33, 40–42]. Using ^{14}C- and ^{13}C-labeled substrates, it was shown that *L. pneumophila* catabolizes glucose via the ED pathway [43, 44]. Besides glucose, metabolites such as α-ketoglutarate, pyruvate, acetate, or glycerol are also utilized by *L. pneumophila*, at least upon growing in media [29, 31, 45]. Isotopologue experiments using ^{13}C-glycerol and other carbon tracers revealed a bipartite metabolism, where glycerol and carbohydrates like glucose are mainly fed into anabolic processes, while serine serves as major energy supply [46]. Finally, complex polysaccharides might also contribute to *L. pneumophila* nutrition. Chitin, cellulose, glycogen, and starch were found to be substrates for secreted enzymes, although none of these were essential for intracellular growth of *L. pneumophila* [47–49].

1.4 Analysis of L. pneumophila Metabolism by Pathogen Vacuole Proteomics

The analysis of bacterial metabolism by pathogen vacuole proteomics offers the advantage that host and pathogen factors implicated in nutrient acquisition and degradation can be analyzed in the same isolated replication compartment. Mass spectrometry is the method of choice to determine the proteome of a pathogen vacuole, since this is a powerful tool to identify and also quantify a vast number of proteins from one sample at the same time [50].

LCV proteomes from different host cells—*D. discoideum* amoebae [24] or RAW 264.7 macrophages [10]—might also yield insights into the compartmentalisation of the intracellular metabolism of *L. pneumophila* (*see* Fig. 1). Metabolic processes occur in the host cytoplasm, the LCV membrane, the LCV lumen or the bacterium. Accordingly, different host and pathogen factors contribute to the intracellular bacterial metabolism, including

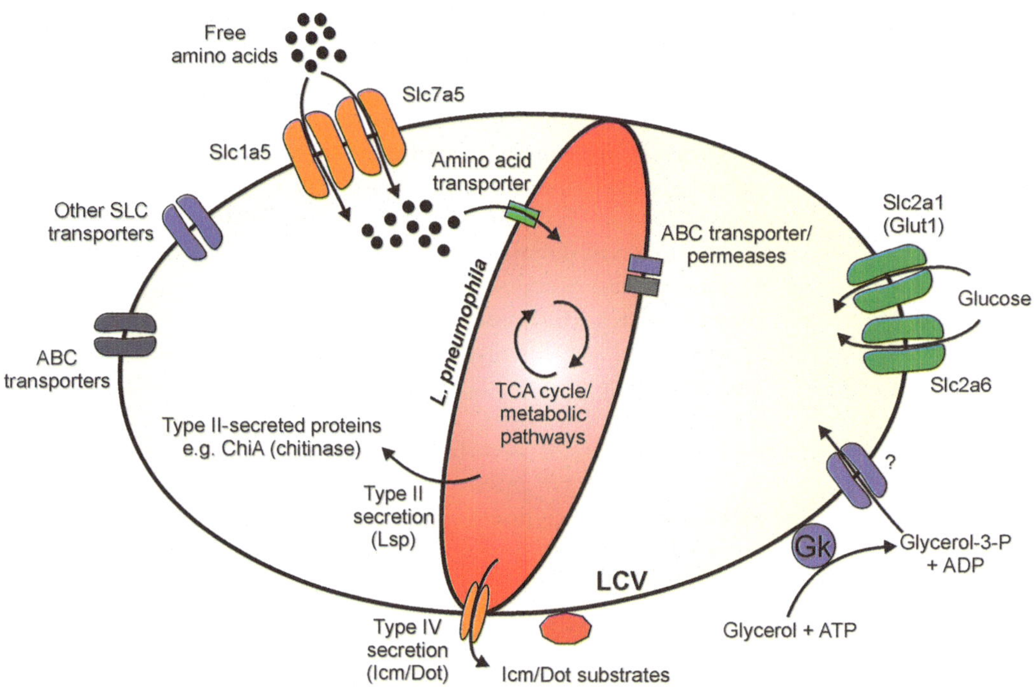

Fig. 1 Analysis of *L. pneumophila* metabolism by pathogen vacuole proteomics. Proteomes of LCVs isolated from infected *D. discoideum* amoebae or RAW 264.7 macrophages reveal the possible compartmentalization of the intracellular metabolism of *L. pneumophila*. The metabolic processes occur in the host cytoplasm, the LCV membrane, the LCV lumen, or the bacterium. Host cell transporters in the LCV membrane can facilitate the transport of different carbon sources or micronutrients into the LCV lumen. These include Slc2a1 and Slc2a6 (glucose), Slc1a5 and Slc7a5 (amino acids), or other Slc family transporters as well as ABC transporters. Host cytoplasmic enzymes, e.g., glycerol kinase Gk, are possibly "hijacked" by the pathogen for metabolic processes. Pathogen factors implicated in intracellular metabolism include type II-secreted hydrolases (ChiA, CelA, GamA) presumably active in the LCV lumen and type IV-secreted effectors that play a role in the formation of the replication-permissive LCV, as well as bacteria-associated transporters and metabolic enzymes implicated in the uptake and catabolism of substrates

eukaryotic transporters and enzymes, as well as bacterial effector proteins and enzymes.

The LCV proteome revealed a number of host and pathogen factors likely implicated in intracellular bacterial metabolism [10, 24]. Host factors identified on LCVs from macrophages included 20 members of the Slc transporter family, which facilitate different transport processes in eukaryotic cells. Among these, some proteins likely catalyze glucose transport (Slc2a1, Slc2a6) or (neutral) amino acid transport (Slc7a5, Slc1a5). Interestingly, Slc1a5 is highly upregulated in *L. pneumophila*-infected macrophages and essential for intracellular bacterial replication [37]. LCVs from infected *D. discoideum* also harbored different types of ABC transporters. Whereas transporters play an obvious role for intracellular replication of *L. pneumophila*, cytoplasmic host cell enzymes might be "hijacked" by the bacteria. For example, glycerol kinase catalyzes the phosphorylation of glycerol to glycerol-3-phosphate, which could then be imported into the LCV and used as a carbon source.

In the LCV lumen bacterial factors likely play a dominant role for *L. pneumophila* metabolism (*see* Fig. 1). In addition to the Icm/Dot T4SS, *L. pneumophila* also possesses the Lsp type II secretion system (T2SS) that is essential for intracellular replication in amoebae and macrophages [51, 52]. Through this T2SS the pathogen secretes an endoglucanase (CelA), a glycoamylase (GamA), and a chitinase (ChiA) [47–49], which might be implicated in the extracellular as well as the intracellular metabolism of *L. pneumophila*.

Proteins localizing to the bacterial cytoplasm obviously also play a vital role for the intracellular metabolism of *L. pneumophila*. A number of bacterial amino acid transporters, ABC transporters and permeases have indeed been identified in the LCV proteome and are expected to facilitate the uptake of nutrients [10]. Furthermore, enzymes of central metabolic pathways such as glycolysis or the TCA cycle provide clues about metabolic pathways active during intracellular growth. In the course of an infection different substrates might be preferentially used, and accordingly, the metabolic pathways might adapt to reflect the new nutrient availability. Finally, LCVs purified from infected *D. discoideum* or macrophages harbored several components of the Icm/Dot T4SS (19 or 9, respectively), as well as many Icm/Dot substrates (60 or 29, respectively) [10]. Since the LCVs were isolated 1 h post infection, these effector proteins are likely required for LCV formation early during infection. As the effector proteins govern the formation of the replication-permissive pathogen vacuole, they also play an essential role for intracellular metabolism of *L. pneumophila*. Collectively, MS-based proteomics of purified pathogen vacuoles represents an excellent starting point for a further hypothesis-driven analysis of intracellular bacterial metabolism.

2 Materials

2.1 Chemicals

1. Unless otherwise indicated, all chemicals are from Sigma-Aldrich. Deionized, distilled H_2O is used in all recipes and protocol steps.

2.2 *Legionella pneumophila*: Strain and Medium

1. The virulent *L. pneumophila* Philadelphia-1 strain JR32 harboring plasmid pSW001 is used [24, 53]. The bacteria constitutively produce the red fluorescent protein DsRed-Express under the control of the P_{tac} ($\Delta lacI^{q}$) promoter. Plasmid pSW001 also contains a chloramphenicol (Cm) resistance cassette allowing the selection of plasmid-bearing bacteria. *L. pneumophila* is a biosafety level 2 microorganism. The corresponding biosafety regulations need to be observed.
2. For extracellular growth of *L. pneumophila* ACES buffered yeast extract medium (AYE) is used [54].

   AYE medium: 10 g/L ACES, 10 g/L yeast extract, 3.3 mM L-cysteine, 0.6 mM $Fe(NO_3)_3$. Dissolve 10 g ACES (*N*-(2-acetamido)-2-aminoethane sulfonic acid) and 10 g Bacto™ yeast extract (Difco, *see* **Note 1**) in 950 mL H_2O. Add 0.4 g L-cysteine and 0.25 g $Fe(NO_3)_3$ each separately in 10 mL H_2O using two 15 mL reaction tubes and filter-sterilize both solutions. Then add the L-cysteine, followed by the iron solution. This has to be done drop-wise while stirring the medium to prevent precipitation. Adjust the pH to 6.9 using 10 M KOH (*see* **Note 2**), fill up with H_2O to 1 L and add 5 μg/L Cm to yield AYE/Cm. To remove precipitates (note that the medium will appear very dark), passage the medium through a layer of seven glass fiber filters using a suction filter and repeat the procedure once using new filter plates. At last, sterilize the medium by using a 0.2 μm filter cartouche and store it at 4 °C in the dark (*see* **Note 3**).
3. The common solid growth medium for *L. pneumophila* is buffered charcoal yeast extract (CYE) agar.

   CYE agar: 10 g/L ACES, 10 g/L yeast extract, 2 g/L activated charcoal powder, 15 g/L agar, 3.3 mM L-cysteine, 0.6 mM $Fe(NO_3)_3$. Dissolve 10 g ACES and 10 g yeast extract (Difco, *see* **Note 1**) in 950 mL H_2O, adjust pH to 6.9 using 10 M KOH and fill up to 1 L (*see* **Note 2**). Add 2 g activated charcoal powder (puriss. p.a.) and 15 g agar in a 1 L Schott bottle, followed by the ACES/yeast extract solution. Add a stir bar and autoclave the medium. After sterilization, let the solution cool down to 50 °C while stirring. Slowly add filter-sterilized L-cysteine and then $Fe(NO_3)_3$ solution (*see* **step 1**). Add 5 μg/L Cm to yield CYE/Cm, pour plates, let them dry for 1 day at room temperature and store them at 4 °C afterward.

4. Chloramphenicol (Cm) stock solution: 30 mg/mL Cm in ethanol. Filter-sterilize, make aliquots and store at −20 °C.

2.3 Dictyostelium discoideum: Cells and Medium

1. *D. discoideum* strain AX3 harboring plasmid pCalnexin-GFP is used [55, 56]. The amoeba strain constitutively produces calnexin-GFP under the control of the P_{act15} promoter (*see* **Note 4**).
2. For growth of *D. discoideum* modified HL5 medium is used [57].

   HL5 Medium: 0.5% yeast extract, 0.5% proteose peptone, 0.5% thiotone peptone, 56 mM glucose monohydrate, 1.3 mM Na_2HPO_4, 2.5 mM KH_2PO_4. Dissolve 5 g BBL™ yeast extract (Becton Dickinson, *see* **Note 1**), 5 g Bacto ™ proteose peptone (Becton Dickinson, *see* **Note 1**), 5 g BBL™ thiotone peptone (Becton Dickinson, *see* **Note 1**), 11 g D-glucose monohydrate, 0.35 g Na_2HPO_4 and 0.35 g KH_2PO_4 in 950 mL H_2O. Adjust pH to 6.5 using 1 M HCl and fill up to 1 L with H_2O. Autoclave the medium and store at 4 °C. During cell culture add 10 μg/mL geneticin (G418) to select for calnexin-GFP producing amoebae.
3. Sörensen phosphate buffer with $CaCl_2$ (SorC) [58]: 15 mM KH_2PO_4, 2 mM Na_2HPO_4, 50 μM $CaCl_2$. SorC is used as a washing and flow cytometry buffer for *D. discoideum*. Dissolve 2 g KH_2PO_4 and 0.29 g Na_2HPO_4 in 950 mL H_2O and adjust pH to 6.0 using 1 M HCl or 1 M KOH. Then add 1 mL of a 50 mM $CaCl_2$ stock solution, fill up to 1 L with H_2O, autoclave and store at room temperature.
4. Antibiotic stock solutions: 20 mg/mL geneticin (G418) in H_2O, 10,000 U/mL penicillin (Pen) in H_2O, 0.1 mg/mL streptomycin (Strep) in H_2O. Filter-sterilize and store stock solutions at −20 °C.

2.4 RAW 264.7 Macrophages: Cell Line and Medium

1. Abelson murine leukemia virus transformed macrophage cell line RAW 264.7 (ATCC, TIB-71™) is used. RAW 264.7 macrophages have to be handled at biosafety level 2. The corresponding biosafety regulations need to be observed.
2. For cultivation of RAW 264.7 macrophages, we use commercially available RPMI 1640 medium (Gibco, *see* **Note 5**) containing 2 mM L-glutamine and 10% heat-inactivated fetal calf serum (FCS). The medium is stored at 4 °C in the dark.
3. Phosphate buffered saline (PBS) buffer: 137 mM NaCl, 2.7 mM KCl, 10 mM Na_2HPO_4, 1.8 mM KH_2PO_4. PBS is used as a washing and flow cytometry buffer for macrophages. For a 10× stock solution dissolve 80 g NaCl, 2 g KCl, 14.4 g

Na_2HPO_4 and 2.4 g KH_2PO_4 in 950 mL H_2O and adjust pH to 7.4 using 1 M HCl. Fill up to 1 L with H_2O, autoclave and store at room temperature.

4. Antibiotic stock solutions: 10,000 U/mL Pen in H_2O, 10 mg/mL Strep in H_2O. Filter-sterilize and store at −20 °C.

2.5 Cell Lysis and LCV Purification

1. Homogenization (HS) buffer: 20 mM *N*-2-hydroxyethylpiperazine-*N*′-2-ethanesulfonic acid (HEPES), 250 mM sucrose, 0.5 mM ethyleneglycoltetraacetic acid (EGTA). HS buffer is used to lyse cells and stabilize LCVs. Adjust the pH to 7.2 with 1 M KOH, filter-sterilize using a 0.2 μm filter (do not autoclave), and store at 4 °C for up to 6 months. Freshly add a complete EDTA-free protease inhibitor cocktail tablet (Roche, cat. no. 11-836-170-001) according to the manufacturer's instructions (1 tablet/10 mL solution), and use the buffer ice cold.
2. FCS at a final concentration of 2% (v/v) is used as blocking reagent prior to treatment with first antibody.
3. Affinity-purified polyclonal rabbit anti-SidC serum (NeoMPS, [56]) is used as primary antibody. Store at 4 °C.
4. Magnet-activated cell sorting (MACS) goat anti-rabbit IgG micro beads (Miltenyi Biotec, cat. no. 130-048-602) is used as secondary antibody. Store at 4 °C. MACS MultiStand (Miltenyi Biotec, cat. No. 130-042-303) and MACS-MS columns (Miltenyi Biotec, cat. no. 130-042-201) are used to enrich intact LCVs.
5. 10% and 35% Histodenz solutions are used for density gradient centrifugation. Dissolve 10 or 35 g Histodenz in 100 mL PBS. Store at 4 °C.
6. Sterile, round microscopy coverslips are coated with sterile 0.01% poly-L-lysine solution. Store at 4 °C.
7. Microscopic samples are fixed with a 4% (w/v) paraformaldehyde (PFA) solution. Weigh 40 g PFA and dissolve in 1 L PBS, heating the solution to 50 °C while constantly stirring. Adjust pH to 7.3, make aliquots and store them at −20 °C. Perform all work under a chemical fume hood, and wear gloves and a mask, as PFA is toxic.

2.6 Additional Equipment and Consumption Items

1. Incubators: 23 °C (*D. discoideum* cell culture), 25 °C (*D. discoideum* infection), 37 °C with 5% CO_2 (macrophage cell culture and infection), 37 °C (*L. pneumophila* on CYE agar plates), 37 °C with rotation wheel or shaker (*L. pneumophila* in broth).
2. 75 cm^2 tissue culture flasks, 24-well flat-bottomed tissue culture plates.

3. 15 mL test tubes, 1.5 mL microcentrifuge tubes.
4. Plastic cell scraper, 3 mL Luer-lock syringes.
5. Stainless steel ball homogenizer (8 μm clearance, 0.5 mL chamber; Isobiotec).
6. Overhead rotation wheel at 4 °C.
7. Spectrophotometer.
8. Microscopy glass slides, suitable mounting medium (e.g., Vectashield; Vectalabs).
9. Glass Pasteur pipettes (230 mm, unplugged).

3 Methods

3.1 Cell Culture of D. discoideum

1. Inoculate 1×10^4 *D. discoideum* cells/mL in 10 mL HL5 medium in a 75 cm^2 tissue culture flask and incubate at 23 °C (*see* **Note 6**). After 2–3 days or when cells reach 80% confluence, subculture the cells by removing old medium, adding 10 mL new HL5 medium and tapping off the cells. Dilute the cells 1:100 in 10 mL HL5, transfer the cells to a new cell culture flask and incubate again at 23 °C. To select for the plasmid encoding calnexin-GFP add 20 μg/mL G418. To maintain sterility, optionally add 100 U/mL Pen and 0.1 mg/mL Strep.

3.2 Cell Culture of RAW 264.7 Macrophages

1. Grow RAW 264.7 macrophages in 10 mL RPMI 1640 medium containing 2 mM L-glutamine and 10% heat-inactivated FCS in a 75 cm^2 tissue culture flask at 37 °C and 5% CO_2. After 2–3 days or when cells reach 80% confluence, subculture the cells by removing old medium, adding 10 mL new RPMI 1640 medium and remove cells mechanically from the culture flask with a cell scraper. Dilute the cells 1:10 in 10 mL RPMI 1640 medium and transfer them to a new cell culture flask. Further incubate the cells at 37 °C and 5% CO_2. To maintain sterility, optionally add 100 U/mL Pen and 0.1 mg/mL Strep.

3.3 Growth of L. pneumophila

1. Four days before an experiment, streak out *L. pneumophila* harboring plasmid pSW001 from frozen glycerol stocks onto CYE/Cm plates, and incubate at 37 °C for 3 days. One day before an experiment prepare liquid cultures of *L. pneumophila* in AYE/Cm. To this end, suspend bacteria from CYE plates in 1 mL AYE and determine the cell count using a spectrophotometer. Inoculate 3 mL AYE/Cm medium with *L. pneumophila* at an optical density (OD_{600}) of 0.1 in a 15 mL plastic test tube, and incubate the bacteria at 37 °C on a rotation wheel for 21–22 h. The OD_{600} used for infections should be between 3.0 and 3.3, as bacteria are most virulent in the post-exponential growth phase (*see* **Note 7**).

3.4 Infection of Phagocytes with *L. pneumophila*

1. Seed out 1×10^7 *D. discoideum* or RAW 264.7 macrophages in 10 mL HL5 or RPMI 1640 medium, respectively, in a 75 cm^2 tissue culture flask 1 day before the experiment. Do not use Pen/Strep anymore from this point onward. Incubate the cells at 23 °C (*D. discoideum*) or 37 °C/5% CO_2 (macrophages) overnight. The phagocytes should have reached 80% confluence (~2×10^7 cells per flask), allowing for maximum uptake efficiency of *L. pneumophila*. Use three flasks per infection and sample (equals a total of ~6×10^7 cells).
2. Infect phagocytes with virulent *L. pneumophila* (*see* Subheading 3.3) at a multiplicity of infection (MOI) of around 50 (*see* **Note 8**).
3. Centrifuge the flasks at 25 °C and 500 × *g* for 10 min to synchronize infection.
4. Incubate the infected cells at 25 °C (*D. discoideum*) or 37 °C/5% CO_2 (RAW 264.7 macrophages) for 1 h (*see* **Note 9**; Fig. 2).

3.5 Homogenization of Infected Phagocytes

1. For homogenization perform all steps on ice.
2. After incubation for 1 h, place the flasks containing the infected cells on ice and remove the supernatant. Wash the cells 2–3 times by carefully adding 10 mL cold SorC (*D. discoideum*) or cold PBS (RAW 264.7 macrophages). Gently agitate and take off the supernatant. This step removes medium and nonphagocytosed bacteria (*see* **Note 9**).
3. Suspend 2×10^7 infected cells (i.e., 1 tissue culture flask) in 3 mL ice-cold HS buffer containing Roche complete EDTA-free protease inhibitor by releasing the cells from the bottom with a plastic cell scraper. Combine suspensions from all three flasks (total 9 mL) in a 15 mL test tube and put on ice. Set aside 180 μL for fluorescence microscopy.
4. Prepare the ball homogenizer (8 μm exclusion size). Wash all parts thoroughly with 70% ethanol and then extensively with H_2O before use. Put the homogenizer on ice and flush it once with cold HS buffer using a 3 mL disposable syringe to get rid of air within the apparatus.
5. Transfer the suspension from **step 3** into a 3 mL disposable Luer-lock syringe and mount the syringe onto the prechilled ball homogenizer. Press the suspension through the homogenizer into a second syringe and passage the suspension back and forth nine times (*see* **Note 10**). The homogenate will become clearer with every passage. Transfer the homogenate into a new 15 mL test tube and repeat this step until all 9 mL of a given cell suspension sample are homogenized. Set aside 180 μL homo-genate for fluorescence microscopy (*see* Fig. 2).

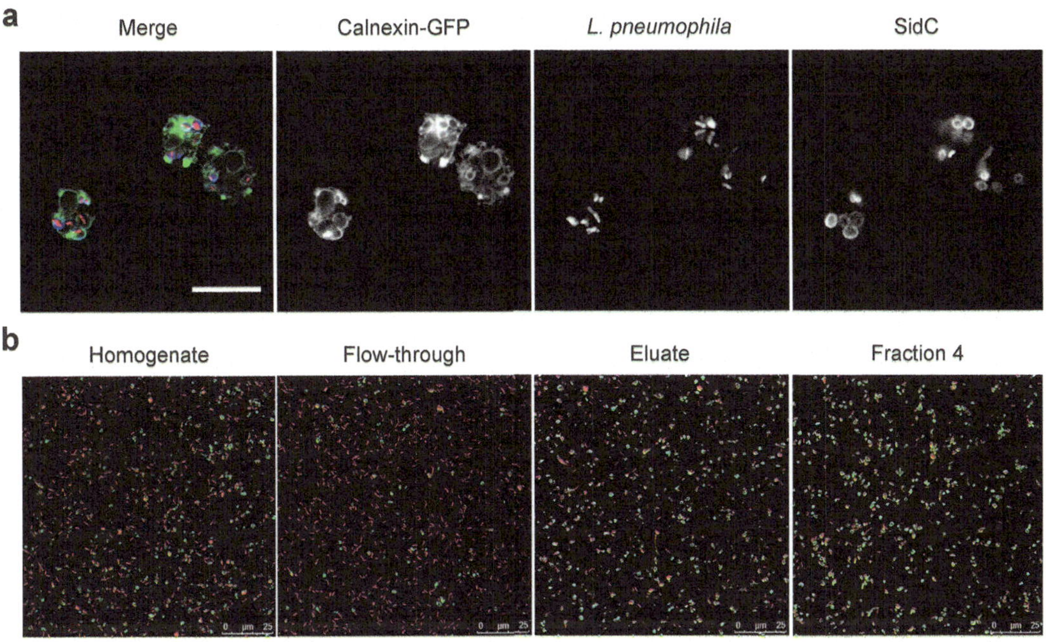

Fig. 2 (**a**) Intact calnexin-GFP producing *D. discoideum* amoebae infected with DsRed-labeled *L. pneumophila*. Cells were fixed with methanol and stained with a polyclonal anti-SidC antibody followed by a Cy5-coupled goat anti-rabbit antibody showing the recruitment of calnexin and SidC to LCVs. Bar, 10 μm. (**b**) Immunomagnetic separation of LCVs from calnexin-GFP producing *D. discoideum* infected with DsRed-labeled *L. pneumophila*. The images show homogenate of infected *D. discoideum* cells, the flow-through and eluate of a magnetic cell separation (MACS) column loaded with homogenate, and fraction 4, which contains enriched intact LCVs after Histodenz density gradient centrifugation

6. If more than one sample has to be homogenized, dismantle and thoroughly wash the ball homogenizer before proceeding with the next sample (*see* **step 4**).

3.6 Enrichment of LCVs by Immunomagnetic Separation

1. Add to the homogenate a final concentration of 2% (v/v) of FCS as blocking reagent, and incubate for 30 min at 4 °C on an overhead spinning wheel turning at 10–20 rpm (*see* **Note 11**).
2. As primary antibody we use an affinity-purified polyclonal rabbit antiserum against the *L. pneumophila* effector protein SidC [56, 59]. Vortex the serum prior to use, add at a dilution of 1:3000 to the homogenate and further incubate the suspension on the overhead spinning wheel for 1 h at 4 °C. Antibodies against other *L. pneumophila* markers can possibly be used, if the antigen specifically decorates the LCV membrane.
3. Centrifuge the homogenate at 2700 × *g* and 4 °C for 15 min. Remove the supernatant completely and suspend the pellet in 1.5 mL HS buffer. Transfer the suspension to a new 15 mL test tube.

4. Add the MACS goat anti-rabbit IgG microbeads. Vortex the magnetic beads, as they tend to sink to the bottom of the flask, and add 60 μL of the magnetic bead slurry to 1.5 mL homogenate. Incubate for 30 min at 4 °C on an overhead spinning wheel turning at 10–20 rpm.
5. Place a magnetic MACS-MS separation column into the magnetic field of the MACS separator, and equilibrate the column by gravity flow with 0.5 mL ice-cold HS buffer.
6. Load the 1.5 mL homogenate (corresponding to 6×10^7 phagocytes) onto the equilibrated MACS-MS separation column and let it flow through. Set aside 30 μL of this flow-through for fluorescence microscopy (*see* Fig. 2).
7. Wash the loaded MACS columns three times with 0.5 mL ice-cold HS buffer. Again, this is done by gravity flow (*see* **Note 12**). Take a 30 μL sample for fluorescence microscopy.
8. Remove the washed MACS columns from the MACS separator, and elute the LCVs with 0.5 mL ice-cold HS buffer into a 1.5 mL microcentrifuge tube. Do this by firmly pressing down with the plunger that is provided with the MACS columns. This will substantially increase the yield of purified LCVs. Take a 10 μL sample of the eluate for fluorescence microscopy (*see* Fig. 2).

3.7 Purification of LCVs by Histodenz Density Gradient Centrifugation

1. The Histodenz gradient will need 1 h preparation time, so best set it up while the homogenates are incubating with the first antibody (Subheading 3.6).
2. The linear gradient consists of 10–35% Histodenz in a total volume of 11.5 mL PBS. Fill 5.75 mL of a 35% Histodenz/PBS solution (Subheading 2.5) into a 15 mL capped centrifuge tube. Then carefully and slowly add 5.75 mL of a 10% Histodenz/PBS solution (Subheading 2.5). This should be done drop-wise and without mixing the solutions. If done correctly, two distinct phases of 35% Histodenz at the bottom of the tube and a 10% Histodenz phase at the top will form. Cap the tube.
3. Gently lay down the tube horizontally and let it rest for 1 h. Avoid any shocks to the tube during this time. The two Histodenz phases will slowly mix and form the linear 10–35% gradient.
4. Slowly return the tube into a vertical position and load 0.5 mL eluate of the magnetically enriched LCV suspension (Subheading 3.6) onto the Histodenz cushion. Centrifuge the cushion at $3350 \times g$ and 4 °C for 1 h. To minimize forces acting on the gradient during deceleration, adjust the brakes to the lowest setting possible.
5. The gradient is divided into 8 fractions. Using a glass Pasteur pipette, carefully push down to the bottom of the 15 mL test

tube holding the gradient and take 1.5 mL of sample. Carefully remove the pipette from the gradient and collect the sample in a 1.5 mL microcentrifuge tube. This is fraction 1. Repeat this until you have collected all eight fractions. Put the fractions on ice and take 100 μL of each for fluorescence microscopy. Fraction 4 is expected to contain the highest concentration of purified LCVs (*see* Fig. 2).

6. The purified LCVs can now be analyzed by fluorescence microscopy (Subheading 3.8), Western blot and/or mass spectrometry.

3.8 Analysis of Purified LCVs by Fluorescence Microscopy

1. Place sterile poly-L-lysine-coated microscopy coverslips (Subheading 2.5) into the wells of a 24-well flat-bottomed tissue culture plate.
2. Put the microscopy samples taken during the process of purifying LCVs (Subheadings 3.5–3.7) onto the coated coverslips in the wells. Add 1 mL ice-cold HS buffer and centrifuge at 600 × *g* and 4 °C for 10 min.
3. Carefully remove the supernatant, and fix the samples by adding 0.5 mL 4% PFA (Subheading 2.5) per well. Incubate at 25 °C for 20 min. Then wash the coverslips twice with 0.5 mL/well SorC (*D. discoideum*) or PBS (RAW 264.7 macrophages), each time carefully removing the supernatant.
4. Samples from infected *D. discoideum* producing calnexin-GFP can be directly mounted on microscopy glass slides using Vectashield (or another suitable mounting medium) and analyzed by microscopy afterward.
5. Samples from infected RAW 264.7 macrophages have to be stained for SidC (or any other suitable LCV marker) before microscopy. To this end, add 0.5 mL of 1% BSA (w/v) in PBS (blocking solution) to the PFA-fixed samples and incubate at 25 °C for 20 min. Take off the supernatant and add 30 μL of primary antibody (affinity-purified rabbit anti-SidC diluted 1:1000 in blocking solution). Incubate for 1 h at 25 °C in a humidity chamber, and wash twice with PBS. Then add 30 μL of secondary antibody (e.g., anti-rabbit-Cy5 diluted 1:200 in blocking solution) and further incubate for 1 h at 25 °C in a humidity chamber. Again wash twice with PBS and mount the coverslips on microscopy glass slides using Vectashield or another suitable mounting medium.
6. Analyze morphology, integrity, and quantity of LCVs per view field with an epifluorescence microscope equipped with suitable filters (*see* Fig. 2).
7. Alternatively, the purified LCVs can be further subjected to proteome analysis using LC-MS.

4 Notes

1. Source, quality, and composition of all media components should not be changed. For example, the composition of yeast extract, especially regarding amino acids, influences growth and therefore virulence of *L. pneumophila*. We use Bacto™ yeast extract from Difco for AYE/CYE medium and BBL™ yeast extract, Bacto™ proteose peptone and BBL™ thiotone peptone from Becton Dickinson for HL5 medium. To maintain reproducibility, always use the same suppliers and batches and carefully test growth and physiology when using new ones.
2. Carefully adjust the pH. Even minor changes in the pH can influence growth of *L. pneumophila*, and the bacteria might be less infectious in stationary growth phase.
3. Cysteine is light sensitive and rapidly oxidizes to cystine, which is unavailable for the bacteria [60]. Storing the medium at 4 °C in the dark slows down that process.
4. *D. discoideum* grows on bacteria as its natural food source. The AX3 strain used here is adapted to grow axenically in synthetic medium, i.e., without bacteria. The membrane protein calnexin is a marker of the ER and the LCV [56, 61].
5. RPMI 1640 medium contains 2 g/L sodium bicarbonate as a buffer and therefore requires a CO_2 environment during cell culture to maintain physiological pH values. The medium contains phenol red as a pH indicator and appears pink-red. Most eukaryotic cells grow best at a pH around 7.0. Cell growth produces organic acids that acidify the medium. This results in a color change of the medium from pink to orange and yellow, indicating the need to change the medium (or presence of bacterial contamination). Phenol red interferes with fluorescence assays, and therefore, RPMI 1640 medium should be replaced with PBS when performing such assays.
6. *D. discoideum* is light sensitive and frequent alterations between light and dark should be avoided. Furthermore, temperatures over 25 °C should be avoided, as the amoebae will react with a heat shock and eventually die. Never use temperatures higher than 25 °C when infecting *D. discoideum* with *L. pneumophila*.
7. When reaching post-exponential/early stationary growth phase at $OD_{600} = 3.0$, *L. pneumophila* is virulent. The bacteria are somewhat smaller than in exponential growth phase, flagellated and highly motile at this point. The population homogeneity, cell morphology, and motility can be determined by light microscopy. To determine the viability of the bacteria, plate out

an aliquot of the culture on CYE/Cm agar plates. For infections, the OD_{600} should not be below 3.0, as otherwise the infection efficiency is likely suboptimal.

8. An optical density of 3.0 equals ~2×10^9 bacteria/mL, but the correlation between the OD_{600} and the cell number is dependent on the spectrophotometer used and should be determined experimentally beforehand.
9. The infected cells can be incubated longer than 1 h, and intact LCVs can still be successfully isolated. Infection times ranging from 15 min up to 14 h have been tested [24]. Upon extension of the infection time, infected cells will lift off the bottom of the tissue flasks, which requires to adjust the washing step before homogenization of phagocytes. In this case, release the cells from the bottom by using a cell scraper, transfer the cell suspension to a 15 mL reaction tube, centrifuge at $500 \times g$ and 4 °C for 10 min, remove the supernatant and suspend the cells in 10 mL cold SorC (*D. discoideum*) or PBS (RAW 264.7 macrophages). Repeat centrifugation and finally suspend cells in cold HS buffer containing EDTA-free protease inhibitor.
10. Use an odd number of strokes to homogenize the suspension of *L. pneumophila*-infected phagocytes. This ensures that unbroken cells remain in the first syringe and do not end up in the second syringe containing the homogenate to be used further.
11. The blocking reagent should be optimized depending on the primary antibody that is used. BSA or other reagents might be more suitable for other antibodies.
12. Washing the loaded MACS columns will take more time, if LCVs of a high density are purified, as this will clog the column matrix. Also, washing homogenates of RAW 264.7 macrophages takes longer than homogenates of *D. discoideum*. The washing step can be accelerated by pipetting the HS buffer up and down on top of the column. Do not let the column run dry.

Acknowledgments

This work received financial support from the Swiss National Science Foundation (SNF; 31003A_153200), the German Research Foundation (DFG; SPP1316, SPP1617) and the "Bundesministerium für Bildung und Forschung" (BMBF; "Medical Infection Genomics" project 0315834C, and "ERA-NET Infect-ERA" EUGENPATH project 031A401A).

References

1. Hoffmann C, Harrison CF, Hilbi H (2014) The natural alternative: protozoa as cellular models for *Legionella* infection. Cell Microbiol 16(1):15–26
2. Gao LY, Harb OS, Abu Kwaik Y (1997) Utilization of similar mechanisms by *Legionella pneumophila* to parasitize two evolutionarily distant host cells, mammalian macrophages and protozoa. Infect Immun 65 (11):4738–4746
3. Hilbi H, Haas A (2012) Secretive bacterial pathogens and the secretory pathway. Traffic 13(9):1187–1197
4. Isberg RR, O'Connor TJ, Heidtman M (2009) The *Legionella pneumophila* replication vacuole: making a cosy niche inside host cells. Nat Rev Microbiol 7(1):13–24
5. Hubber A, Roy CR (2010) Modulation of host cell function by *Legionella pneumophila* type IV effectors. Annu Rev Cell Dev Biol 26:261–283
6. Zhu W, Banga S, Tan Y, Zheng C, Stephenson R, Gately J, Luo ZQ (2011) Comprehensive identification of protein substrates of the Dot/Icm type IV transporter of *Legionella pneumophila*. PLoS One 6(3):e17638
7. Finsel I, Ragaz C, Hoffmann C, Harrison CF, Weber S, van Rahden VA, Johannes L, Hilbi H (2013) The *Legionella* effector RidL inhibits retrograde trafficking to promote intracellular replication. Cell Host Microbe 14(1):38–50
8. Haneburger I, Hilbi H (2013) Phosphoinositide lipids and the *Legionella* pathogen vacuole. Curr Top Microbiol Immunol 376:155–173
9. Rothmeier E, Pfaffinger G, Hoffmann C, Harrison CF, Grabmayr H, Repnik U, Hannemann M, Wölke S, Bausch A, Griffiths G, Müller-Taubenberger A, Itzen A, Hilbi H (2013) Activation of Ran GTPase by a *Legionella* effector promotes microtubule polymerization, pathogen vacuole motility and infection. PLoS Pathog 9(9):e1003598
10. Hoffmann C, Finsel I, Otto A, Pfaffinger G, Rothmeier E, Hecker M, Becher D, Hilbi H (2014) Functional analysis of novel Rab GTPases identified in the proteome of purified *Legionellaa*-containing vacuoles from macrophages. Cell Microbiol 16(7):1034–1052
11. Eisenreich W, Dandekar T, Heesemann J, Goebel W (2010) Carbon metabolism of intracellular bacterial pathogens and possible links to virulence. Nat Rev Microbiol 8(6):401–412
12. Mills SD, Finlay BB (1998) Isolation and characterization of *Salmonella typhimurium* and *Yersinia pseudotuberculosis*-containing phagosomes from infected mouse macrophages: *Y. pseudotuberculosis* traffics to terminal lysosomes where they are degraded. Eur J Cell Biol 77(1):35–47
13. Vorwerk S, Krieger V, Deiwick J, Hensel M, Hansmeier N (2015) Proteomes of host cell membranes modified by intracellular activities of *Salmonella enterica*. Mol Cell Proteomics 14 (1):81–92
14. Matsumoto A (1981) Isolation and electron microscopic observations of intracytoplasmic inclusions containing *Chlamydia psittaci*. J Bacteriol 145(1):605–612
15. Sturgill-Koszycki S, Haddix PL, Russell DG (1997) The interaction between *Mycobacterium* and the macrophage analyzed by two-dimensional polyacrylamide gel electrophoresis. Electrophoresis 18(14):2558–2565
16. Herweg JA, Hansmeier N, Otto A, Geffken AC, Subbarayal P, Prusty BK, Becher D, Hensel M, Schaible UE, Rudel T, Hilbi H (2015) Purification and proteomics of pathogen-modified vacuoles and membranes. Front Cell Infect Microbiol 5:48
17. Urwyler S, Finsel I, Ragaz C, Hilbi H (2010) Isolation of *Legionella*-containing vacuoles by immuno-magnetic separation. Curr Protoc Cell Biol 3:34
18. Hoffmann C, Finsel I, Hilbi H (2013) Pathogen vacuole purification from *Legionella*-infected amoeba and macrophages. Methods Mol Biol 954:309–321
19. Finsel I, Hoffmann C, Hilbi H (2013) Immunomagnetic purification of fluorescent *Legionella*-containing vacuoles. Methods Mol Biol 983:431–443
20. Luo ZQ, Isberg RR (2004) Multiple substrates of the *Legionella pneumophila* Dot/Icm system identified by interbacterial protein transfer. Proc Natl Acad Sci U S A 101(3):841–846
21. Brombacher E, Urwyler S, Ragaz C, Weber SS, Kami K, Overduin M, Hilbi H (2009) Rab1 guanine nucleotide exchange factor SidM is a major phosphatidylinositol 4-phosphate-binding effector protein of *Legionella pneumophila*. J Biol Chem 284(8):4846–4856
22. Weber S, Dolinsky S, Hilbi H (2013) Interactions of *Legionella* effector proteins with host phosphoinositide lipids. Methods Mol Biol 954:367–380
23. Dolinsky S, Haneburger I, Cichy A, Hannemann M, Itzen A, Hilbi H (2014) The *Legionella longbeachae* Icm/Dot substrate SidC selectively binds phosphatidylinositol 4-phosphate with nanomolar affinity and promotes pathogen vacuole-endoplasmic

reticulum interactions. Infect Immun 82 (10):4021–4033

24. Urwyler S, Nyfeler Y, Ragaz C, Lee H, Mueller LN, Aebersold R, Hilbi H (2009) Proteome analysis of *Legionella* vacuoles purified by magnetic immunoseparation reveals secretory and endosomal GTPases. Traffic 10(1):76–87
25. Hilbi H, Hoffmann C, Harrison CF (2011) *Legionella* spp. outdoors: colonization, communication and persistence. Environ Microbiol Rep 3(3):286–296
26. Rohmer L, Hocquet D, Miller SI (2011) Are pathogenic bacteria just looking for food? Metabolism and microbial pathogenesis. Trends Microbiol 19(7):341–348
27. Abu Kwaik Y, Bumann D (2013) Microbial quest for food in vivo: 'nutritional virulence' as an emerging paradigm. Cell Microbiol 15 (6):882–890
28. Manske C, Hilbi H (2014) Metabolism of the vacuolar pathogen *Legionella* and implications for virulence. Front Cell Infect Microbiol 4:125
29. Pine L, George JR, Reeves MW, Harrell WK (1979) Development of a chemically defined liquid medium for growth of *Legionella pneumophila*. J Clin Microbiol 9(5):615–626
30. Ristroph JD, Hedlund KW, Gowda S (1981) Chemically defined medium for *Legionella pneumophila* growth. J Clin Microbiol 13 (1):115–119
31. Tesh MJ, Morse SA, Miller RD (1983) Intermediary metabolism in *Legionella pneumophila*: utilization of amino acids and other compounds as energy sources. J Bacteriol 154 (3):1104–1109
32. Cazalet C, Rusniok C, Brüggemann H, Zidane N, Magnier A, Ma L, Tichit M, Jarraud S, Bouchier C, Vandenesch F, Kunst F, Etienne J, Glaser P, Buchrieser C (2004) Evidence in the *Legionella pneumophila* genome for exploitation of host cell functions and high genome plasticity. Nat Genet 36 (11):1165–1173
33. Chien M, Morozova I, Shi S, Sheng H, Chen J, Gomez SM, Asamani G, Hill K, Nuara J, Feder M, Rineer J, Greenberg JJ, Steshenko V, Park SH, Zhao B, Teplitskaya E, Edwards JR, Pampou S, Georghiou A, Chou IC, Iannuccilli W, Ulz ME, Kim DH, Geringer-Sameth A, Goldsberry C, Morozov P, Fischer SG, Segal G, Qu X, Rzhetsky A, Zhang P, Cayanis E, De Jong PJ, Ju J, Kalachikov S, Shuman HA, Russo JJ (2004) The genomic sequence of the accidental pathogen *Legionella pneumophila*. Science 305(5692):1966–1968
34. Sauer JD, Bachman MA, Swanson MS (2005) The phagosomal transporter A couples threonine acquisition to differentiation and replication of *Legionella pneumophila* in macrophages. Proc Natl Acad Sci U S A 102 (28):9924–9929
35. Fonseca MV, Sauer JD, Crepin S, Byrne B, Swanson MS (2014) The *phtC-phtD* locus equips *Legionella pneumophila* for thymidine salvage and replication in macrophages. Infect Immun 82(2):720–730
36. Faucher SP, Mueller CA, Shuman HA (2011) *Legionella pneumophila* transcriptome during intracellular multiplication in human macrophages. Front Microbiol 2:60
37. Wieland H, Ullrich S, Lang F, Neumeister B (2005) Intracellular multiplication of *Legionella pneumophila* depends on host cell amino acid transporter SLC1A5. Mol Microbiol 55 (5):1528–1537
38. Price CT, Al-Quadan T, Santic M, Rosenshine I, Abu Kwaik Y (2011) Host proteasomal degradation generates amino acids essential for intracellular bacterial growth. Science 334(6062):1553–1557
39. Schunder E, Gillmaier N, Kutzner E, Eisenreich W, Herrmann V, Lautner M, Heuner K (2014) Amino acid uptake and metabolism of *Legionella pneumophila* hosted by *Acanthamoeba castellanii*. J Biol Chem 289 (30):21040–21054
40. D'Auria G, Jimenez-Hernandez N, Peris-Bondia F, Moya A, Latorre A (2010) *Legionella pneumophila* pangenome reveals strain-specific virulence factors. BMC Genomics 11(1):181
41. Schroeder GN, Petty NK, Mousnier A, Harding CR, Vogrin AJ, Wee B, Fry NK, Harrison TG, Newton HJ, Thomson NR, Beatson SA, Dougan G, Hartland EL, Frankel G (2010) *Legionella pneumophila* strain 130b possesses a unique combination of type IV secretion systems and novel Dot/Icm secretion system effector proteins. J Bacteriol 192(22):6001–6016
42. Cazalet C, Gomez-Valero L, Rusniok C, Lomma M, Dervins-Ravault D, Newton HJ, Sansom FM, Jarraud S, Zidane N, Ma L, Bouchier C, Etienne J, Hartland EL, Buchrieser C (2010) Analysis of the *Legionella longbeachae* genome and transcriptome uncovers unique strategies to cause Legionnaires' disease. PLoS Genet 6(2):e1000851
43. Eylert E, Herrmann V, Jules M, Gillmaier N, Lautner M, Buchrieser C, Eisenreich W, Heuner K (2010) Isotopologue profiling of *Legionella pneumophila*: role of serine and glucose as carbon substrates. J Biol Chem 285 (29):22232–22243

44. Harada E, Iida K, Shiota S, Nakayama H, Yoshida S (2010) Glucose metabolism in *Legionella pneumophila*: dependence on the Entner-Doudoroff pathway and connection with intracellular bacterial growth. J Bacteriol 192 (11):2892–2899
45. Weiss E, Peacock MG, Williams JC (1980) Glucose and glutamate metabolism of *Legionella pneumophila*. Curr Microbiol 4:1–6
46. Häuslein I, Manske C, Goebel W, Eisenreich W, Hilbi H (2016) Pathway analysis using (13) C-glycerol and other carbon tracers reveals a bipartite metabolism of *Legionella pneumophila*. Mol Microbiol 100(2):229–246
47. DebRoy S, Dao J, Soderberg M, Rossier O, Cianciotto NP (2006) *Legionella pneumophila* type II secretome reveals unique exoproteins and a chitinase that promotes bacterial persistence in the lung. Proc Natl Acad Sci U S A 103 (50):19146–19151
48. Pearce MM, Cianciotto NP (2009) *Legionella pneumophila* secretes an endoglucanase that belongs to the family-5 of glycosyl hydrolases and is dependent upon type II secretion. FEMS Microbiol Lett 300(2):256–264
49. Herrmann V, Eidner A, Rydzewski K, Bladel I, Jules M, Buchrieser C, Eisenreich W, Heuner K (2011) GamA is a eukaryotic-like glucoamylase responsible for glycogen- and starch-degrading activity of *Legionella pneumophila*. Int J Med Microbiol 301(2):133–139
50. Otto A, Becher D, Schmidt F (2014) Quantitative proteomics in the field of microbiology. Proteomics 14(4–5):547–565
51. Hales LM, Shuman HA (1999) *Legionella pneumophila* contains a type II general secretion pathway required for growth in amoebae as well as for secretion of the Msp protease. Infect Immun 67(7):3662–3666
52. Liles MR, Edelstein PH, Cianciotto NP (1999) The prepilin peptidase is required for protein secretion by and the virulence of the intracellular pathogen *Legionella pneumophila*. Mol Microbiol 31(3):959–970
53. Mampel J, Spirig T, Weber SS, Haagensen JAJ, Molin S, Hilbi H (2006) Planktonic replication is essential for biofilm formation by *Legionella pneumophila* in a complex medium under static and dynamic flow conditions. Appl Environ Microbiol 72(4):2885–2895
54. Horwitz MA (1983) The Legionnaires' disease bacterium (*Legionella pneumophila*) inhibits lysosome-phagosome fusion in human monocytes. J Exp Med 158:2108–2126
55. Müller-Taubenberger A, Lupas AN, Li H, Ecke M, Simmeth E, Gerisch G (2001) Calreticulin and calnexin in the endoplasmic reticulum are important for phagocytosis. EMBO J 20(23):6772–6782
56. Weber SS, Ragaz C, Reus K, Nyfeler Y, Hilbi H (2006) *Legionella pneumophila* exploits PI(4)P to anchor secreted effector proteins to the replicative vacuole. PLoS Pathog 2(5):e46
57. Watts DJ, Ashworth JM (1970) Growth of myxameobae of the cellular slime mould *Dictyostelium discoideum* in axenic culture. Biochem J 119(2):171–174
58. Malchow D, Nagele B, Schwarz H, Gerisch G (1972) Membrane-bound cyclic AMP phosphodiesterase in chemotactically responding cells of *Dictyostelium discoideum*. Eur J Biochem 28(1):136–142
59. Ragaz C, Pietsch H, Urwyler S, Tiaden A, Weber SS, Hilbi H (2008) The *Legionella pneumophila* phosphatidylinositol-4 phosphate-binding type IV substrate SidC recruits endoplasmic reticulum vesicles to a replication-permissive vacuole. Cell Microbiol 10 (12):2416–2433
60. Ewann F, Hoffman PS (2006) Cysteine metabolism in *Legionella pneumophila:* characterization of an L-cystine-utilizing mutant. Appl Environ Microbiol 72(6):3993–4000
61. Lu H, Clarke M (2005) Dynamic properties of *Legionella*-containing phagosomes in *Dictyostelium* amoebae. Cell Microbiol 7(7):995–1007

Part III

Protein Quantification

Chapter 7

Detection and Identification of Low-Abundant Proteins Using HPE Gels, Fluorescent Stains, and MALDI-ToF-ToF-MS

Martin Moche, Dirk Albrecht, Reiner Westermeier, and Knut Büttner

Abstract

Two-dimensional electrophoresis as a complementary approach to gel-free proteomic methods possesses the ability to separate physiologically important isoforms of proteins in an unbiased manner. Frequently, those isoforms are low-abundant regulators, and therefore, detection and identification of low-abundant proteins is highly necessary to exploit this advantage. We describe an experimental sequence of classical operations to process gels but optimized them, in order to identify each detectable protein spot on gel.

Key words High performance electrophoresis, Low-abundant proteins, Fluorescent stains, Scanning conditions, Tryptic digestion, MALDI-ToF-MS

1 Introduction

Proteomic techniques based on two-dimensional (2D) gels have several constraints compared to gel-free proteomics. The most important disadvantages are the limited separation power of 2D gels in terms of protein amount and resolution, and the physicochemical features of gels that exclude the majority of hydrophobic proteins and very large and very small proteins too. However, the major advantage of 2D gels is an unbiased view on complex cellular processes, because gels are separating intact proteins, i.e., the real biological players [1]. This is in contrast to almost all gel-free techniques that are separating peptides of digested proteins. After an electrophoretic run in a 2D gel single proteins will be migrated into discrete spots [2, 3] inclusive of the separation of their isoforms, and will be stained, imaged, cut and digested for mass-spectrometrical (MS) identification. Frequently, Peptide Mass Fingerprint (PMFP) and mass-spectrometrical fragmentation (MS/MS) of only two protein-specific peptides is considered to be sufficient for a successful identification of a protein. In untargeted discovery experiments, the sequence coverage of the respective proteins by MS is rather low, mostly below 50% [4]. This is true

Dörte Becher (ed.), *Microbial Proteomics: Methods and Protocols*, Methods in Molecular Biology, vol. 1841,
https://doi.org/10.1007/978-1-4939-8695-8_7,

both for gel-free and gel-based techniques. But by using gel-free procedures it may have fatal consequences, if the biologically important part of the protein that makes the difference, e.g., a posttranslationally modified peptide, will not be detectable in MS. In that case, the differentiation of biologically important isoforms of proteins cannot be performed by the investigator, although it is essential for understanding of the biological meaning. Especially the quantification of protein isoforms is affected by this disadvantage.

In contrast, in 2D gels two differently positioned spots can be observed in such cases, because protein isoforms have a different p*I* and/or MW regularly. For their identification and quantification it is not necessary to detect the modified peptide in MS. The discrete spots of the isoforms are sufficient to differentiate between and to quantify them. This technical difference is very important for investigations in relatively unexplored bacteria. Usually a number of unknown isoforms for particular proteins exists in such microorganisms [5]. The peptide-based bias of gel-free techniques endangers the recognition of those isoforms without hints from earlier research at all. Therefore, by using solely gel-free techniques important biological differences might be ignored by the investigator. But it is impossible to ignore different spots on a 2D gel, even if they are completely unexpected!

In order to exploit this advantage, there is a strong need to enable detection and identification of low-abundant proteins in 2D gels. The isoforms of physiologically important regulators belong to protein species which are expressed in quantities that are some orders of magnitude below the enzymes of main metabolic pathways or other highly abundant intracellular proteins. In the following chapter we describe an experimental procedure to detect and identify nearly each visible protein in 2D gels [6].

Our procedure accounts for some expensive technical equipment which is hardly available in each laboratory. Simplification of our method is partially possible in terms of disclaiming automatically performed operations of some robots. But if doing this, one should always pay attention to the much higher susceptibility to malfunction of manually performed operations. Especially to avoid any contamination with keratin from a variety of sources, e.g., hair, skin, textiles (particularly wool), is challenging if working manually. Already traces of keratin will impede detection of low-abundant proteins. Therefore, automatic processing is preferred whenever possible.

2 Materials

Ultrapure water and analytical grade reagents have to be used for all solutions. Always wear powder-free, disposable gloves made from nitrile to avoid any contamination with keratin or polyethylene glycol.

2.1 Isoelectric Focusing (IEF)

1. Rehydration buffer: 8 M urea, 2 M thiourea, 1% (w/v) CHAPS, 0.35% (w/v) Dithiothreitol (DTT), 0.5% (w/v) Pharmalytes 3–10 (*see* **Note 1**). Add 2.4 g urea, 760 mg thiourea, 50 mg CHAPS, 0.15 mg DTT, and 26 μL Pharmalyte 3–10 to a 15-mL tube and fill up to 5 mL with water (*see* **Note 2**). The rehydration buffer can be stored in suitable aliquots at −20 °C (*see* **Note 3**).
2. IPG: Precast, dehydrated, frozen strips of gels polymerized onto a plastic support and containing an immobilized pH-gradient (IPG) [7], at least 17 cm long.
3. Rehydration trays made of acrylic glass with slots in dimensions slightly longer and broader than IPG strips.
4. Parafilm.
5. Electrode wicks.
6. Electrophoresis system for IEF, consisting of power supply, thermostatic circulator and electrophoresis unit (*see* **Note 4**).
7. Mineral oil in medical purity to cover the IPG strips.

2.2 High Performance Electrophoresis (HPE)

1. HPE™ Tower System (SERVA Electrophoresis GmbH, Heidelberg, Germany).
2. HPE™-2D-Large-Gel flatbed NF 12.5% Kit (SERVA).
3. PaperPool (SERVA).
4. Fixation and acidification solution: 1% (w/v) citric acid, 15% (v/v) ethanol in H_2O.

2.3 Fluorescent Staining

1. SERVA Purple™ dye (SERVA) (*see* **Notes 5** and **6**).
2. Fluorescence staining trays made of nontransparent plastic (300 × 210 × 50 mm).
3. Fixation and acidification solution: 1% (w/v) citric acid, 15% (v/v) ethanol in H_2O.
4. Staining buffer: Dissolve 6.2 g boric acid and 3.85 g sodium hydroxide in 1 L H_2O.
5. Washing solution: 15% (v/v) ethanol in H_2O.

2.4 Scanning

1. High end laser scanner (*see* **Note 7**). We are operating Typhoon 9400 multi-mode imager (GE Healthcare, Little Chalfont, UK).

2.5 Image Analysis and Spot Picking

1. Gel-image analysis software. Different types and suppliers are available and can be used. We are operating Delta2D (Decodon GmbH, Greifswald, Germany).
2. Automatic spot cutting system able to import files (picklists) from the image analysis software. We are operating Ettan Spot Picker (GE Healthcare, Little Chalfont, UK) (*see* **Note 8**).

3. Marker stickers made from white paper (circle of 6 mm diameter).
4. 96-well barcoded microplates made from low-binding plastics, and tapered wells (*see* **Note 9**).
5. High end laser scanner (*see* **Note** 7). We are operating Typhoon 9400 multi-mode imager (GE Healthcare, Little Chalfont, UK).

2.6 Tryptic Digestion and MALDI Spotting

1. Automatic system to digest proteins in gel plugs, to extract peptide mixtures out of gel plugs, and to spot these mixtures onto a MALDI target. These three procedures can be performed separately in different machines or combined with one apparatus. We are operating Ettan Spot Handling Workstation (GE Healthcare, Little Chalfont, UK) that performs combined protocols.
2. Washing solution: 50 mM NH_4HCO_3 in 50% (v/v) methanol.
3. Drying solution: 75% (v/v) acetonitrile in H_2O.
4. Trypsin solutions: (*see* **Notes 10** and **11**).
   (a) Dissolve 20 μg trypsin in 5 mL 20 mM NH_4HCO_3 solution for highly abundant proteins.
   (b) Dissolve 20 μg trypsin in 10 mL 20 mM NH_4HCO_3 solution for medium-abundant proteins.
   (c) Dissolve 20 μg trypsin in 20 mL 20 mM NH_4HCO_3 solution for low-abundant proteins.
5. Extraction solution: 50% (v/v) acetonitrile, 0.1% (v/v) trifluoroacetic acid (TFA) in H_2O.
6. Spotting solution: Prepare a stock solution of 50% (v/v) acetonitrile, 0.5% (v/v) trifluoroacetic acid (TFA) in H_2O. Dissolve 3.3 mg α-cyano-4-hydroxycinnamic acid in 1.5 mL stock solution freshly and store not longer than one week (*see* **Note 12**).
7. 96-well barcoded microplates made from low-binding plastics, and tapered wells (*see* **Note 9**).
8. MALDI target: 384 Well Insert (dimension: 123 × 81 mm) (*see* **Note 13**).

2.7 Protein Identification

1. High End MALDI TOF mass spectrometer. We are operating AB SCIEX TOF/TOF 5800 Analyzer (AB Sciex, Foster City, CA, USA).
2. MALDI target: 384 Well Insert (dimension: 123 × 81 mm) (*see* **Note 13**).
3. GPS Explorer version 3.6 (AB Sciex, Foster City, CA, USA) (*see* **Note 14**).
4. Mascot version 2.4.1 (Matrix Science Ltd., London, UK).

3 Methods

3.1 Isoelectric Focusing with Rehydration Loading

1. Dissolve a sample of 200 μg total protein in 400 μL rehydration buffer (*see* **Notes 15** and **16**).
2. Dispense the dissolved protein sample into a slot of the rehydration tray and place an IPG strip into the slot with the gel side down facing the dispensed solution.
3. Thoroughly seal the rehydration tray with Parafilm to avoid evaporation and prevent crystallization of urea (*see* **Note 17**).
4. Incubate the trays at room temperature for 16 h.
5. Set the IEF unit to 20 °C.
6. Remove the rehydrated IPG strips from rehydration tray and place them—gel surface up—into the IPG trays of the IEF unit.
7. Position electrode wicks in required lengths and soak them with H_2O to become moist but not wet. Place the wicks between the IPG strips and the electrodes (*see* **Note 18**).
8. Place the anode and the cathode on top of the electrode wicks.
9. Cover the IPG with mineral oil completely to avoid any evaporation and prevent crystallization of urea.
10. Start the IEF with a suitable program:

    Limit the current to 50 mA per IPG. A successful example is in the first phase an application of a gradient to 500 V within 500 Vh, in second phase keeping 500 V until a total of 2500 Vh is reached, in third phase the application of a gradient to 3500 V until a total of 10,000 Vh is reached and a final keeping of 3500 V until a total of 35,000 Vh is reached (*see* **Note 19**).
11. Stop the power supply and place the focused IPG strips in a box for freezing, avoid contact of the gel surface with the box (plasticizer contaminations). They can be stored at −20 °C for several weeks.

3.2 High Performance Electrophoresis (HPE)

1. Prepare and process the gels according to instructions of the manufacturer (*see* **Notes 20** and **21**).
2. When the separation is finished place the gels into plastic trays, cover them with 400 mL fixation solution, and incubate for at least 1 h.

3.3 Fluorescent Staining with SERVA Purple

1. For pH equilibration, incubate the gel in 300 mL staining buffer (without the dye) for 15 min with gentle shaking; change the solution and incubate for another 15 min.
2. Add 1 mL SERVA Purple dye to 250 mL staining buffer and incubate the gel immediately for 1 h with gentle shaking (*see* **Note 22**).

3. Discard the staining solution and incubate the gel in 400 mL washing solution for 30 min with gentle shaking.
4. Discard the washing solution and incubate the gel in 400 mL fixation solution for at least 30 min with gentle shaking.

3.4 Scanning

1. Detection: Place the HPE gel on the laser scanner gel side down and cover it with a scan frame. Scan at 100 μm resolution using the green laser (532 nm) with long pass filter (560 nm). Adjust the PMT voltage to ensure that the most intense spot is below saturation by using the software tools of the scanner.

3.5 Image Analysis and Spot Picking

1. Load gel images into a software for gel-image analysis. We are using the software Delta2D.
2. Perform automatic spot detection and edit the spots as required using the spot editing tool in the dual view window, or edit the spots of interest manually.
3. Open the image of the gel to be cut in "Quantitation View". Sort the spots by normalized spot volumes rising from the lowest to the most intense spot.
4. In the tab "Labels", choose "label the unlabeled spots with numbers". The spots should be numbered now according to rising intensity.
5. Paste white marker stickers inside the spot picking tray so that they are aligned centrally to the right and to the left side of the gel. The marker must be placed outside the gel area used for protein separation but inside the central area used by the spot picker for marker detection (*see* **Note 23**) (Fig. 1).
6. Scan the gels again using the laser scanner and import the images into Delta2D (*see* **Note 24**). This and the following steps must be performed to ensure correct cutting in automatic mode of spots below the detection level of the cutting tool (Fig. 1).
7. Use the warping tool to align the new image with the original one correctly.
8. Transfer the labels from the original file to the new image (that one containing the two marker spots).
9. Label the center of the left marker spot with IR1 and the center of the right marker spot with IR2.
10. Export a picklist into a data format the spot picker can deal with. In Delta2D go to "Dual View", and click the tab "Export Picklist". The original picklist is a .txt file.
11. To achieve maximum identification rates the spots must be digested with different trypsin concentrations depending on spot size and density (*see* **Note 10**). Because digestion conditions cannot be changed within one microplate, it is necessary

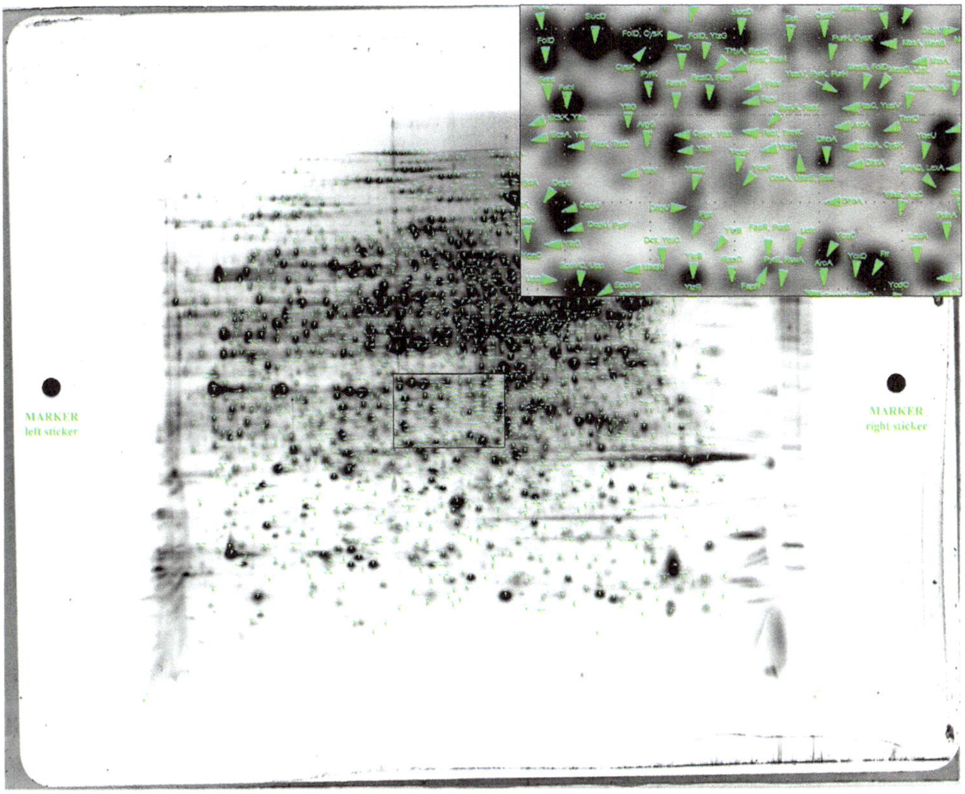

Fig. 1 Annotated fluorescent image from a high end laser scanner of a HPE gel. The marker stickers were added at the left and at the right side of the gel to enable the spot picking robot to recalculate the x/y positions of the protein spots from the imported picking list. The image is imported into the spot picker and the gel plugs are cut in automatic mode. The image is electronically enhanced for detection of low-abundant proteins. Spots were processed and identified as described. Findings could be successfully used to annotate nearly each spot of the gel. In order to demonstrate density and accuracy of the results, a central sector of the gel is enlarged and shown in the upper right part of the figure

to distribute the spots of similar protein amount into the same microplates. In order to determine at which amount the plates have to be switched, visualize the image with labels. Open the histogram and reduce the contrast to the lowest value that corresponds to the original image. Click the tab "Export" in the "Dual View" and "Export presentations slide...". The resulting PowerPoint file can be searched for numbers.

12. Open the histogram again and increase the contrast to visualize low-abundant proteins in best possible quality (Fig. 1). Export the view as described in **step 11**. The proteins have to be divided into three different classes (*see* **Note 25**). Open the exported picklist and split it into these three different picklists accordingly. Keep exactly the format of the original picklist.
13. Place the gel into the gel picking tray. The marker stickers must be positioned inside the area used for marker detection.

14. Adjust the gel according to the marker positions (Fig. 1). Press the button "Manual detection" of the spot picker. The visual system will be placed over IR1 (marker sticker 1). If the position is correctly the signal will switch to green. Press the button "Accept marker". After correct recognition of IR1 this is done for IR2 (marker sticker 2) automatically.
15. Load the first picklist into the spot picker.
16. Place enough 96-well microplates into the spot picker to distribute all spots in the picklist.
17. Select the suitable cutting tool of 1.5 mm diameter in order to remove the HPE gel from plastic support. Edit the cutting parameter for the cutting tool. The horizontal jazz movement of the picking head has to be 1.5 mm for HPE gels.
18. Start spot picking process.
19. Repeat the **steps 15–18** with each of the other two picklists.

3.6 Tryptic Digestion and MALDI-Spotting

1. Place the microplates with the cut spots into the correct position in the Spot Handling Platform. Fill empty microplates to catch extracted peptides into equivalent positions in the Spot Handling Platform. Add the required number of MALDI targets into necessary positions in the Spot Handling Platform.
2. Start the program for digestion and spotting of the Spot Handling Platform (Fig. 2). All following operations are performed automatically.
3. All pipetting procedures are carried out by an eight-channel-needle-tool from stainless steel that is always pipetting a complete microplate before starting next operation. The only exception occurs during MALDI-spotting and will be described at **step 26**.
4. Those wells in the microplates containing a cut spot include an additional cover layer of 150 μL H_2O originating from the picking process.
5. Remove the cover layer from the wells.
6. Add 100 μL washing solution to each well.
7. Incubate for 30 min at room temperature.
8. Remove washing solution.
9. Repeat **steps 6–8** twice.
10. Add 100 μL drying solution to each well.
11. Incubate for 30 min at room temperature.
12. Remove drying solution. The cut spots are predried now.
13. Place the microplate containing the predried spots into the dryer of the Spot Handling Platform.

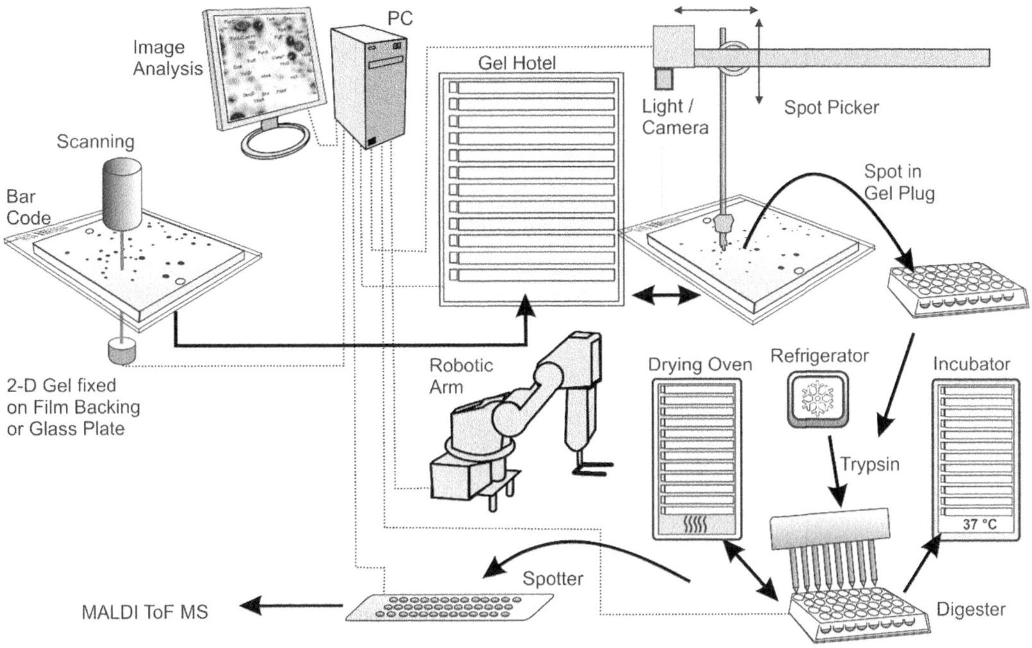

Fig. 2 Scheme of operations. The first few steps, 2DE, staining, scanning, image analysis, and spot cutting of and in the HPE gels are performed outside the Spot Handling Platform using separate equipment. For the following steps, the cut gel plugs are processed inside the Spot Handling Platform as described. Spotted MALDI targets will be removed from the Spot Handling Platform and placed into a MALDI instrument

14. Leave them for 30 min there. The dryer will blow air with 40 °C onto the predried spots. The spots will be dried completely after this procedure.
15. Add 10 μL trypsin solution to each well. Decide for the suitable type of this solution (A, B, or C) (*see* **Notes 10** and **25**).
16. Place the microplate containing trypsinated spots into the incubator of the Spot Handling Platform.
17. Incubate for 2 h at 37 °C.
18. Add 60 μL extraction solution to each well.
19. Incubate for 30 min at room temperature.
20. Remove the extraction solutions and transfer them into the wells of a new microplate.
21. Add 40 μL extraction solution again to each well of the microplate containing trypsinated spots.
22. Incubate again for 30 min at room temperature.
23. Remove again extraction solutions and transfer them into the wells of the new microplate already containing those 60 μL primarily removed extractions solutions.
24. Place the microplate containing per well 100 μL removed and joined extractions solutions into the dryer of the Spot

Handling Platform, and allow the dryer to blow air with 40 °C for approximately 3 h until the liquid is completely evaporated.

25. Pipetting operations using the eight-channel-needle-tool are finished now. Next operations are performed by a single-needle-tool from stainless steel. These procedures do not follow the previous scheme to pipette a complete microplate before starting next operation. Now the complete sequence of next operations (**steps 26–28**) will be performed one well after another well. The sequence for the successive well will only start after the sequence for the previous well is finished.
26. Add 1.5 μL spotting solution into a single well of the completely dried microplate.
27. Dissolve the peptide mixture by 5× strokes of the spotting solution in the needle.
28. Spot 1.1 μL of the peptide containing solution onto the MALDI target.

3.7 Protein Identification

1. Place the MALDI target into the MALDI instrument (*see* **Note 26**).
2. Spectra should be recorded in an analytical window from 900 to 3700 Da.
3. The focus mass should be at 1700 Da inside this mass range.
4. Accumulate 25 subspectra per main spectrum, the peptide mass fingerprint (PMFP).
5. Perform 100 laser shots per subspectrum.
6. Use a random search pattern for screening activity of the laser.
7. Use the autoproteolytic fragments of trypsin to calibrate the spectra (*see* **Note 11**). The tryptic fragment causing the monoisotopic peak $(M + H)^+$ m/z at 2211.104 is especially suitable, but the signal-to-noise ratio (S/N) has to be 40 or higher. In those cases use that peak to perform an internal calibration as single point calibration.
8. Standard mass deviation has to be below 0.15 Da.
9. In a few cases (below 1%) automatic calibration fails. Those spectra have to be manually calibrated.
10. Select the five most intense peaks of the previous PMFP for subsequent MS/MS analysis.
11. Accumulate 25 subspectra per main MS/MS spectrum.
12. Perform 255 laser shots per subspectrum.
13. Use a random search pattern for screening activity of the laser.
14. Use the monoisotopic peaks of Arginine $(M + H)^+$ m/z at 175.119 or Lysine $(M + H)^+$ m/z at 147.107 for an internal

calibration as single point calibration. The signal-to-noise ratio (S/N) has to be 5 at least.

15. Create a peak list by using the software GPS Explorer™ version 3.6 (*see* **Note 14**).
16. Use for PMFP these settings:
    - Mass range 900–3700 Da.
    - Density 20 peaks per 200 Da.
    - Minimal S/N ratio 15 peaks per spot.
    - Maximal S/N ratio 65 peaks per spot.
17. Use for MS/MS these settings:
    - Mass range from 60 to Precursor—20 Da.
    - Density 25 peaks per 200 Da.
    - Minimal S/N ratio ten peaks per precursor.
    - Maximal S/N ratio 65 peaks per precursor.
18. Analyze the peak list via Mascot search engine version 2.4.1 using a sequence database suitable for your organisms.

4 Notes

1. Pharmalytes is a trademark and a particular mixture of carrier ampholytes. Different mixtures of carrier ampholytes from different suppliers are available. They differ not only by their trade names, but also by their chemical structures! Common to all of them is only their amphoteric nature. But there are large differences between their other chemical features (molecular size, functional groups, etc.). These differences heavily influence some of the separation features, compatibility with buffer components, and compatibility with subsequent staining methods (background, adsorption of dye).
2. The molarity of the solution is close to saturation. Therefore, dissolving urea requires some time. It is difficult to estimate the volume reduction of the urea crystals by solubilization. Add water to the desired volume, observe volume reduction, and repeat the procedure carefully until all chemicals will be dissolved. The solution must not be heated to facilitate this process because at temperatures above 30 °C urea will cause artificial carbamylation of proteins leading to series of artificial spots (isoforms) in IEF.
3. Before usage, ensure that the urea is completely dissolved after thawing. Again, the solution must not be heated above 30 °C (*see* **Note 2**).

4. Different combinations of power supplies, electrophoretic chambers and thermostatic circulators are principally possible. It is recommended to use a complete system from one of the suppliers because the electrical conditions of the IEF process should be balanced between the performance of the power supply and to the cooling effectiveness. Highly advantageous are those IEF units that are offering the independent electrical control of the IEF process for each individual IPG strip.
5. A broad variety of fluorescent dyes with different features is available from different suppliers. Always take care to decide for the dye with the most suitable features for your experiment. In our case highest sensitivity combined with compatibility to the HPE gels was important.
6. SERVA Purple™ is a synthetic fluorescent dye that labels lysine, arginine, and histidine. The dye should be stored at −20 °C. The dye has a limited stability in the basic staining solution. The limit of detection is approximately 50 pg protein. The dye can be excited at a wavelength of 405 or 500 nm but also at 543 nm, 532 nm, or 488 nm. The signal (610 nm) can be detected with a 610 nm band pass filter or a 560 nm long pass filter.
7. There is a variety of laser scanners available with low or medium performance. But to detect all proteins of lowest abundance on gel high end laser scanners are unavoidably necessary. The requirements are fulfilled from several scanner types of the Typhoon series (GE Healthcare) and the former FLA series (Fuji).
8. Different spot picker or spot cutter systems are available from different suppliers. Some of them have an incorporated source for excitation of fluorescent stains. In our experience this feature shows very limited benefits, because the sensitivity of the corresponding image is far from that of high end laser scanners. Much more important and for our work flow indispensable is the ability of the spot picker to import and to use files (picklists) created with image analysis software and founded onto images from high end laser scanners. Their incomparable image quality is the best basis to process spot cutting (*see* **Note 24** and Fig. 1).
9. Shape and material of microplates are very important for identification. Most of the tremendous variety of microplates are made for enzymatic or immunological tests, e.g., ELISA, and have several features adversely for protein digests. The shape of the wells must guarantee the central position of cut gel plugs inside the well during numerous pipetting procedures. Sticking of the plugs to the needles of the robot would cause artificial transfer into wrong positions. Low-protein binding material reduces the losses of low concentrated peptides by adsorption

to the walls. Barcodes are necessary for automatic processing by the robot during digestion and MALDI spotting (*see* Fig. 2).

10. Trypsin is a hydrolytic enzyme. The dynamic range of that enzymatic reaction is some orders of magnitude lower than the dynamic range of amount of intracellular proteins. Therefore, it is not possible to digest all the different proteins under only one condition (*see* ref. 6). Outside the optimal ratio of substrate and enzyme the efficiency of hydrolytic reaction decreases considerably. It is possible to have both—too much trypsin in the digest (biochemical inhibition of substrate) or a too low amount of this enzyme. The results are highly similar: no identification because of insufficient digestion.

11. There are several methods existing to calibrate MALDI spectra. Common to all internal methods of calibration (using peaks of known mass inside the acquired spectrum), that are more accurate than external calibrations, is the need for a suitable ratio of peaks used to calibrate and those to be determined. We have best experience with peptide peaks generated by the autoproteolytic activity of trypsin. In order to generate these peaks, it is necessary to use a type of trypsin that is not modified to protect against its autoproteolytic reaction.

12. One and a half mL spotting solution ready to use is sufficient for two completely spotted MALDI targets (768 spots). This 1.5 mL should be prepared even for a lower number of solutions to be spotted, because a minimum liquid level for the operation of the pipetting needle is needed. Larger volumes can always be prepared if needed.

13. Targets with the dimensions mentioned above are supplied by different manufacturers and can be used in MALDI instruments of these manufacturers. For automatic processing of the targets by robots during MALDI spotting (*see* Fig. 2) it might be useful or even necessary to design and build a carrier frame for these targets by your own.

14. There are several types of software existing to process raw data of MALDI instruments (*see* ref. 8). Protein Pilot can be used instead of GPS Explorer. For MALDI instruments from other manufacturers their recommended software should be used.

15. The correct volume depends on the type of IPG strips used. If the volume is not given in the instructions of the supplier it must be determined. Too small volumes lead to partially dry IPG strips with fatal consequences in electrophoresis. Too large volumes cause losses of protein because a part of the proteins will not enter the gel by capillary force during rehydration.

16. If the protein is not a pellet, which is directly dissolved in rehydration buffer, but a solution containing a different buffer composition, it is very important that the concentrated protein

solution does not exceed 10% of the final volume (*see* **Note 15**) in order to maintain the required concentrations of the rehydration buffer ingredients.

17. It has to be avoided, that the Parafilm becomes leaky: Evaporation during the long rehydration process can cause partial drying of the IPG strips with fatal consequences in electrophoresis. Therefore, we recommend to apply a double layer of Parafilm to avoid any leaks. Some instructions recommend the use of mineral oils for this purpose. But that makes the IPG strips very slippery and uncomfortable in handling after rehydration.
18. The electrode strips do not only have the function to supply the IPG strips with moisture during the long electrophoresis, but also to serve as reservoir for all nonamphoteric compounds, like salt ions, that are migrating out of the gel. When a sample contains an unusual high amount of ions, apply a double layer of electrode strips.
19. Focusing programs can differ considerably. It depends on the type and origin of proteins (which organisms), and from the IEF system used.
20. A broad variety of protocols for gels of second dimension is available. High performance electrophoresis is different from all these classical gels because HPE is a flatbed process, whereas classical gels run vertically. HPE offers numerous advantages and is highly recommended for the detection of low-abundant proteins (*see* ref. 6).
21. Dispose of iodoacetamide containing buffers in halogen-containing waste.
22. ServaPurple is not stable for longer than five hours in the alkaline staining buffer; hence it is important to add the dye to the staining buffer shortly before use, and incubation should not exceed 3 h. The staining is reversible and can be removed using ammonium carbonate.
23. Sometimes the markers do not stick permanently to the gel. These problems can be bypassed via cutting two holes in size of a 1 mL pipette into the gel instead of sticking markers. These holes will be recognized and can be used to align the gel therefore. Alternatively, it is possible to attach marker stickers directly on the plastic backing (*see* Fig. 1). It sticks much better onto a plastic film than onto a gel.
24. It is indispensable to import the images of high end laser scanners because no picking or cutting tool has comparable detection abilities (*see* **Note 8**). This means, if the aim of the work is the identification of low-abundant proteins, that tool has to cut spots not detectable by the cutting or picking tool itself. The instrument has to work only on basis of the imported file (*see* Fig. 1).

25. Fluorescent dyes are invisible to the naked eye and their signals have to be detected electronically by photomultipliers. Their dynamic range is much larger than the ability of the human eye to differentiate between colors and their shading. Therefore, it is impossible to show all protein spots to the observer in the same quality in the same image (*see* ref. 9 and Fig. 1). Usually one has to decide to see the highly abundant spots below saturation leading to invisible low-abundant spots or vice versa. But all information is available in the scans. If necessary one can use copies of the image with different contrast levels:
    (a) Most intense and large spots that cover the whole 2 mm diameter of the picking head.
    (b) Majority of spots, which are visible on the gel image without in silico enhancement of the contrast.
    (c) Spots that are invisible in the PowerPoint image, but become visible after in silico enhancement of the contrast of gel images (*see* ref. 6).
26. This instrument is designed for high throughput measurement. Therefore, it is able to acquire the raw data, to calibrate the spectra and to analyze the data using the TOF/TOF™ Series Explorer™ Software V4.1.0., automatically.

References

1. Hecker M, Antelmann H, Büttner K, Bernhardt J (2008) Gel-based proteomics of Gram-positive bacteria: a powerful tool to address physiological questions. Proteomics 8(23-24):4958–4975
2. O'Farrell PH (1975) High resolution two-dimensional electrophoresis of proteins. J Biol Chem 250(10):4007–4021
3. Klose J (1975) Protein mapping by combined isoelectric focusing and electrophoresis of mouse tissues. A novel approach to testing for induced point mutations in mammals. Humangenetik 26(3):231–243
4. Muntel J, Fromion V, Goelzer A, Maabeta S, Mader U, Büttner K, Hecker M, Becher D (2014) Comprehensive absolute quantification of the cytosolic proteome of *Bacillus subtilis* by data independent, parallel fragmentation in liquid chromatography/mass spectrometry (LC/MS(E)). Mol Cell Proteomics 13 (4):1008–1019
5. Markert S, Arndt C, Felbeck H, Becher D, Sievert SM, Hugler M, Albrecht D, Robidart J, Bench S, Feldman RA, Hecker M, Schweder T (2007) Physiological proteomics of the uncultured endosymbiont of *Riftia pachyptila*. Science 315(5809):247–250
6. Moche M, Albrecht D, Maass S, Hecker M, Westermeier R, Büttner K (2013) The new horizon in 2D electrophoresis: new technology to increase resolution and sensitivity. Electrophoresis 34(11):1510–1518
7. Bjellqvist B, Ek K, Righetti PG, Gianazza E, Gorg A, Westermeier R, Postel W (1982) Isoelectric focusing in immobilized pH gradients: principle, methodology and some applications. J Biochem Biophys Methods 6(4):317–339
8. Hessling B, Büttner K, Hecker M, Becher D (2013) Global relative quantification with liquid chromatography-matrix-assisted laser desorption ionization time-of-flight (LC-MALDI-TOF)—cross-validation with LTQ-Orbitrap proves reliability and reveals complementary ionization preferences. Mol Cell Proteomics 12 (10):2911–2920
9. Griebel A, Obermaier C, Westermeier R, Moche M, Büttner K (2013) Simplification and improvement of protein detection in two-dimensional electrophoresis gels with SERVA HPE lightning red. Arch Physiol Biochem 119(3):94–99

Chapter 8

Applications of Difference Gel Electrophoresis (DIGE) in the Study of Microorganisms

Kathleen Trautwein and Ralf Rabus

Abstract

OMICs-based investigations of microorganisms are becoming more and more widespread in the upcoming era of systems and synthetic biology. Here, proteomics plays a key role and two-dimensional difference gel electrophoresis (2D DIGE) remains the "gold-standard" for globally determining protein abundance changes on a quantitative and statistically confident level—in particular also for laboratories not having full-cycle proteomic facilities at their disposal. In this contribution we summarize our methodological procedures and experiences with 2D DIGE accumulated over the past 15 years.

Key words 2D DIGE, 2DE, Fluorescence labeling, Image analysis, Quantitative proteome profiling, Differential proteomics, Microorganisms

1 Introduction

1.1 Prospects of Differential Proteomics in Microbiology

Microbial research is wide-ranged by encompassing biotechnological applications, biomedicine, basic research interests and the relevance of microorganisms for ecosystem functioning. Modern DNA-sequencing technologies allow affordably determining the genomes of essentially any microbial species of interest. Likewise, bioinformatic and proteomic tools have reached a level of robustness and user-friendliness that increasingly enables the general (micro)biologist to hands-on access these technologies [1]. With the advent of systems and synthetic biology as the great challenges and promising opportunities in life sciences for the next decades, proteomics will retain its central role in OMICS-based technologies. Remarkably, this trend is also already consolidating in the field of environmental microbiology [2, 3] and was addressed in the PROTEOMICS Special Issue 13(18–19):2697–2930. Considering that soluble proteins (intra- and/or extracellular) mediate a substantial share of the metabolic activities of microbes, global protein abundance profiling by the "gold-standard" 2D DIGE should attract increasing attention and application. Even more so,

Dörte Becher (ed.), *Microbial Proteomics: Methods and Protocols*, Methods in Molecular Biology, vol. 1841, https://doi.org/10.1007/978-1-4939-8695-8_8,

as 2D DIGE is becoming affordable also for small or nonspecialist laboratories, enabling them to discriminate context-relevant proteins against the global proteome background and thereby to direct subsequent protein identification in a very target-oriented and economic manner.

1.2 General Background on Two-Dimensional Gel Electrophoresis (2DE)

2DE provides the highest resolution among gel-based protein separation techniques. Moreover, it is a robust method with low variability, offering also the possibility of multiplexing [4]. First-dimension separation is based on isoelectric focusing of proteins in immobilized pH gradients (IPG), the development of which proved decisive for reproducibility and ease of use in 2DE [5, 6]. Resolution and sensitivity can be further enhanced by application of overlapping, narrow-range pH-gradients, e.g., [7]. Following equilibration of IPG strips from the nonionic IEF to the ionic SDS PAGE system, second dimension separation is based on molecular weight differences. Noteworthy, proteins extracted from the analyzed sample remain intact, whereas gel-free separation techniques typically require cleavage of proteins into multiple peptides. This property of 2DE provides subsequent MS-based analysis with the highest degree of sample decomplexation as well as complete primary sequence information of the intact proteins.

1.3 Two-Dimensional Difference Gel Electrophoresis (2D DIGE)

DIGE is based on pre-electrophoretic labeling of samples with three different fluorophores (cyanine dyes [CyDyes] Cy2, Cy3 and Cy5) and was originally developed by Ünlü et al. [8]. Key to the functionality of CyDyes in 2D DIGE are the discrete absorption and emission spectra as well as the so-called charge and mass matching: i.e., (1) a positive charge compensates for the one lost during amide bond formation of CyDyes with the amino group in the side chain of lysine (preventing undesired p*I* shifts), and (2) equal masses of ~500 Da lead to almost identical mass shifts of the labeled proteins. Multiplexing (coseparating differentially labeled samples in the same 2DE gel) reduces gel-to-gel variation, allowing for the determination of protein abundance changes with high statistical confidence. For consistency and intuitive visualization of protein abundance changes (green: increased; red: decreased), we generally use Cy3 for the test state(s), Cy5 for the reference state and Cy2 for the internal standard. The latter comprises aliquots of the reference and all test states, and forms the basis for normalization and inter-gel comparison. Since the internal standard contains every protein present in the analyzed samples, spot matching can be reliably performed across all gels, independently of gel-to-gel electrophoretic variations. A pioneering evaluation of 2D DIGE in the study of microorganisms was reported by Gade et al. [9].

2 Materials

The materials and methods described in the following represent the protocols and laboratory infrastructure established and optimized over the last 15 years in our group. They are essentially based on the 2DE protocols of Görg et al. [5, 6] and the 2D DIGE protocol reported by Gade et al. [9]. Chemicals, instruments and software from other than the named providers may yield comparable results.

Double-distilled, sterile-filtered deionized water ($\geq$18.2 MΩ resistance) is used for preparation of all solutions and for the final rinsing of 2DE lab equipment. Use chemicals of highest quality and purity.

2.1 Cell Harvest, Cell Lysis, and Protein Concentration Determination

1. Cold (4 °C) washing buffer: 100 mM Tris–HCl, 5 mM $MgCl_2$; final pH 7.5. Autoclaved for sterilization.
2. Cell lysis buffer (*see* **Note 1**): 30 mM Tris, 7 M urea, 2 M thiourea, and 4% (w/v) CHAPS; final pH 8.5. To prepare 100 mL cell lysis buffer, first 42 g urea and then 15.2 g thiourea (add each in small portions) are completely dissolved in 35 mL H_2O. Use a water bath at 33 °C to facilitate dissolution (*see* **Note 2**). Add 4 g CHAPS and 3 mL Tris (from 1 M stock solution). Adjust to pH and final volume. Store in 1 mL aliquots at −20 °C.
3. PlusOne® Sample Grinding Kit (GE Healthcare, Uppsala, Sweden).
4. 1.5 mL polypropylene microcentrifuge tubes (Beckman Coulter, Krefeld, Germany).
5. pH-Fix test strips pH 7.5–9.5 (Macherey-Nagel, Düren, Germany).
6. Protein assay (Bio-Rad, Munich, Germany).
7. Fatty acid-free bovine serum albumin fraction V (e.g., Roche, Penzberg, Germany).
8. UV-quartz cuvette.
9. Liquid nitrogen.
10. High performance centrifuge (e.g., Avanti J-26XP with rotor JA-10; Beckman Coulter).
11. Centrifugation beakers with screw caps.
12. Refrigerated benchtop centrifuge (e.g., Eppendorf 5417 R; Eppendorf, Hamburg, Germany).
13. Ultracentrifuge (e.g., Optima Max-E with fixed-angle rotor TLA-110 and Delrin tube adapters for 1.5 mL microcentrifuge tubes; Beckman Coulter).
14. Spectrophotometer (e.g., UVmini-1240; Shimadzu, Duisburg, Germany).

2.2 Fluorescence Minimal Labeling

1. 5 nmol CyDye DIGE Fluor Minimal Labelling Kit (GE Healthcare).
2. 99.8% anhydrous *N*,*N*-dimethylformamide (DMF) (*see* **Note 3**).
3. 10 mM solution of L-lysine monohydrate, sterile-filtered. Stored in aliquots at −20 °C.
4. Mini centrifuge (e.g., MCF-2360; LMS Co., Tokyo, Japan).
5. Ice bath with lid.
6. Vortex shaker.

2.3 Sample Preparation for IEF

1. Carrier ampholytes with a pH range matching the used IPG strip (e.g., IPG buffer pH 3–11 NL; GE Healthcare).
2. 2× 2DE sample buffer: 7 M urea, 2 M thiourea, 2% (v/v) carrier ampholytes, 2% (w/v) DTT, 4% (w/v) CHAPS. To prepare 25 mL 2DE sample buffer, first 10.5 g urea and then 3.8 g thiourea (add each in small portions) are completely dissolved in 8.7 mL H_2O. Use a water bath at 33 °C to facilitate dissolution (*see* **Note 2**). Add 0.5 mL carrier ampholytes, 0.5 g DTT, and 1 g CHAPS and adjust to final volume. Store in 0.5 mL aliquots at −20 °C.
3. Rehydration buffer: 6 M urea, 2 M thiourea, 0.4% (w/v) DTT, 2% (w/v) CHAPS, and 0.5% (v/v) carrier ampholytes. Prepare and store analogous to the 2× 2DE sample buffer (see above). We prefer using DeStreak rehydration solution (GE Healthcare) to avoid horizontal smearing in the alkaline range of the 2DE gel (add also 0.5% (v/v) carrier ampholytes prior to use).

2.4 Active Rehydration: Loading and IEF

1. 24-cm long IPG strips (e.g., GE Healthcare; Serva Electrophoresis, Heidelberg, Germany) with a defined and immobilized pH gradient (*see* **Note 4**).
2. Gel-side down electrode wicks (e.g., Bio-Rad).
3. Mineral oil (e.g., Serva HPE IPG Overlay; Serva Electrophoresis).
4. Strip holder cleaning solution (e.g., Bio-Rad).
5. Protean i12 IEF system with 24 cm focusing tray (Bio-Rad).
6. Two pairs of tweezers.

2.5 Equilibration

1. Resolving gel buffer: 1.5 M Tris, 0.4% (w/v) SDS; adjust the pH to 8.8 using 4 M HCl. Store up to 2 weeks at 4 °C.
2. Basic equilibration buffer: 6 M urea, 30% (w/v) glycerol, 2% (w/v) SDS, 33.4 mL/L resolving gel buffer. Prepare at least 500 mL 1 day before the equilibration step due to slow dissolution of urea and SDS. Can be stored up to 2 weeks at room temperature.

3. Bromophenol blue solution: Dissolve 125 mg bromophenol blue (0.25%, w/v) in 50 mL resolving gel buffer containing 0.1 mg/mL sodium azide. Store at 4 °C.
4. DTT solution: Dissolve 2 g DTT (stored at 4 °C) in 200 mL basic equilibration buffer. Add 1 mL 0.25% (w/v) bromophenol blue solution. Prepare freshly before use.
5. Iodoacetamide solution: Dissolve 8 g iodoacetamide (stored at 4 °C) in 200 mL basic equilibration buffer. Add 1 mL 0.25% (w/v) bromophenol blue solution. Prepare freshly before use.
6. Glass tubes (diameter: 1.6 cm, length: 27 cm; Glasgerätebau Ochs, Bovenden, Germany) and plastic stoppers.
7. Two pairs of tweezers.
8. Electronic pipette (e.g., Easypet 3; Eppendorf).
9. Glass pipettes (20 mL).
10. Horizontal shaker.

2.6 SDS-PAGE

1. Lint-free paper.
2. Preparation of 14 SDS-PAGE gels (size 27 × 21 cm; 1.5 mm thick): (a) 415 mL 30% (w/v) acrylamide–bis solution (37.5:1; Serva Electrophoresis); (b) 247 mL slab gel buffer (1.5 M Tris: 1.08 M Tris and 0.42 M Tris–HCl; final pH 8.7–9.0; store at 4 °C); (c) 325 mL H_2O; (d) 10.3 mL 10% (w/v) SDS; (e) 0.51 mL *N*,*N*,*N*′,*N*′-tetramethylethane-1,2-diamine (TEMED), and (g) 2.5 mL 10% (w/v) ammonium persulfate (freshly prepared).
3. 1× running buffer (>10 L): 25 mM Tris, 192 mM glycine, and 0.1% (w/v) SDS. Prepare a tenfold concentrate of the running buffer and dilute accordingly. Store at room temperature.
4. Prepare 100 mL of a 40% (w/v) glycerol solution dissolved in 1× running buffer and add 400 μL 0.25% (w/v) bromophenol blue solution for coloration.
5. Prepare 0.1% (w/v) SDS solution and store in a spray bottle at room temperature.
6. Agarose sealing solution. Dissolve 0.5% (w/v) agarose and some crystals of bromophenol blue in 1× running buffer. Aliquot 1 mL portions of the hot solution in 1.5 mL microcentrifuge tubes.
7. Ettan Dalt*twelve* large vertical electrophoresis system (GE Healthcare).
8. Ettan Dalt*twelve* gel caster for 14 SDS-PAGE gels (GE Healthcare).
9. Fourteen low-fluorescence glass plate kits (size 27 × 21 cm; e.g., GE Healthcare).

10. Vacuum filtration unit.
11. Magnetic stirrer.
12. Glass pipettes (5 and 10 mL).
13. Funnel for gel casting (e.g., GE Healthcare).
14. Scalpel.
15. Two pairs of tweezers.
16. Thermomixer (e.g., from Eppendorf).
17. 5 mL pipette and tips (e.g., from Eppendorf).

2.7 Digital Acquisition of Gel Images and Software-Based Image Analysis

1. Typhoon 9400 three-mode laser fluorescence scanner (e.g., GE Healthcare) or high-performance CCD camera system (e.g., Intas Science Imaging Instruments GmbH, Göttingen, Germany).
2. ImageQuant software (GE Healthcare).
3. DeCyder 2D software package (GE Healthcare) or alternative.

3 Methods

An important basis for a successful 2D DIGE study is the establishment of a stringent working timeline (exemplified in Fig. 1) and a well-conceived experimental design to assure correct fluorescence labeling (Fig. 2).

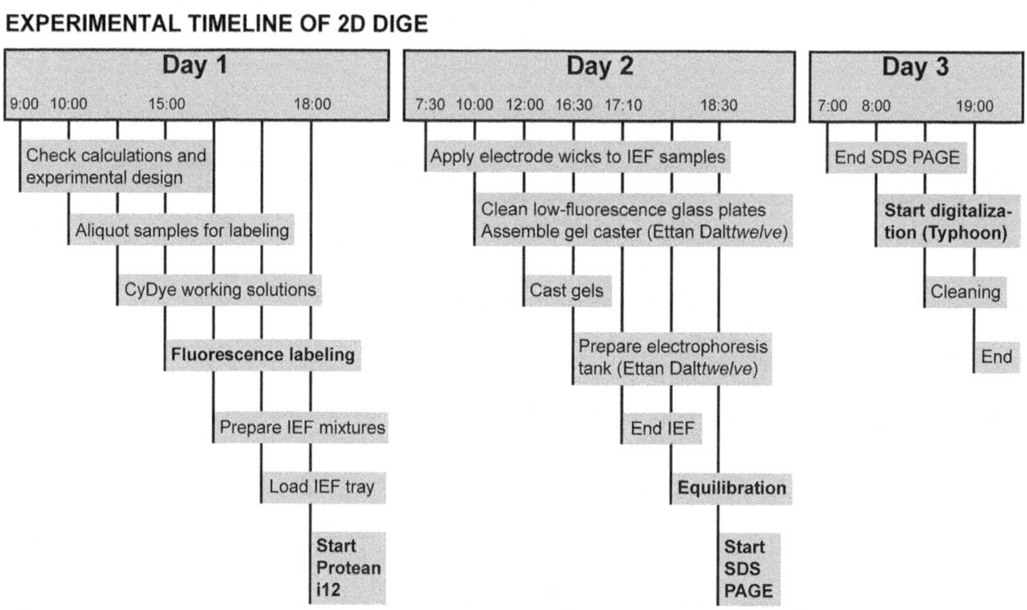

Fig. 1 A typical timeline for 2D DIGE experiments without image analysis. The preparation of protein samples is performed in advance and not shown. With complex DIGE work packages it can be helpful to prepare and freeze the aliquots for labeling before day 1. If no time point is indicated, the respective task has to be completed prior to the subsequent step

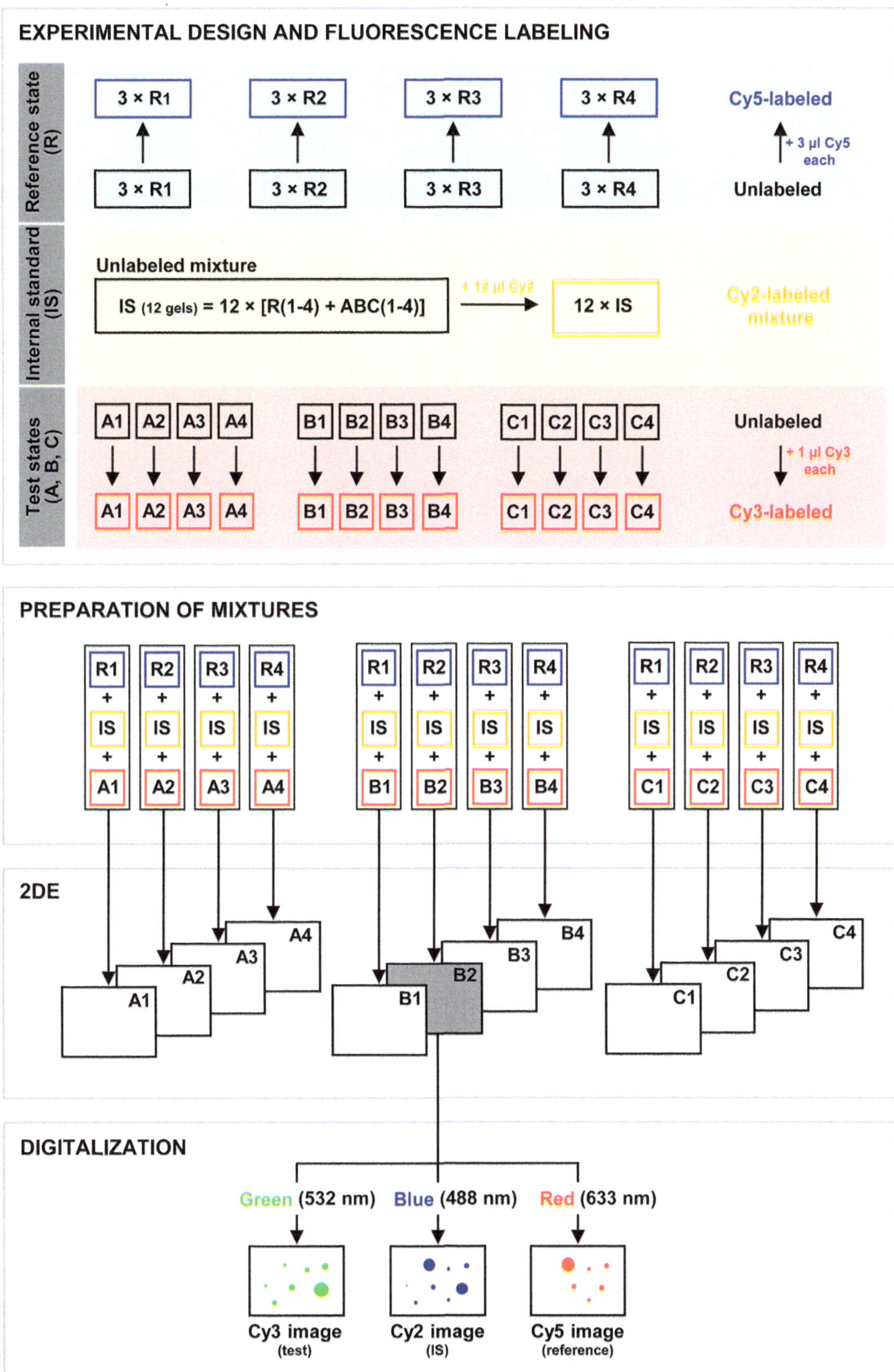

Fig. 2 Design and workflow of a typical 2D DIGE experiment divided in four major working steps. (1) *Experimental design and fluorescence labeling.* Samples derive from four independent cultures (biological

3.1 Preparation of Cell Lysate for 2D DIGE

1. Harvest bacterial cells by centrifugation (e.g., 10,000 × *g*, 20 min, 4 °C; *see* **Note 5**) and wash cell pellets twice with cold Tris washing buffer. Keep cell pellets on ice during harvesting. Cell pellets (about 100 mg wet weight each; *see* **Note 6**) should be frozen immediately in liquid nitrogen and stored at −80 °C until cell lysis.
2. Cell pellets are quickly resuspended in 1 mL cell lysis buffer (*see* **Note 6**). Cell breakage is performed at room temperature with the PlusOne Sample Grinding Kit (GE Healthcare). Resuspend the cell pellet with a pipette first in 0.5 mL cell lysis buffer, before adding another 0.5 mL of the latter. The cell suspension can be difficult to pipette. Try to transfer about 200 μL to the grinding tube and grind for 1 min using a timer. Move the grinding pistil fast up and down (be careful not to use strong force). Centrifuge for 2–3 min at room temperature and transfer the supernatant into a polypropylene microcentrifuge tube (for ultracentrifugation). Repeat the procedure until the entire cell suspension has been processed (*see* **Note 7**).
3. Separate proteins from cell debris by ultracentrifugation (100,000 × *g*, 1 h, 17 °C). Determine the pH of the supernatant using a 10 μL sample pipetted onto a pH test strip of pH 7.5–9.5. Use the cell lysis buffer (pH 8.5) for positive color control. Adjust the pH of the protein extract to 8.4–8.6 using 1 M NaOH or 1 M HCl (mix by inverting the tube only). Protein extracts are frozen in liquid nitrogen and stored in aliquots of 200–250 μL at −80 °C. Use one aliquot of 30 μL for determination of the protein concentration with the Bio-Rad colorimetric assay and bovine serum albumin (BSA) as calibration standard.
4. Determination of exact protein concentrations is crucial for quantifying small changes in protein abundance. The exact concentration of the BSA standard solution is measured in a quartz cuvette at 280 nm against water as blank (BSA stock solution of 1 mg/mL has an absorption of 0.667). Dilute the BSA stock solution (stored in aliquots at −20 °C) with water to obtain 4 mL of a 20 μg/mL working solution. Use the latter to

Fig. 2 (continued) replicates) of different treatment groups (e.g., cells grown with different substrates). Each 50 μg protein sample is labeled with 1 μL (200 pmol) CyDye. Four 150 μg aliquots of the reference state (R1–R4) are labeled each with 3 μL Cy5. Twelve 50 μg aliquots of the test states (A1–A4, B1–B4, C1–C4) are labeled each with 1 μL Cy3. The pooled internal standard (IS) contains all protein samples from the work package and a 12-fold preparation of the IS is labeled with 12 μL Cy2. (2) *Preparation of mixtures.* The labeled proteins are mixed to yield 50 μg protein for the IS, reference and test state. (3) *2DE.* The protein mixtures are separated by isoelectric focusing (first dimension) and SDS PAGE (second dimension). (4) *Digitalization.* 2D DIGE gels are digitalized with a three-mode fluorescence laser scanner (Typhoon), acquiring three CyDye-specific images per gel. Modified from Rabus and Trautwein [10]

prepare triplicates of BSA standards within a concentration range from 0–10 μg/mL. Prepare in triplicates four different dilutions of the protein sample between 1000- and 10,000-fold. Determine the protein concentration according to the manufacturer's instructions. Use the values of at least two different dilutions to calculate the protein concentration in the samples destined for 2D DIGE. If values deviate strongly, repeat the protein assay.

3.2 Experimental Design

1. Carefully prepare an exact pipetting scheme for preparation of the internal standard, reference and each test state (*see* **Note 8**).
2. If the protein concentration exceeds 10 μg/μL, dilute the samples with cell lysis buffer (for fluorescence labeling the protein concentration should range between 1 and 10 μg/μL).
3. Prepare the internal standard from equal amounts of all protein samples (reference and test states) to yield a total of 50 μg protein per gel. For twelve 2D DIGE gels, prepare a 26-fold master mix of the internal standard, mix gently and aliquot in 12-fold portions (the second aliquot serves as backup only).
4. Prepare aliquots of 50 μg protein per gel from the reference and each test state (prepare an extra aliquot as backup).
5. Freeze all aliquots in liquid nitrogen and store them at −80 °C or use them immediately for fluorescence labeling.

3.3 Fluorescence Minimal Labeling

1. Due to the light sensitivity of the CyDyes, the labeling reaction should be carried out without any delay and at weak illumination.
2. The three CyDyes are reconstituted in DMF according to the manufacturer's instructions. 5 nmol of each CyDye are dissolved in 5 μL of DMF to prepare the CyDye stock solutions (1 nmol/μL). The latter is stored in two aliquots of 2.45 μL each at −20 °C. Each aliquot is sufficient for the preparation of twelve 2D DIGE gels (*see* **Note 9**).
3. Before the labeling, 10 μL DMF is added to yield the CyDye working solution (200 pmol/μL). Mix by vortexing and spin down the solution using a mini centrifuge and keep CyDyes on ice in the dark.
4. Set a timer to 30 min and start the labeling (200 pmol CyDye per 50 μg protein) by adding 1 μL CyDye to each protein sample (e.g., 12 μL to the 12 × internal standard, 3 μL to each 3 × reference state, 1 μL to each 1 × test state; see Fig. 2), mix gently and spin down shortly in a mini centrifuge. Incubate on ice in the dark for 30 min. Remember time of CyDye addition and chronological order of reaction tubes to assure an equal labeling time for all samples.

5. Stop the labeling reaction by addition of 1 μL 10 mM lysine per 50 μg protein, mix gently and spin down shortly in mini centrifuge. Incubate at least 10 min on ice in the dark.

3.4 Active Rehydration: Loading and IEF

1. Prepare 12 microcentrifugation tubes (1.5 mL) labeled from number 1–12.
2. Add to each tube the required 2× 2DE sample buffer (volume added is equal to the sum of the three labeled samples) and rehydration buffer/DeStreak to obtain a final volume of 450 μL (*see* **Note 10**). Add 50 μg of the Cy2-, Cy3-, or Cy5-labeled protein extracts in each tube. Vortex the protein mixtures and spin them down shortly.
3. Distribute 450 μL protein mixture evenly in one lane of the 24 cm focusing tray of the Protean i12. Use two tweezers to remove the plastic cover from the IPG strip and to place the strip gel-side down into the focusing lane. Carefully remove air bubbles and cover the lane with 4 mL mineral oil to prevent evaporation and urea or protein precipitation. Proceed with the next sample accordingly until all lanes are occupied. Position the electrodes and insert the focusing tray into the Protean i12. Add at least another 2 mL mineral oil per lane to cover the IPG strips completely. Place one IPG strip retainer close to each electrode to keep the strips in position during IEF.
4. The following settings are applied in our laboratory for active rehydration of IPG strips, sample entry and IEF (*see* **Note 11**) with the Protean i12 (50 μA/strip, 25 °C): (step 1) 50 V for 13 h; (step 2) 200 V for 1 h; (step 3) 1000 V for 1 h; (step 4) gradual increase from 1000 to 10,000 V in 2 h; (step 5) 10,000 V for 60,000 Vhs (increase Vhs for narrow-range pH gradients; *see* **Note 12**). Add an additional step to keep the focused IPG strips at 1500 V until you end the run.
5. Wet 24 gel-side down electrode wicks with water and place them between the electrode and IPG strips in **step 2** (*see* **Note 13**).
6. Remove focused IPG strips with tweezers and rinse shortly with water to remove excess oil. Place each IPG strip in a long glass tube for subsequent equilibration (*see* **Note 14**).
7. Clean the focusing tray only with a nonionic detergent.

3.5 Equilibration

1. Add 15 mL of DTT-containing basic equilibration buffer to each tube and close with stopper immediately. Incubate IPG strips for 15 min on a horizontal shaker at 20 rpm.
2. Remove stoppers and pour out the buffer completely from the glass tubes. Prevent the IPG strip from dropping out by placing a (glove-protected) finger in front of the glass tube opening.

3. Add 15 mL of iodoacetamide-containing basic equilibration buffer to each tube and close with new stopper immediately. Incubate IPG strips again for 15 min on a horizontal shaker at 20 rpm.
4. The equilibrated IPG strips (*see* **Note 15**) are directly used for SDS PAGE (*see* Subheading 3.6).

3.6 SDS-PAGE

Casting Gels

1. Reclean the inner face of 14 low-fluorescence glass plate pairs and the spacers using lint-free wipes first with denatured ethanol (100%) and then with water. Remove dust particles as good as possible and assemble the glass plate sandwich.
2. Assemble the Ettan Dalt*twelve* gel-casting chamber according to the manufacturer's instructions.
3. Prepare the gel polymer as follows (*see* Subheading 2.6 for exact volumes): (a) filter the Tris slab gel buffer using a vacuum manifold; (b) add water; (c) remove the filter assembly from the bottle and add acrylamide–bis solution; (d) close the bottle with a rubber stopper, stir slowly and degas the solution under vacuum for 10 min; (e) add 10% SDS, TEMED and freshly prepared ammonium persulfate (avoid reintroducing air bubbles).
4. Pour the gel polymer carefully into a funnel (avoid air bubbles) and fill the gel-casting chamber slowly from bottom to top. Replace the gel polymer in the tubing of the gel caster with the blue-colored glycerol solution. The gel polymer should stop 2–3 cm below the shorter glass plate to insert the IPG strip later.
5. Cover the top of the gels with a layer (some millimeters) of sprayed 0.1% SDS. Close the chamber with the lid and leave to polymerize for at least 3 h (at 18–20 °C).

SDS-PAGE

6. Mix 1 L of 10 × running buffer and 9 L of water in the Ettan Dalt*twelve* electrophoresis chamber (switch on the pump). Remove 2.5 L of the 1 × running buffer (to be added back later) before inserting the gel cassettes into the electrophoresis chamber.
7. Take out the gel cassettes from the gel caster and clean the bottom from excess polymer using a scalpel. Rinse glass plates and recess for IPG strips thoroughly. Visually inspect the quality of the gels (for trapped air bubbles or dust particles, polymerization defects, smooth and straight top rim of the gels). Select the twelve best gels and insert the wet glass plate pairs into the electrophoresis chamber.

8. Add 1× running buffer until the recess for IPG strips on top of the gels is completely filled.
9. Dissolve 12 aliquots of 1 mL agarose sealing solution in a thermomixer (first 95 °C until dissolution, then 65–70 °C).
10. Briefly rinse the equilibrated IPG strip with water and 1× running buffer. Place the IPG strip between the two glass plates on top of the SDS polyacrylamide gels. The acidic end of the IPG strip is placed on the left side, with the gel-side facing the experimenter; gently push the IPG strip down (e.g., with a separator-sheet from the gel caster accessories). Quickly and evenly distribute hot agarose sealing solution to fix the position of the IPG strip in its recess. Remove displaced agarose sealing solution from the electrophoresis chamber with a 5 mL pipette.
11. Fill the electrophoresis chamber with the remaining 1× running buffer and start the electrophoresis at 30 W constant power and 25 °C (run time 10–12 h). Stop the SDS-PAGE, when the bromophenol blue marker (from the agarose sealing solution) has reached the bottom of the gels.
12. Rinse the outer faces of the glass plate pairs very thoroughly and store the 2D DIGE gels temporarily at 4 °C in the dark (*see* **Note 16**).

3.7 Acquisition of Digital Gel Images

1. Start immediately after SDS-PAGE is completed with the digitalization of the first two 2D DIGE gels (still sandwiched between glass plates!).
2. Dry the glass surface of the 2D DIGE gels completely with lint-free paper and remove any smear or dust particles. Use the gel alignment guide of the Typhoon scanner accessories to place the two 2D DIGE gels on the glass surface of the scanner (*see* **Note 17**).
3. In the Typhoon scanner control software adjust the laser settings to Blue2 (488 nm) for Cy2, Green (532 nm) for Cy3, and Red (633 nm) for Cy5. The emission filters are set to 520 nm (bandpass 40) for Cy2, 580 nm (bandpass 30) for Cy3 and 670 nm (bandpass 30) for Cy5. Activate "press sample" and "DIGE file naming format," set the focal plane to +3 mm and choose the correct orientation of the gels on the scanner surface from the menu.
4. Perform a prescan with a pixel size of 1000 μm resolution to adjust the photomultiplier voltage (typically between 480 and 550 V) for every laser and gel to optimal pixel intensities of 50,000–80,000 ($\geq$100,000 the signal is saturated and unusable for quantification). Pixel intensities are measured with ImageQuant software. Repeat prescan, if necessary.

5. Perform the main scan at 100 μm resolution. The scan time is approximately 80 min for two gels. Control the pixel intensities after each scan with ImageQuant software.
6. Following successful digitalization, discard the gels (*see* **Note 18**) and clean the glass plates.

3.8 Software-Based Image Analysis (DeCyder 2D)

The basic principles of image analysis by DeCyder 2D software are displayed in Fig. 3.

1. Image analysis with DeCyder 2D software is performed as described by the manufacturer (*see* **Note 19**).

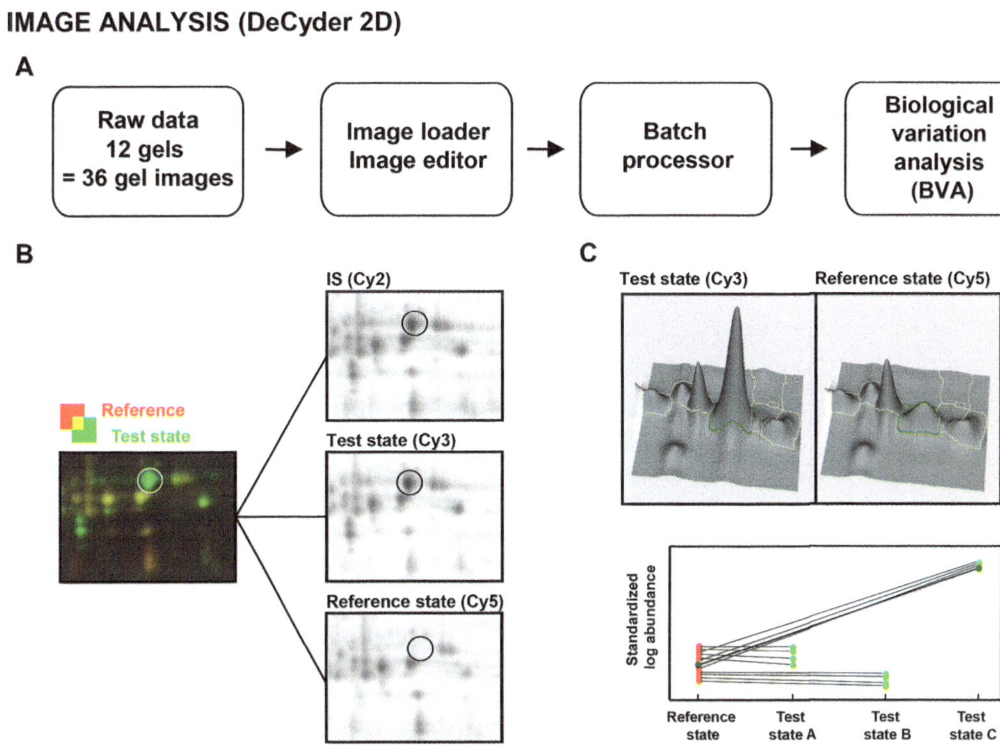

Fig. 3 Basic principle of image analysis by DeCyder 2D software. (**a**) Raw images are loaded into DeCyder 2D using the image loader tool. The image editor tool can be used to remove unnecessary gel parts (e.g., the spacer of the glass cassettes). The batch processor automatically performs spot detection and matching, exclusion of nonproteinaceous spots and applies statistics to detect changes in protein abundances across all gel images in the work package. The biological variation analysis (BVA) module of DeCyder 2D is used to manually correct the spot matching and to assess differentially abundant protein spots. (**b**) Selected part from a 2D DIGE gel shown as false-color overlay (Cy3 *versus* Cy5) or as CyDye-specific grey-colored images. Proteins with increased abundance appear green, whereas those with decreased abundance appear red. Unchanged abundances are reflected by the yellow color. The marked protein spot is present in the IS and test state, but absent in the reference state. (**c**) The BVA module visualizes a selected protein spot (dark green spot boundary) and its surrounding. The small circles in the diagram at the bottom show the abundance changes of this spot in the selected (shown in dark green) and across all other gel images. Modified from Rabus and Trautwein [10]

2. Use the DeCyder image loader tool to select and incorporate all gel images into a new or existing project.
3. Always crop the gel images to similar size to exclude nonessential image parts (in ImageQuant or directly with the DeCyder image editor tool).
4. Use the batch processor to automatically analyze all gel images of the work package (*see* Fig. 3 in ref. 10). Add all images by "add DIA batch item". Select the DeCyder spot detection algorithm 6.0 and enter "3000" for the estimated number of spots. Adjust parameters for exclusion of nonproteinaceous signals. The following exclusion filter settings proved applicable to many different samples in our laboratory: slope >1, area <200, height <190 and volume <30,000 or 60,000. Check the "include in BVA batch list" box and click "ok" to proceed. Define the experimental groups and sort gel images according to their CyDye label. Select one representative Cy2-labeled image as master for spot matching and click "ok". In the next window, define the calculated statistics: select "average ratio" and "Student's *t*-test," and if more than two different test states are included in the analysis, also "one-way ANOVA". Save and start batch processing by "run batch".
5. Open the biological variation analysis (BVA) module and load the BVA file created by the batch processor. Use this program to manually control all differentially abundant protein spots for correct spot matching and to select protein spots for MS-based identification with an average ratio of $\leq$–1.5 and $\geq$1.5 (*see* **Note 20**), an ANOVA P value of <0.05, a *t*-test value of $\leq 10^{-4}$ and present in at least 75% of all analyzed gels.

3.9 Identification of Differentially Abundant Proteins

1. Prepare separate 2DE gels (300 μg protein load each; *see* **Note 18**) and stain them with colloidal Coomassie Brilliant Blue (cCBB) G250 according to the protocol reported by Neuhoff et al. [11]. Acquire digitalized gel images, mark protein spots of interest and excise them from the 2DE gel (manually or automatic).
2. Digest excised proteins with trypsin, analyze resulting peptides by MALDI-TOF-MS(/MS) and identify proteins by peptide mass fingerprinting (*see* **Note 21**).

4 Notes

1. The cell lysis buffer does not contain DTT and carrier ampholytes to prevent thiol and primary amino groups from reacting with the NHS esters of the CyDyes during fluorescence labeling. At pH 8.5 optimal labeling conditions are achieved.

2. Prevent carbamylation of proteins by keeping urea-containing buffers below 37 °C. Carbamylation changes the isoelectric point of proteins, since cyanate formed from urea at elevated temperatures reacts with amino groups (i.e., N-terminus as well as side chains of lysine and arginine). This can cause electrophoretic artifacts (i.e., appearance of multiple spots of the same protein).
3. Always use fresh and anhydrous DMF (<3 months old). Immediately before CyDye reconstitution, penetrate the airtight seal of the DMF bottle with a thin needle and take out 100–200 μL using a disposable syringe.
4. Choose the appropriate pH gradient for the IPG strips depending on your required analytical window (e.g., "zoom-in" pH 3–4 or broad-range pH 3–11). Choose a nonlinear (NL) gradient, if you need a higher pH resolution in the center of the pH gradient.
5. Skilled, reproducible cultivation and cell harvesting practices are absolutely necessary. For 2D DIGE analyses of substrate-dependent catabolic subproteomes adapt the cells to the desired substrate (by passaging five times) and harvest the cells during active growth (e.g., at respective half-maximal optical densities). This avoids detection of unwanted and interfering growth phase-dependent changes ("false positives"). Always try to minimize introduction of artifacts due to cultivation handling. Use only cultures with a similar cultivation history (i.e., derived from the same stock culture, similar number of passages). Cultivate and harvest three to four biological replicates per sample state (due to the low technical variance of 2D DIGE, technical replicates are generally not necessary with the model organisms investigated in our group; *see* **Note 20**).
6. If the amount of cell material is restricted, also cell pellets of 50 mg wet weight or even less can be sufficient. In this case, reduce the volume of the cell lysis buffer to about 0.5 mL.
7. Control the performance of cell breakage with a microscope. This is especially important in case you are working with an organism for the first time.
8. Based on our experience up to ten different states can be analyzed per 2D DIGE work package (i.e., having the same internal standard). Automatic spot detection and matching in DeCyder 2D is based on the internal standard. With >10 analyzed states in the internal standard, the automatic spot detection algorithm may not recognize low abundant proteins. Plan the experiment and pipetting scheme carefully (e.g., using a commercially available spreadsheet software). Calculate the required volumes for obtaining 50 μg total protein per gel and label. Using color-coded labels (matching the CyDye labeling)

on the lids of the microcentrifugation tubes helps not to mix up samples.

9. This is a deviation from the manufacturer's recommendation of using 400 pmol per CyDye for labeling 50 μg protein. To our experience, the same labeling results are obtained using only 200 pmol per CyDye. This allows preparation of 24 instead of only twelve 2D DIGE gels per 5 nmol CyDye and decreases the costs.
10. Make sure that you added the carrier ampholytes. Broad-range carrier ampholyte solutions (e.g., pH 3–11) will also work for other IPG strips within this pH range (e.g., pH 4–7).
11. The protein sample enters the gel matrix of the IPG strip during active rehydration at low voltage. Alternatively, the IPG strips can be rehydrated with 450 μL rehydration buffer/DeStreak (with carrier ampholytes) in the absence of the protein sample in separate rehydration trays for 10–12 h (passive rehydration). The volume of the rehydration buffer depends on the length of the IPG strip. Proteins are then actively loaded onto the IPG strips from plastic cups sitting on top of the gel surface in close vicinity to the anode (cup-loading). See the Protean i12 guidelines for details on this method. In case of bad pH resolution or smearing, changing the loading procedure of the IPG strips can improve the quality of the 2DE gels.
12. For narrow-range pH gradients of only one pH unit (e.g., pH 3–4) increase to 90,000 Vhs in **step 5**.
13. Electrode wicks trap ions (including proteins with p*I*s beyond the applied pH range) and thereby help to reduce the electric current within the IPG strip. A low electric current (<50 μA/strip) is necessary to achieve the applied high voltages in IEF. At higher ionic strengths, the Protean i12 restricts the voltage to lower values and depending on the ionic strength in your sample, this can result in substantially elongated IEF times.
14. The focused and oil-covered IPG strips can be stored at −20 °C for 3 months. Do not rinse the IPG strips with water before freezing. In general, however, this should be avoided.
15. The DTT added to the equilibration buffer reduces disulfide bonds in proteins. This is a reversible process, which is prevented by subsequent alkylation of reduced thiol groups by iodoacetamide during the second equilibration step.
16. Keep the gels at 4 °C until scanning to limit diffusion of protein spots.
17. Different companies are currently developing camera-based digitalization systems for 2D DIGE. We recently tested a CCD camera system (Intas Advanced 2D Imager; Intas Science

Imaging Instruments GmbH) in comparison to the old "gold standard" Typhoon fluorescence scanner. Comparable performances were observed for protein abundance changes, linear range and limit of the detection [12].

18. In some cases, it might be useful to keep 2D DIGE gels for subsequent colloidal Coomassie Brilliant Blue (cCBB) staining. However, low abundant proteins may not be visible after staining with cCBB, since one 2D DIGE gel contains only 150 μg total protein (three times 50 μg). Due to the simultaneous presence of internal standard, reference and test state, spot excision for MS-based analysis can result in ambiguous protein identifications. It is therefore recommended to routinely prepare separate, cCBB-stained 2DE gels containing 300 μg total protein of only one sample.
19. Alternative software for analysis of 2D DIGE gel images is, e.g., Delta2D software (Decodon, Greifswald, Germany) or SameSpots (TotalLab Ltd., Newcastle upon Tyne, UK).
20. The significance threshold in protein abundance changes of >|1.5| has been determined experimentally and depends on the biological (genetic) variation of individual biological replicates (~20%, equals to average ratio of |1.2|) [13]. Technical variation due to the labeling and electrophoretic differences is generally negligible due to using an internal standard and coseparation of different samples on a single 2D DIGE gel.
21. The mass spectrometric (MS) analysis provides information on the molecular weight of peptides. Protein identification is achieved by comparison of the determined mass spectrometric data to a protein sequence database (peptide mass fingerprint, PMF). The latter is built from the (high-quality) genome sequence of the study organism and contains the molecular weight of each peptide generated during an in silico tryptic digestion of all translated proteins [1].

Acknowledgment

Proteomic research (incl. DIGE) in our laboratory was supported by the German Federal Ministry of Education and Research BMBF (REGX-project), the Max Planck Society, and the Deutsche Forschungsgemeinschaft (SPP 1319, SFB TRR 51, GRK 1885).

References

1. Wöhlbrand L, Trautwein K, Rabus R (2013) Proteomic tools for environmental microbiology—a roadmap from sample preparation to protein identification and quantification. Proteomics 13(18–19):2700–2730
2. Rabus R, Trautwein K, Wöhlbrand L (2014) Towards habitat-oriented systems biology of "*Aromatoleum aromaticum*" EbN1: chemical sensing, catabolic network modulation and growth control in anaerobic aromatic

compound degradation. Appl Microbiol Biotechnol 98(8):3371–3388

3. Rabus R (2014) Fifteen years of physiological proteo(geno)mics with (marine) environmental bacteria. Arch Physiol Biochem 120 (5):173–187
4. Rabilloud T, Chevallet M, Luche S, Lelong C (2010) Two-dimensional gel electrophoresis in proteomics: past, present and future. J Proteome 73(11):2064–2077
5. Görg A, Obermaier C, Boguth G, Harder A, Scheibe B, Wildgruber R, Weiss W (2000) The current state of two-dimensional electrophoresis with immobilized pH gradients. Electrophoresis 21(6):1037–1053
6. Görg A, Weiss W, Dunn MJ (2004) Current two-dimensional electrophoresis technology for proteomics. Proteomics 4(12):3665–3685
7. Hoving S, Voshol H, van Oostrum J (2000) Towards high performance two-dimensional gel electrophoresis using ultrazoom gels. Electrophoresis 21(13):2617–2621
8. Ünlü M, Morgan ME, Minden JS (1997) Difference gel electrophoresis: a single gel method for detecting changes in protein extracts. Electrophoresis 18(11):2071–2077
9. Gade D, Thiermann J, Markowsky D, Rabus R (2003) Evaluation of two-dimensional difference gel electrophoresis for protein profiling. Soluble proteins of the marine bacterium Pirellula sp. strain 1. J Mol Microbiol Biotechnol 5(4):240–251
10. Rabus R, Trautwein K (2010) Proteogenomics to study the anaerobic degradation of aromatic compounds and hydrocarbons. In: Timmis KN (ed) Handbook of hydrocarbon and lipid microbiology. Springer, Berlin, pp 4385–4405
11. Neuhoff V, Arold N, Taube D, Ehrhardt W (1988) Improved staining of proteins in polyacrylamide gels including isoelectric focusing gels with clear background at nanogram sensitivity using Coomassie Brilliant Blue G-250 and R-250. Electrophoresis 9(6):255–262
12. Strijkstra A, Trautwein K, Roesler S, Feenders C, Danzer D, Riemenschneider U, Blasius B, Rabus R (2016) High performance CCD camera system for digitalization of 2D DIGE gels. Proteomics 16(14):1975–1979
13. Zech H, Echtermeyer C, Wöhlbrand L, Blasius B, Rabus R (2011) Biological versus technical variability in 2-D DIGE experiments with environmental bacteria. Proteomics 11 (16):3380–3389

Chapter 9

Proteomic Signatures in *Staphylococcus aureus*

Susanne Engelmann and Stephan Fuchs

Abstract

Despite all its apparent limitations proteome analysis based on two-dimensional protein gels combined with mass spectrometry is still the method of choice to study global protein synthesis activity in bacterial cells. Alterations in global protein synthesis play an important role during adaptation of bacteria to changing environmental conditions which are rather the role than the exception in their natural habitats. The protein synthesis pattern in response to a certain stimulus is highly specific and reflects the new challenges the bacterium has to meet. Here we present the techniques to analyze global protein synthesis in bacteria as exemplified by *Staphylococcus aureus* which is an important human pathogen and one main cause of nosocomial infections with severe outcome.

Key words 2D PAGE, [^{35}S]-L-methionine pulse labeling, Proteomic signatures, *Staphylococcus aureus*, Expression database

1 Introduction

Bacteria are frequently exposed to changing environmental conditions and adapt by rapidly reprogramming global gene expression at different levels including transcriptional, translational and post-translational events. Gel based proteomics in combination with mass spectrometry is a powerful technique to study effects on global protein synthesis and accumulation in bacteria triggered by specific stimuli [1]. Under well-defined laboratory conditions each specific stimulus induces and represses the synthesis of a definite set of proteins also referred as proteomic signature of the respective condition [2]. Due to their specificity proteomic signatures derived from several well-defined laboratory conditions are excellent tools to decipher the adaptation process of bacterial cells under conditions typically found in their natural habitats where they are confronted with a multitude of unknown signals [3]. In addition, proteomic signatures have been successfully applied to understand the mode of action of new antibacterial compounds [4, 5].

Dörte Becher (ed.), *Microbial Proteomics: Methods and Protocols*, Methods in Molecular Biology, vol. 1841, https://doi.org/10.1007/978-1-4939-8695-8_9, © Springer Science+Business Media, LLC, part of Springer Nature 2018

Staphylococcus aureus is a Gram-positive bacterium that normally colonizes the human nose without any symptoms. However, the number and severity of infections caused by this pathogen is steadily increasing during the last years and new therapeutic tools and an improved prevention of these infections is urgently needed. Adaptation of *S. aureus* to its host during colonization and infection is poorly understood so far and thus demands special attention in the future.

To visualize rapid changes in protein synthesis of *S. aureus* in response to specific environmental changes, metabolic pulse labeling of proteins using [^{35}S]-methionine was applied and proteins were subsequently separated by two-dimensional (2D) protein gel electrophoresis (Fig. 1). The protein synthesis patterns obtained for the different conditions were compared and proteins with altered synthesis profiles have been detected (Fig. 2a) [3, 6–11]. The comparison of different signatures is essential to differentiate between stimulus specific and growth rate dependent effects. To compare protein expression profiles between different experiments, an integrative identification-based approach has been introduced previously [10]. The identity of proteins was determined by allocating each protein spot detected under the different conditions to a

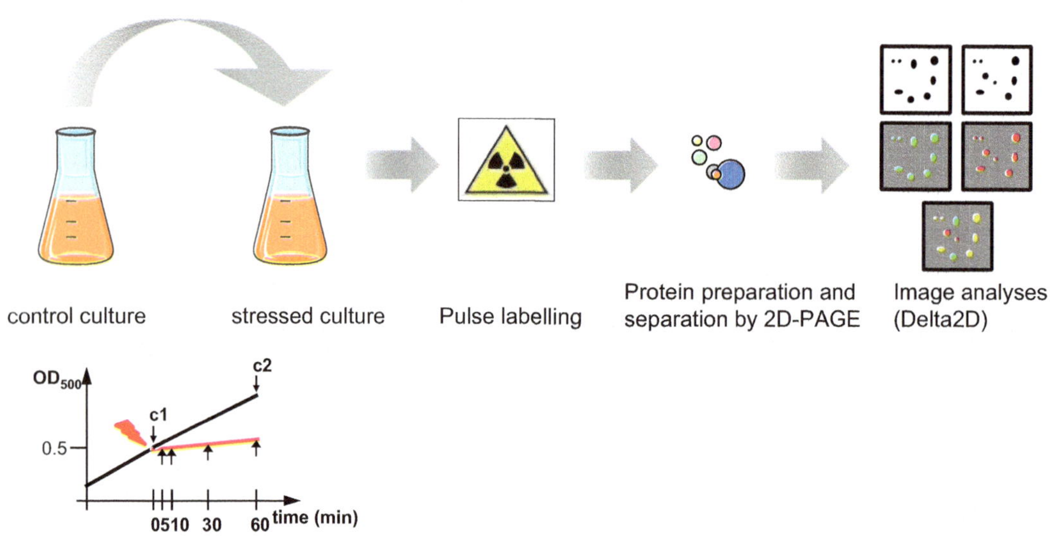

Fig. 1 Typical workflow of an in vivo protein pulse labeling experiment using [^{35}S]-L-methionine to determine protein synthesis rates. Bacterial cells are cultivated in a synthetic medium without methionine to the exponential growth phase (OD_{540} = 0.5). Cells are exposed to the respective conditions and proteins are labeled with [^{35}S]-L-methionine for 5 min at different time points after imposition to stress. Cytoplasmic proteins are prepared and separated on 2D-PAGE. Radioactively labeled proteins are detected using phosphor screens and a Typhoon imager. The obtained images are compared using DELTA-2D software (Decodon, Greifswald, Germany). Protein spots are quantified and synthesis ratios for each protein spot in response to the respective stress condition are calculated

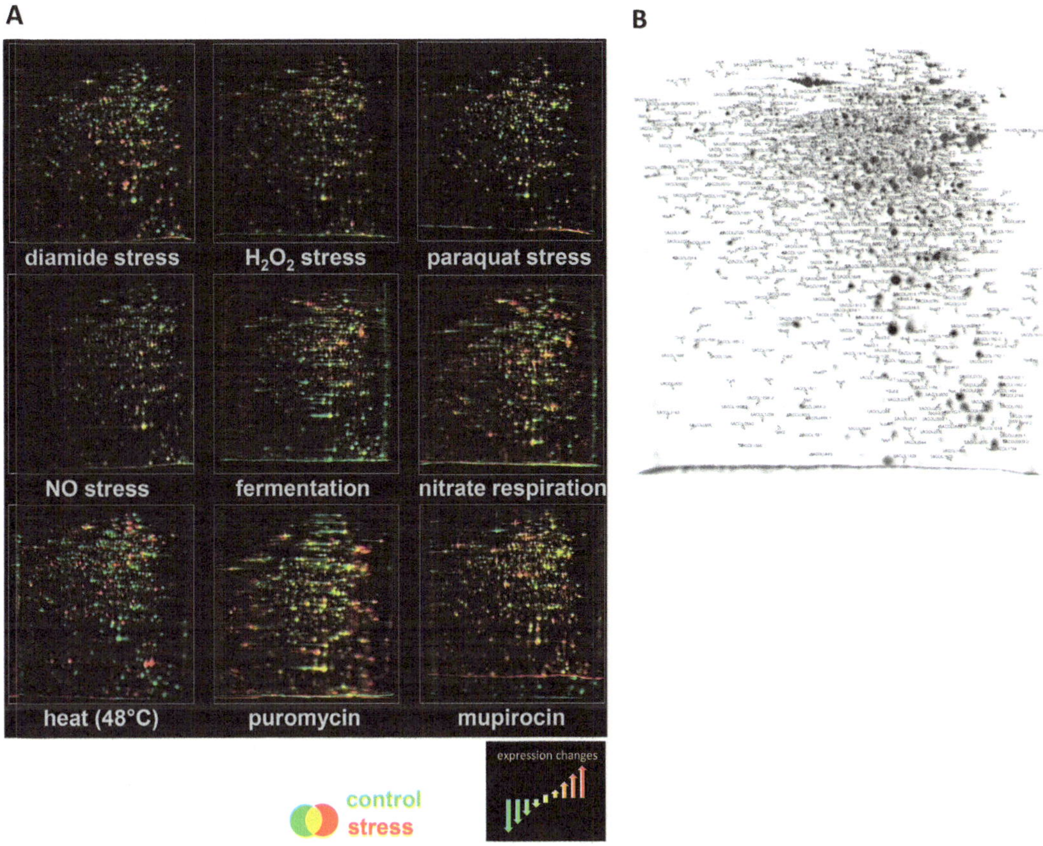

Fig. 2 2D-Reference map of *S. aureus* COL and proteomic signatures in response to different growth conditions in *S. aureus* COL. (**a**) *S. aureus* COL was grown in synthetic medium to an optical density of 0.5 and exposed to the different growth conditions. Global protein synthesis in response to the different stimuli and under control conditions was analysed by [^{35}S]-L-methionine pulse labeling. Cytoplasmic proteins were separated on 2D gels and detected using a Typhoon scanner. Image analyses were done using Delta2D software (Decodon, Greifswald, Germany). The protein pattern under control conditions is shown in green and those obtained under the different stress conditions are shown in red. (**b**) 2D reference map of *S. aureus* COL

corresponding protein spot on a 2D reference gel (Fig. 2b) [1, 3, 10]. The reference gel was established by separating cytoplasmic proteins on 2D protein gels based on the same protocols as used for the pulse labeled proteins (*see* also [11]). All detected proteins spots were cut from the gel and identified by MALDI MS analysis (Fig. 2b). A database system was created for storage, filtering and visualisation of these large protein expression data sets. The resulting online resource called *Aureolib* is publicly available (http://www.aureolib.de) and currently contains 4692 time-dependent expression profiles for 521 *S. aureus* proteins obtained under nine infection-relevant conditions [3].

2 Materials

2.1 Culture Conditions

1. For the Basic medium of the synthetic growth medium (2×): For 1 L 3.54 g Na_2HPO_4, 2.72 g KH_2PO_4, 0.4 g $MgSO_4 \cdot 7H_2O$, 1 g NH_4Cl, and 1 g NaCl are dissolved in deionized water and autoclaved for 20 min.
2. 1 M citrate: prepared in deionized water and autoclaved for 20 min.
3. 20% (w/v) glucose: prepared in deionized water and autoclaved for 10 min.
4. MOPS-buffer (25×): 209.26 g MOPS are dissolved in deionized water and filled to 1 L. The pH is adjusted to 7.0 and the buffer is autoclaved for 10 min.
5. L-Amino acid mixtures (50×): Mixtures of 1–4 different amino acids containing 50 mM of each amino acid are prepared: (a) alanine, valine, leucine, isoleucine; (b) aspartate, glutamate; (c) serine, threonine, cysteine; (d) arginine, lysine, proline; (e) phenylalanine, tryptophane, histidine; (f) glycine. Amino acids are dissolved in deionized water (*see* **Note 1**). The mixtures are sterilized by filtration (0.45 μm).
6. Vitamins (1000×): All vitamin solutions are prepared separately: 0.036 mM cyanocobalmine, 0.29 mM p-aminobenzoate, 0.04 mM biotin, 0.81 mM nicotinic acid, 0.21 mM Ca-D-pantothenic acid, 0.62 mM pyridoxamine hydrochloride, 0.29 mM thiaminium dichloride, 0.26 mM riboflavin. The vitamins are dissolved in deionized water and sterilized by filtration (0.45 μm).
7. Trace elements (1000×): All trace element solutions are prepared separately: 0.51 mM $ZnCl_2$, 0.5 mM $MnCl_2$, 0.097 mM H_3BO_3, 1.46 mM $CoCl_2$, 0.015 mM $CuCl_2$, 0.1 mM $NiCl_2$, 0.148 mM Na_2MoO_4. The trace elements are dissolved in deionized water and sterilized by filtration (0.45 μm).
8. 0.5 mM $FeCl_3$: dissolved in 2 N HCl.
9. 2 N NaOH in deionized water.
10. TE Buffer: 10 mM Tris, 1 mM EDTA, pH 7.5, autoclaved.
11. For 100 mL synthetic medium 4 mL 25× MOPS buffer, 14 μL 1 M citrate, 740 μL 20% glucose, 2 mL of each amino acid mixture (50×), 100 μL of each vitamin solution (1000×), 100 μL of each trace element solution (1000×), 150 μL FeCl3 and 150 μL 2 N NaOH are added to 50 mL 2× basic medium. Afterwards the medium is made up to 100 mL with sterilized, deionized water.

12. 15 mCi/mL L-[^{35}S] methionine.
13. Stop solution: 15 mg L-methionine and 10 mg chloramphenicol are dissolved in 10 mL 0.1 M Tris–HCl, pH 7.5.

2.2 Cell Lysis

1. Glass beads (Ø 0.10–0.11 mm; Braun, Göttingen, Germany).
2. Homogeniser Precellys® 24 (Thermo Electron Cooperation).

2.3 Two-Dimensional Polyacrylamide Gel Electrophoresis (2D PAGE)

1. Rehydration solution: 2.4 g urea, 0.76 g thiourea, 50 mg Chaps (e.g., ROTH, Karlsruhe, Germany), and 15 mg DTT (e.g., SERVA Electrophoresis GmbH, Heidelberg, Germany) are dissolved in 5 mL water (deionized water additionally purified by using the water purification system "Synergy" from Millipore). Subsequently, 52 μL Servalyte 3-10 (SERVA Electrophoresis GmbH, Heidelberg, Germany) is added.
2. IPG strips: linear pH range 4–7, 18 cm (SERVA Electrophoresis GmbH, Heidelberg, Germany).
3. Rehydration chamber.
4. Multiphor II unit (GE Healthcare, Little Chalfont, UK).
5. Electrode strips (GE Healthcare, Little Chalfont, UK).
6. 1.5 M Tris–HCl, pH 8.8 (adjusted at 18 °C), stored at room temperature (*see* **Note 2**).
7. 10% (w/v) SDS dissolved in deionized water, stored at room temperature.
8. 10% (w/) Ammonium persulfate (APS), dissolved in deionized water, stored at 4 °C no longer than 1 week.
9. 0.5 M Tris–HCl, pH 6.8 (adjusted at 18 °C), stored at room temperature (*see* **Note 2**).
10. Upper buffer (4×): 0.4% (w/v) SDS is dissolved in 0.5 M Tris–HCl pH 6.8.
11. Equilibration solution (100 mL): 36 g urea are dissolved in a mixture of 10 mL 0.5 M Tris–HCl (pH 6.8), 30 mL glycerol, and 40 mL 10% SDS. For solution **A** 175 mg DTT (SERVA Electrophoresis GmbH, Heidelberg, Germany) are added to 50 mL equlibration solution. For solution **B** 2.25 g iodoacetamide (toxic!) and traces of bromphenol blue are added to 50 mL equilibration solution.
12. Running buffer (10×): 150 g Tris, 720 g glycine, and 50 g SDS are dissolved in deionized water. When all components are completely in solution the buffer is filled to 5 L with deionized water.
13. Protean® plus Dodeca™ cell gel system, power supply, glass plates (Biorad-Laborities, Hercules, USA).

14. For preparing 12 slab gels 335.3 mL 40% acrylamide, 179 mL 2% bis-acrylamide, 272.54 mL 1.5 M Tris–HCL (pH 8.8), 11.48 mL 10% SDS, and 300 mL deionized water are mixed while stirring on a magnetic stirrer. 2.8 mL 10% APS and 0.55 mL TEMED are added to this solution and mixed by stirring (*see* **Note 3**). Generation of bubbles should be avoided.

2.4 Protein Detection

1. Storage Phosphor Screens 20 × 25 cm (Molecular Dynamics, Krefeld, Germany).
2. Typhoon 9400 variable mode imager (Amersham Biosciences, Freiburg, Germany).

2.5 In-Gel Digestion

1. Spot cutter (Proteome Work™) with a picker head of 2 mm (GE Healthcare, Little Chalfont, UK).
2. Ettan Spot Handling Workstation (GE Healthcare, Little Chalfont, UK).
3. 50 mM ammonium bicarbonate–50% (v/v) methanol.
4. 75% (v/v) acetonitrile.
5. Trypsin solution: 1 mg/mL mg trypsin dissolved (Promega, Madison, WI, USA) in 20 mM ammonium carbonate (freshly prepared).
6. 50% (v/v) acetonitrile–0.1% (w/v) trifluoroacetic acid (TFA).
7. Matrix solution 50% (v/v) acetonitrile–0.5% (w/v) TFA saturated with α-cyano-4-hydroxycinnamic acid (CHCA).
8. 0.5% (w/v) trifluoroacetic acid–50% (v/v) acetonitrile.

2.6 Protein Identification by Mass Spectrometry

1. Proteome-Analyzer 4800 (Applied Biosystems, Foster City, CA, USA).
2. 4800 Explorer™ software
3. Genome sequence of *S. aureus* COL in FASTA format (www.ncbi.nlm.nih.gov).

3 Methods

3.1 Bacterial Growth Conditions and Preparation of Cytoplasmic Protein Extracts Labeled with [^{35}S]-L-Methionine (Pulse-Labeling Reaction)

1. For radioactive pulse labeling experiments bacterial strains are cultivated in 100 mL synthetic medium without methionine in a 500 mL Erlenmeyer flask. The synthetic medium is inoculated with exponentially growing cells of the specific *S. aureus* strain grown over night in the same medium to an initial OD_{500} of 0.07–0.1.
2. Cells are grown at 37 °C to an optical density of 0.5.
3. 10 mL of the culture volume are transferred to each new 50 mL Erlenmeyer flask. One sample serves as a control and the other

samples are exposed to the respective stimuli. Protein synthesis at different time points after exposition to stress or under control conditions was measured by adding 10 μL (150 μCi) L-[^{35}S]-methionine to each culture. The labeling reaction is stopped after 5 min by adding 1 mL stop solution and by transferring the Erlenmeyer flask to ice.

4. Cells are pelleted by centrifugation (8000 × *g*) for 5 min at 4 °C.
5. Cell pellets are washed twice with 1 mL ice cold TE buffer, resuspended in 500 μL TE buffer and transferred into screw top tubes containing 500 μL glass beads.
6. Cells are disrupted by homogenization with glass beads using the homogeniser Precellys® 24 (Thermo Electron Cooperation) for 30 s at 6.5 m/s.
7. After homogenisation, the sample is centrifuged at 21,000 × *g* for 25 min at 4 °C. The supernatant is transferred to a new tube and the centrifugation process is repeated for 30 min.
8. Protein concentration is determined using ROTI-Nanoquant (Roth, Karlsruhe Germany) (*see* **Note 4**).
9. The protein solution is stored frozen at −20 °C.

3.2 2D PAGE

1. For Isoelectric focussing (IEF) protein samples (80–100 μg of radioactively labeled proteins) (*see* **Note 5**) are made up to 40 μL with deionized water. Subsequently 360 μL rehydration buffer are added and the solution is mixed by shaking at room temperature for 30 min. The rehydration mix is centrifuged for 5 min at 21,000 × *g* (20 °C) to remove insoluble proteins.
2. The supernatant is equally dispensed in one slot of the rehydration chamber. Position the IPG strip with the gel side down and lower the strip onto the solution. To help coat the entire IPG strip, gently lift and lower the strip and slide it back and forth along the surface of the solution. Avoid large bubbles under the IPG strip! Rehydration occurs over night for at least 15 h (no longer than 24 h).
3. The isoelectric focusing is performed with the Multiphor II unit. The Dry strip aligner and the electrodes should be cleaned with soft tissues and all cables should be connected. The temperature of the thermostatic circulator should be set to 20 °C. Two electrode strips of a length of 110 mm are placed on a clean flat surface and soaked with distilled water. Excess of water is removed by blotting with filter (*see* **Note 6**).
4. In order to remove the rehydrated IPG strips from its slots the ends of the strip are grabbed with tweezers and the strip is lifted out of the tray. Any contact with the gel surface should be avoided! The strips are positioned in the DryStrip Aligner in

adjacent grooves and all strips are aligned so that the anodic gel edges are lined up. The electrode strips are placed across the cathodic and anodic ends of the aligned IPG strips. The electrode strips must at least partially contact the gel surface of each IPG strip. When each electrode is aligned with the electrode strip, they are pressed down to contact the electrode strips. The IPG strips are covered with mineral oil to provide electric insulation.

5. The isoelectric focusing is performed at 20 °C by using the following voltage profile: Step 1: 500 V (gradient), 2 mA, 5 W, 1 Vh. Step 2: 3.5 kV (gradient), 2 mA, 5 W, 3 kVh. Step 3: 3.5 kV, 2 mA, 5 W, 23.5 kVh. As IEF proceeds, the current has to be checked! (Low current is normal for IPG gels.) If no current is flowing, the contact between electrodes and electrode strips should be checked.
6. After IEF proceed to the second-dimension immediately or store the IPG strips at −20 °C in foil.
7. For SDS Polyacrylamide Gel Electrophoresis (PAGE) running buffer (1×) is added to the gel system and the system is adjusted to 18 °C:
8. The separation of the focused proteins according to their molecular weight is done in 12.5% acrylamide and 2.6% bis-acrylamide polyacrylamide gels using the Tris–glycine system (*see* **Note 7**).
9. Prepare the solution for 12 slab gels. The gels are poured immediately by filling the gel cassette about 1 cm below the top of the glass plates. The gels are overlayed immediately after pouring with a thin layer (1 mL) of water-saturated n-butanol or water immediately after pouring the gels to minimize the gel exposure time to oxygen and to create a flat gel surface. Polymerization takes at least 4 h. Each gel should be inspected and the top of surface of each gel should be straight and flat. The butanol is removed and the gel surface is rinsed with deionized water. The butanol should be completely removed!
10. Prepare the stacking gel solution. An appropriate amount of the stacking gel solution is used to quickly rinse the top of each gel and is overlaid with a thin layer (1 mL) of water. The stacking gel should be polymerized for 1 h (and no longer than 2 h).
11. Prepare equilibration buffer A and B. Each IPG strip is placed with the gel side up in one slot of an equilibration chamber and 5 mL equilibration buffer A is added to each slot. The IPGs are equilibrated for at least 15 min (and no longer than 20 min) with gently shaking. The equilibration solution A is decanted and 5 mL equilibration solution B is added to each slot. The IPGs are equilibrated again for at least 15 min (and no longer

than 20 min) while shaking. Equilibration solution B is decanted and the IPG strips are placed on filter paper so that they rest on an edge to help drain the equilibration solution.

12. In the meantime gel cassettes containing the slab gels are removed from the Casting chamber and the butanol is removed by rinsing the gel surface with deionized water. (The butanol should be completely removed!) Gel cassettes are placed in a holder with the stacking gel on the top and filled with running buffer.
13. The IPG strips are placed between the plates on the surface of the stacking gel by gently pushing the IPG strip down so that the entire lower edge of the IPG strip is in contact with the top surface of the stacking gel. Prevent air bubbles between the IPG strips and the stacking gel! The IPG strips are fixed with 2% (w/v) agarose in deionized water on both ends.
14. Gels are put in the gel system filled with running buffer adjusted to 18 °C.
15. Electrophoresis is performed at constant power (17–20 W) at 18 °C. A constant temperature during electrophoresis is very important for gel to gel reproducibility.
16. After electrophoresis, gels are removed from their glass plates without the stacking gel in preparation for staining. Each gel should be marked to identify the acidic end of the first dimension preparation.

3.3 Protein Detection

1. For Krypton staining gels are fixed with 200 mL 40% (v/v) ethanol, 10% (v/v) acetic acid in deionized water for one hour and washed with 200 mL deionized water for 5 min.
2. Staining was performed with 200 mL 1× fluorescent dye Krypton™ (Thermo Scientific (Pierce), Waltham, MA, USA) for at least one hour in the dark on a shaker.
3. Carefully remove the staining solution and add 200 mL 5% (v/v) acetic acid in deionized water for 5 min.
4. Gels are scanned using Typhoon 9400 Variable Mode Imager (GE Healthcare, Little Chalfont, UK) with 200 μm and the appropriate laser. The excitation wavelength for Krypton is 532 nm (green) and the emission wavelength 580 nm (*see* **Note 8**). Photomultiplier voltage is adapted to get maximal intensity without saturation effects which can be checked by ImageJ or ImageQuant (*see* **Note 9**).
5. After Krypton staining and scanning, gels are placed on Wattman paper, covered with cellophane sheets and dried on a vacuum dryer at 75 °C for at least 2–4 h (*see* **Note 10**).
6. For Detection of L-[^{35}S]-methionine labeled proteins the dried gels are exposed to "Storage Phosphor Screens" for 2 h to

several days (depending on signal intensity). Storage Phosphor Screens are scanned using Typhoon 9400 Variable Mode Imager (GE Healthcare, Little Chalfont, UK) as described in **step 4** (*see* **Note 8**).

3.4 Protein Identification by Mass Spectrometry

1. For identification of proteins by MALDI-TOF-MS, 80–250 μg unlabeld cytoplasmic protein extracts are separated by 2D PAGE and stained with Kryptone. Detected protein spots are cut from gels using a spot cutter (Proteome Work™) with a picker head of 2 mm and transferred into 96-well microtiter plates.
2. Digestion with trypsin and subsequent spotting of peptide solutions onto the MALDI targets are performed automatically in the Ettan Spot Handling Workstation (GE Healthcare, Little Chalfont, UK) using the following standard procedure [12]: The gel pieces are washed twice 100 μL 50 mM ammonium bicarbonate–50% (v/v) methanol for 30 min and once with 100 μL 75% (v/v) acetonitrile for 10 min. After 17 min of drying 10 μL trypsin solution containing 20 ng/μL trypsin is added and incubated at 37 °C for 120 min. For peptide extraction gel pieces are covered with 60 μL 50% (v/v) acetonitrile–0.1% (w/v) TFA and incubated for 30 min at 37 °C. The supernatant containing the peptides is transferred into a new microtiter plate and the extraction is repeated with 40 μL of the same solution. The supernatants are dried completely at 40 °C for 220 min completely and the peptides are dissolved in 2.2 μL of 0.5% (w/v) TFA–50% (v/v) acetonitrile.
3. 0.7 μL of this solution is directly spotted onto the MALDI target. Then, 0.4 μL of matrix solution is added and mixed with the sample solution by aspirating the mixture five times. Prior to the measurement in the MALDI-TOF instrument the samples are dried on the target for 10–15 min.
4. MALDI-TOF-MS analyses of spotted peptide solutions is carried out on a Proteome-Analyzer 4800 (Applied Biosystems, Foster City, CA, USA). The spectra are recorded in a reflector mode in a mass range from 900 to 3700 Da. For one main spectrum 25 subspectra with 100 spots per subspectrum are accumulated using a random search pattern. If the autolytical fragment of trypsin with the monoisotopic $(M + H)^+$ m/z at 2211.104 reaches a signal to noise ratio (S/N) of at least 10, an internal calibration is automatically performed using the peak for one point calibration. The peptide search tolerance is 50 ppm but the actual standard deviation is between 10 and 20 ppm.
5. Calibration is performed manually for the less than 1% samples for which automatic calibration fails. After calibration the peak

lists are created using the "peak to mascot" script of the 4700 Explorer™ software with the following settings: mass range from 900 to 3700 Da; peak density of 50 peaked per range of 200 Da, minimal area of 100 and maximal 200 peaks per protein spot and minimal S/N ratio of 6. The resulting peak lists are compared with organism specific sequence databases by using the mascot search engine (Matrix Science, London, UK). Peptide mixtures that yielded at least twice a Mowse score of at least 49 (depending on the size and quality of the database) and a sequence coverage of at least 30% are regarded as positive identifications.

6. Proteins that fail to exceed the 30% sequence coverage cutoff are subjected to MALDI-MS/MS. MALDI-TOF-TOF analysis is performed for the three strongest peaks of the TOF spectrum. For one main spectrum 20 subspectra with 125 shots per subspectrum are accumulated using a random search pattern. The internal calibration is automatically performed as one-point calibration if the monoisotopic arginine $(M + H)^+$ m/z at 175.119 or lysine $(M + H)^+$ m/z at 147.107 reaches an S/N of at least 5. The peak lists are created using the "peak to mascot" script of the 4800 Explorer™ software with the following settings: mass range from 60 Da to a mass that was 20 Da lower than the precursor mass; peak density of 5 peaks per 200 Da; minimal area of 100 and maximal 20 peaks per precursor and a minimal S/N ratio of 5. Database searches are performed using the GPS explorer software with the organism specific databases. Proteins with a Mowes score of at least 49 in the reflector mode that is confirmed by subsequent peptide/fragment identifications of the strongest peaks (MS/MS) are regarded as identified. MS/MS analysis is particularly useful for the identification of spots containing more than one protein.

3.5 Protein Quantitation Approaches

1. The 2D gel image analysis is performed with the Delta2D Software (Decodon GmbH, Greifswald, Germany). For each stress experiment, data sets of at least three biological replicates are recommended.
2. Biological (and technical) replicates are grouped together.
3. Gel to gel variations of spot positions are compensated by warping of all images within each stress experiment using the *exact mode*.
4. To ensure a uniform spot detection, a virtual fusion gel is generated based on all gel images. Using the *Union mode*, pixel intensities are combined by a weighted average which reduces the accumulation of artifacts on the resulting fusion gel.

5. The generated fusion gel is used for protein spot detection. Automatically calculated spot borders have to be corrected and optimized manually if necessary.
6. The final spot mask has to be transferred from the fusion gel to each gel image of the corresponding experiment (*see* **Note 11**).
7. For relative quantitation, protein spot intensities should be normalized on each gel to compensate technical variations for example in protein concentration, rehydratation, or staining. Delta2D provides total normalization of spot intensities on the same gel. Alternatively, spot intensities can be normalized using median normalization which might give more robust results.
8. Afterwards synthesis ratios between gel image groups (e.g., stressed samples versus unstressed control samples) are calculated for each protein spot. Presentation of the ratios as binary logarithms is useful to make induction and repression ratios symmetrical.
9. Standardization of relative spot volumes for statistical analyses are performed using z-score [13].
10. Standardized data can be loaded into the TIGR Multiexperiment Viewer which is included in the Delta2D software package [14].
11. For statistical analyses ANOVA is applied to test whether the relative spot intensity of a given protein spot differs between multiple sample groups. Positives are subsequently tested using Student's test (post test). In a previous study, we used a critical alpha of 0.1 and adjusted Bonferroni correction for multiple testing [10].
12. The introduction of an absolute threshold is reasonable to define biological relevance. A ratio of at least twofold is the most frequently used threshold.

3.6 Expression Profiling and Data Mining

1. Hierarchical sample and protein clustering groups the proteins according to their expression profiles and provides first hints on the biological effects induced by the respective stimulus [10].
2. For data interpretation, a lot of information about the proteins can be retrieved from different databases. Protein annotations are referred to public databases such as NCBI and Uniprot [15, 16]. Insights into the transcriptional regulation of the respective genes are provided by Regprecise [17]. The online repository *Aureowiki* combines information about the proteins from various sources, such as prediction of protein localization and functional domains, modifications, hydrophobicity, and conservation in *S. aureus* (http://www.protecs.uni-greifswald.de/aureowiki/).

The Kyoto Encyclopedia of Genes and Genomes (KEGG) provides comprehensive data on metabolic pathways (http://www.genome.jp/kegg/). Tools such as KEGG mapper (http://www.genome.jp/kegg/tool/map_pathway2.html) may help to understand the metabolic networks active under different conditions [18].

STRING-DB provides functional interaction networks based on user-defined gene (or protein) sets and supports the identification of functional clusters of proteins induced or repressed in response to the different growth conditions [19].

3.7 Inter-experimental Analyses of Protein Expression Profiles

1. An identification-based approach to integrate synthesis data of different experiments has been introduced previously [10].

   A very comprehensive 2D reference map (*master gel*) containing as much as possible identified protein spots is a prerequisite for these analyses (*see* Subheading 3.4). To increase the number of identified protein spots, mixed protein extracts derived from cells grown under various growth conditions can be used to generate the master gel [20]. Multiple spots representing the same protein indicate the existence of isoforms of these proteins and have to be clearly differentiated for example by numbering.
2. Image analysis of each experiment is performed as described in Subheading 3.5.
3. Protein identifications have to be manually transferred from the master gel to each experiment-specific fusion gel. Again, gel-to-gel variations have to be compensated by gel warping. Only identifications of unambiguously referable protein spots may be transferred from the master gel to the experiment-specific fusion gels!
4. Spot intensities are quantified for each experiment separately (*see* also **item 8** in Subheading 3.5). Importantly, transferred spot identifications should be considered during manual spot mask curation to avoid artificial protein colocalization caused by inaccurate spot borders.
5. Subsequently, quantification data of the different experiments are merged based on protein spot identification (e.g., using MS Excel). Remember, multiple spots of the same protein are differentiated by numbers.
6. Multifactorial analyses (e.g., two-way ANOVA) are applied to identify significant changes in the synthesis of the proteins between the different experiments (*see* also **item 11** in Subheading 3.5). Hierarchical clustering identifies similar expression patterns or stress-specific responses (*see* also **item 1** in Subheading 3.6).

3.8 Working with the Protein Expression Database Aureolib

To handle the total set of data obtained by integrative and multi-experimental analyses (as described in Subheading 3.6) we created *Aureolib* fulfilling special requirements for (1) data storage and integration, (2) data access, and (3) data visualization. *Aureolib* is based on PHP and MySQL which makes it easily accessible by generic web browsers (http://www.aureolib.de). Currently, the system provides 4692 time dependent expression profiles for 521 *S. aureus* proteins in response to nine infection-related conditions. Additionally, functional annotations and information on transcriptional regulators have been integrated. Here we describe the different options facilitating the access to expression data in *Aureolib* [3].

1. To select specific expression profiles choose: *browse* > *expression filter* (Fig. 3a).
2. Select stimuli that should be considered by filtering (*consider* checkbox).
3. By activating the *significance* checkbox only statistically significant changes will be considered (*see* **Note 12**).
4. Select *induced* (expression induced by the respective stimulus), *repressed* (expression repressed by the respective stimulus), *changed* (expression induced or repressed by the respective stimulus), or *unchanged* (expression NOT induced or

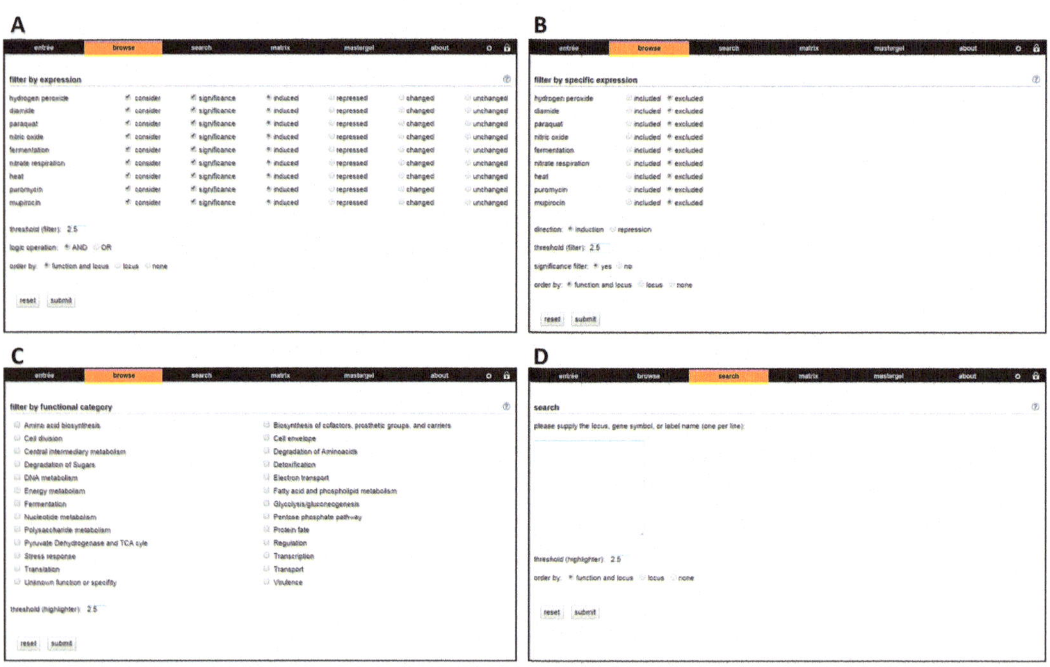

Fig. 3 Search criteria for data retrieval in *Aureolib. Aureolib* provides filter masks to assess specific expression profiles in response to defined stimuli (**a** and **b**), expression profiles of proteins belonging to specific categories (**c**), and expression profiles of specific proteins (**d**)

repressed by the respective stimulus) to define the expression kinetics of interest.

5. *Threshold*: Define a threshold for minimal expression changes.
6. *Logic operation*: Select AND to get proteins with induced or repressed expression profiles in response to all of the selected stimuli or OR to find expression profiles induced or repressed by at least one of the selected stimuli.
7. *Order by*: Select how the expression profiles should be ordered.
8. Expression plots are shown for each protein whose expression profile matches the selected criteria (*see* **Note 13**) (Fig. 4).
9. To select expression profiles of proteins belonging to defined stimulons choose: *browse* > *specific expression filter* (Fig. 3b).
10. *Include* or *exclude* stimuli from the set affecting protein synthesis.

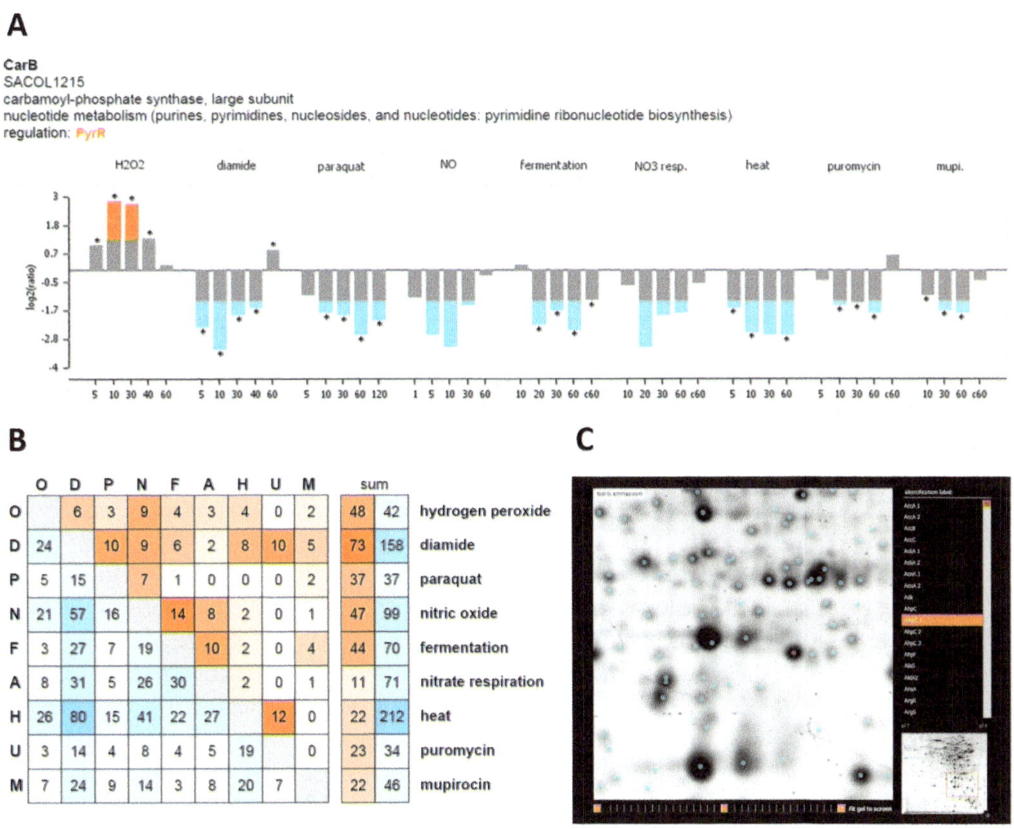

Fig. 4 Data visualization in *Aureolib*. (**a**) Expression data is provided as interexperimental expression plots. Bar charts show expression ratios as binary logarithm for each sample in comparison to the control sample. Induction ratios which pass the user-defined threshold are colored in orange (increased expression) and blue (decreased expression). (**b**) The pairwise comparison matrix summarizes the number of expression profiles with increased (orange area) or decreased (blue area) synthesis rates in response to selected conditions. (**c**) Each identified protein spot can be localized on the digital master gel

11. Proceed with **step 3** in this section.
12. To select expression profiles of proteins belonging to defined functional categories choose: *browse* > *function filter* (Fig. 3c).
13. Select boxes of functional categories of interest.
14. Proceed with **step 3** in this section.
15. To select expression profiles of proteins belonging to defined regulons choose: *browse* > *regulation filter*.
16. Select all regulators or regulatory elements of interest.
17. Proceed with **step 3** in this section.
18. To select expression profiles of specific proteins choose: *search* (Fig. 3d).
19. Paste locus tags or gene symbols of *S. aureus* COL into the text box (one per line).
20. Proceed with **step 3** in this section.
21. For a pairwise comparison of proteomic signatures choose: *matrix*.
22. Define the set of stimuli for pairwise comparison.
23. Proceed with **steps 3** and **5** in this section.

4 Notes

1. The amino acids of one mixture are dissolved successively.
2. For highly reproducible 2D gels the pH of these buffers should be adjusted at a fixed temperature that should be identical to the running temperature of the SDS gels during the second dimension.
3. The total volume of the solution needed depends on the gel sizes, the gel thickness, and the number of the gels cast.
4. Ensure that at least 10 μL of each protein solution is used. Otherwise dilute an appropriate volume of the protein extract prior to measurement. Ensure that the ratio E_{590nm}/E_{450nm} is always the same for each protein extract. Protein extracts should be frozen at −20 °C before measuring the protein concentration. For preparing a calibration curve a protein mix containing three to five different proteins with defined concentrations in TE should be used and handled in the same way as the protein extracts.
5. If the volume containing the desired amount of cytoplasmic proteins exceeds 40 μL the volume should be reduced by using a speed vac.
6. Important: Strips must be damp, not wet!

7. These instructions assume the use of the Protean® plus Dodeca™ cell gel system (Biorad-Laborities, Hercules, USA). They are easily adaptable to other gel systems.
8. For quantitation of protein synthesis and accumulation ensure that no protein spot is saturated.
9. Since fluorescent staining is fading with each scan we avoid the automatic adjustment of PMT voltage provided by the Typhoon scanner. Here the scan process aborts with the first satured pixel to get restarted with a lower PMT voltage. Thus, some gel areas are scanned more often than others.
10. Ensure that the gels are completely dried before interrupting the vacuum! The gel should have lost its softness when you press your finger on it.
11. For further details on 2D gel image analyses *see* [21].
12. Statistical methods are applied as described previously [3].
13. Expression plots show ratios compared to the unstressed control as binary logarithm. Statistical relevance and protein colocalization is indicated.

Acknowledgments

We are very grateful to our former colleagues at the Institute for Microbiology of the Greifswald University who were involved in these experiments. Furthermore, we thank the Decodon GmbH (Greifswald) for providing the Delta2D software. This work was supported by grants from the Bildungsministerium für Bildung und Forschung (InfektionsGenomik) and the Deutsche Forschungsgemeinschaft (SFB-TRR34, FOR 585, GRK840).

References

1. Hecker M, Antelmann H, Büttner K, Bernhardt J (2008) Gel-based proteomics of Gram-positive bacteria: a powerful tool to address physiological questions. Proteomics 8 (23–24):4958–4975
2. VanBogelen RA, Schiller EE, Thomas JD, Neidhardt FC (1999) Diagnosis of cellular states of microbial organisms using proteomics. Electrophoresis 20(11):2149–2159
3. Fuchs S, Zühlke D, Pané-Farré J, Kusch H, Wolf C, Reiss S, Binh le TN, Albrecht D, Riedel K, Hecker M, Engelmann S (2013) Aureolib – a proteome signature library: towards an understanding of *Staphylococcus aureus* pathophysiology. PLoS One 8(8): e70669
4. Bandow JE, Brötz H, Leichert LI, Labischinski H, Hecker M (2003) Proteomic approach to understanding antibiotic action. Antimicrob Agents Chemother 47 (3):948–955
5. Brötz-Oesterhelt H, Bandow JE, Labischinski H (2005) Bacterial proteomics and its role in antibacterial drug discovery. Mass Spec Rev 24 (4):549–565
6. Wolf C, Hochgräfe F, Kusch H, Albrecht D, Hecker M, Engelmann S (2008) Proteomic analysis of antioxidant strategies of *Staphylococcus aureus*: diverse responses to different oxidants. Proteomics 8(15):3139–3153
7. Reiss S, Pané-Farré J, Fuchs S, Francois P, Liebeke M, Schrenzel J, Lindequist U, Lalk M, Wolz C, Hecker M, Engelmann S

(2012) Global analysis of the *Staphylococcus aureus* response to mupirocin. Antimic Agents Chemother 56(2):787–804
8. Fuchs S, Pané-Farré J, Kohler C, Hecker M, Engelmann S (2007) Anaerobic gene expression in *Staphylococcus aureus*. J Bacteriol 189 (11):4275–4289
9. Hochgräfe F, Wolf C, Fuchs S, Liebeke M, Lalk M, Engelmann S, Hecker M (2008) Nitric oxide stress induces different responses but mediates comparable protein thiol protection in *Bacillus subtilis* and *Staphylococcus aureus*. J Bacteriol 190(14):4997–5008
10. Fuchs S, Mehlan H, Kusch H, Teumer A, Zühlke D, Berth M, Wolf C, Dandekar T, Hecker M, Engelmann S, Bernhardt J (2010) Protecs, a comprehensive and powerful storage and analysis system for OMICS data, applied for profiling the anaerobiosis response of *Staphylococcus aureus* COL. Proteomics 10 (16):2982–3000
11. Engelmann S, Hecker M (2008) Proteomic analysis to investigate regulatory networks in *Staphylococcus aureus*. Methods Mol Biol 431:25–45
12. Eymann C, Dreisbach A, Albrecht D, Bernhardt J, Becher D, Gentner S, Tam le T, Büttner K, Buurman G, Scharf C, Venz S, Völker U, Hecker M (2004) A comprehensive proteome map of growing *Bacillus subtilis* cells. Proteomics 4(10):2849–2876
13. Cheadle C, Vawter MP, Freed WJ, Becker KG (2003) Analysis of microarray data using Z score transformation. J Mol Diagnostics 5 (2):73–81
14. Saeed AI, Bhagabati NK, Braisted JC, Liang W, Sharov V, Howe EA, Li J, Thiagarajan M, White JA, Quackenbush J (2006) TM4 microarray software suite. Methods Enzymol 411:134–193
15. Coordinators NR (2014) Database resources of the National Center for Biotechnology Information. Nucleic Acids Res 42(Database issue):7–17
16. UniProt C (2015) UniProt: a hub for protein information. Nucleic Acids Res 43(Database issue):D204–D212
17. Novichkov PS, Kazakov AE, Ravcheev DA, Leyn SA, Kovaleva GY, Sutormin RA, Kazanov MD, Riehl W, Arkin AP, Dubchak I, Rodionov DA (2013) RegPrecise 3.0—a resource for genome-scale exploration of transcriptional regulation in bacteria. BMC Genomics 14:745
18. Kanehisa M, Goto S, Sato Y, Kawashima M, Furumichi M, Tanabe M (2014) Data, information, knowledge and principle: back to metabolism in KEGG. Nucleic Acids Res 42 (Database issue):D199–D205
19. Szklarczyk D, Franceschini A, Wyder S, Forslund K, Heller D, Huerta-Cepas J, Simonovic M, Roth A, Santos A, Tsafou KP, Kuhn M, Bork P, Jensen LJ, von Mering C (2015) STRING v10: protein-protein interaction networks, integrated over the tree of life. Nucleic Acids Res 43(Database issue):D447–D452
20. Kohler C, Wolff S, Albrecht D, Fuchs S, Becher D, Büttner K, Engelmann S, Hecker M (2005) Proteome analyses of *Staphylococcus aureus* in growing and non-growing cells: a physiological approach. Int J Med Microbiol 295(8):547–565
21. Berth M, Moser FM, Kolbe M, Bernhardt J (2007) The state of the art in the analysis of two-dimensional gel electrophoresis images. Appl Microbiol Biotechnol 76(6):1223–1243

Chapter 10

How to Assess Protein Stability: Half-Life Determination of a Regulatory Protein in *Bacillus subtilis*

Alexander Reder, Stephan Michalik, and Ulf Gerth

Abstract

The half-life of a particular protein is highly variable, reaching from minutes to hours, over days and weeks to years or even a whole life time of an organism (e.g., α-crystalline of the mammalian eye). Thus, controlling protein activity by proteolysis is the most dramatic and unambiguous decision by any organism, because reconstitution of the destroyed protein activity requires an "expensive" new synthesis. To distinguish degradation from protein synthesis and accumulation only one method comes into consideration—pulse-chase labeling. In our hands, the most accurate method to track the fate of a single protein is radioactive pulse-chase labeling combined with immunoprecipitation. Besides a detailed description of the standard protocol, the general applicability as well as certain improvements of the method will be discussed here.

Key words ATP-dependent proteolysis, Degradation, Regulatory proteins, Immunoprecipitation, *Bacillus subtilis*

1 Introduction

In comparison with transcriptional regulation by induction or repression and numerous other levels of activity control mechanisms, protein degradation represents the most-dramatic and final step by deactivation of a protein function and marks the end of a proteins life. Especially, the life of regulatory proteins such as transcriptional activators or repressors appears to be rather short, because often they are only needed transiently by the cell. Here, the maintenance of the information flow comes at the cost of a quick protein turnover.

We would like to illustrate this concept and method using the example of the transcriptional regulator MgsR (*m*odulator of *g*eneral *s*tress *r*esponse) of *Bacillus subtilis*. MgsR belongs to the Spx/ArsC (arsenate reductase) protein family and controls a subregulon within the sigmaB (σ^B)-dependent general stress and stationary phase modulon [1, 2].

Dörte Becher (ed.), *Microbial Proteomics: Methods and Protocols*, Methods in Molecular Biology, vol. 1841, https://doi.org/10.1007/978-1-4939-8695-8_10,

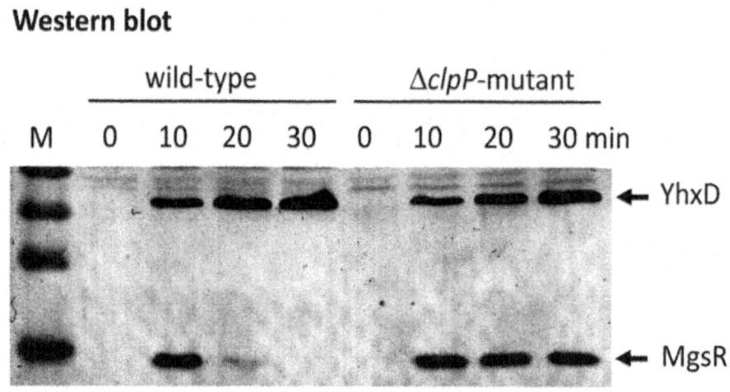

Fig. 1 Western blot kinetics of the stress induced proteins YhxD and the regulator MgsR in response to the addition of 4% (v/v) ethanol to exponentially growing cultures of the *B. subtilis* wild-type 168 and protease mutant Δ*clpP*. Control samples were taken when the cultures reached an optical density OD_{500nm} of 0.4 and ethanol was added immediately to a final con-centration of 4% (v/v). Further samples were taken after 10, 20, and 30 min as indicated in the top row

The Spx protein represents the prototype of a new family of redox-sensitive global regulators in *B. subtilis* [3, 4] and has been shown to interact with the α-subunit of the RNA polymerase and to direct the enzyme to specific promoter sequences [3–6].

The MgsR protein is almost completely absent from cells under normal, unstressed control conditions and its expression is strictly dependent on the stress mediated activation of the alternative sigma factor σ^B. Upon expression, MgsR senses and integrates secondary oxidative stress signals beyond the primary decision of the σ^B activation leading to a hierarchical expression of the MgsR subregulon with more than 50 genes [1, 2].

In order to study protein decay in vivo, Western-blots are only suitable to a limited extend. Regular Western analysis delivers single protein signals as a snapshot that just represent the actual equilibrium between protein synthesis, accumulation and degradation. To illustrate this you find Western blot results for the two proteins YhxD and MgsR in Fig. 1. The expression of both proteins is known to be strictly dependent on stress induction by activation of σ^B [1].

What can be reliably inferred from the Western blot data is that both proteins are indeed not expressed under control conditions and that both proteins become expressed in response to exposition to 4% ethanol stress. The time resolved kinetics for the wild type also shows that the protein YhxD accumulates over time and that the MgsR protein is present only transiently indicating quick proteolytic degradation. The results for YhxD observed in the Δ*clpP* mutant are consistent with the wild-type, whereas the MgsR signal can be detected throughout the observed time window of 30 min in the Δ*clpP* strain. Thus, although it is very likely that the observed effect on MgsR is directly caused by the failure of proteolytic degradation due to the deletion of the ClpP-protease, one cannot be 100% sure of this. The observations for MgsR in the wild type may also be caused by a shift from the cytosolic to the pellet fraction under continuous stress and the results obtained from the Δ*clpP* mutant may for example be attributed to continuous transcription and translation (increased synthesis) of MgsR in the Δ*clpP* background due to for example stabilization or permanent activation of σ^B. Thus, a stable equilibrium between new synthesis, accumulation and degradation could be mistakenly interpreted as an enhanced stabilization against proteolytic decay.

Furthermore, the usage of certain antibiotics as translational inhibitors to stop protein synthesis is no good choice, since inhibition of translation causes several side effects such as oxidative stress or artificial stabilization of proteins by blocking the synthesis of their adapter proteins that are necessary for their proteolysis, etc. In selected and observed cases artificial degradation can be induced or even inhibited. Radioactive pulse-chase labeling with ^{35}S-methionine in defined minimal medium coupled with immunoprecipitation is the method of choice to assess in vivo degradation rates and circumvents all obstacles mentioned above.

A detailed experimental procedure for the half-life determination of the MgsR regulator is given below.

2 Materials

2.1 Instruments

1. Mini-PROTEAN Tetra Cell (Bio-Rad).
2. Storage Phosphor screens (BAS-IP MS 2040 E from GE Healthcare Life Sciences).
3. Typhoon 9400 scanner (GE Healthcare Life Sciences).
4. Heated vacuum gel dryer (e.g., Model 583 and HydroTech™ Pump Gel Drying System from Bio-Rad).
5. Magnetic stand for Eppendorf tubes.
6. Refrigerated centrifuge.
7. Overhead shaker or equivalent equipment.

2.2 Materials

1. Minimal media to reach a maximum of ^{35}S-L-methionine uptake and incorporation by the cells.
2. ^{35}S-L-methionine (e.g., from Hartmann Analytic GmbH) (*see* **Notes 1** and **2**).
3. 10 ml of a 0.3 M stock solution of nonradioactive "cold" L-methionine
4. Lysis Buffer (50 mM Tris–HCl; 5 mM EDTA; 1 mM PMSF; add HPLC grade water to final 100 ml; adjust to pH 7.5, afterward add 12 mg/ml (w/v) of egg white lysozyme.

5. KI-Buffer (50 mM Tris–HCl; 150 mM NaCl; 0,5% (v/v) Triton X-100; 5 mM EDTA; 1 mM PMSF; add HPLC grade water to final 100 ml; adjust to pH 8.0).
6. Primary Antibody.
7. Protein A or G coated Dynabeads (e.g., from Life Technologies) (*see* **Note 3**).
8. *Optional*: total protein extract of a mutant strain lacking the structural gene for the target protein—here a Δ*mgsR* mutant—for improvement of the primary (polyclonal) antibody by blocking unspecific interactions.
9. Resolving gel buffer: 1.5 M Tris–HCl, pH 9.1 at room temperature. Weigh 181.7 g Tris and transfer to a beaker. Add water to a volume of 900 ml. Mix and adjust pH with 10 N HCl to pH 9.1. Fill up to 1 l with double distilled water. Store at 4 °C.
10. Stacking gel buffer: 0.5 M Tris–HCl, pH 6.8. Weigh 60.6 g Tris and prepare a 1 l solution as described in **step 9** of this section. Store at 4 °C.
11. 37.5:1% acrylamide–bisacrylamide mix (e.g., PanReac AppliChem, Germany). Store at 4 °C.
12. Ammonium persulfate: 10% solution in double distilled water. Always prepare fresh.
13. N,N,N,N′-tetramethylethylenediamine (TEMED) (e.g., Sigma-Aldrich). Store at 4 °C.
14. SDS-PAGE running buffer: 0.025 M Tris–HCl, pH 8.3, 0.192 M glycine, 0.1% SDS.
15. SDS sample buffer (5× concentrated): 0.3 M Tris–HCl (pH 6.8), 10% SDS, 25% β-mercaptoethanol, 0.1% bromophenol blue, 45% glycerol. Leave one aliquot at 4 °C for current use and store remaining aliquots at −20 °C.
16. Fixation buffer: 12.5% acetic acid and 50% ethanol in distilled water.

3 Methods

3.1 ^{35}S-L-Methionine Pulse-Chase Labeling

All strains were grown in Belitsky minimal media [7] supplemented with 0.01% (w/v) yeast extract. Here, the addition of yeast extract was necessary due to growth deficiencies of the *clp* mutants [8]. Pulse labeling was carried out in the early exponential growth phase when cells reached an optical density of OD_{500nm} 0.4. Proteins were pulse-labeled by the addition of 25 μCi/ml final activity of ^{35}S-L-methionine (normally you have ~15 μCi/μl of activity in your ^{35}S-methionine stock solution, thus, to reach 25 μCi/ml in a

30 ml cell culture you have to add 50 μl or a total activity of 750 μCi to your culture). The protein synthesis of MgsR was induced by the addition of 4% (v/v) ethanol 2 min after the ^{35}S-L-methionine has been added. The 2 min time period between the addition of the methionine pulse and ethanol stress gives the cells enough time to take up and incorporate the ^{35}S-L-methionine into the tRNA molecules. Ten minutes after exposition to ethanol stress control samples (t_0) were taken immediately after onset of the chase by addition of a 600,000-fold molar excess (400 μl of methionine solution to 30 ml of culture volume) of nonradioactive or "cold" methionine, followed by cell sampling of constant culture volumes of 2 ml after 10, 20, 30, 40, 50, and 120 min. Notice that the existence of just two methionine residues in our protein, MgsR, is sufficient to label the molecule adequately, even the presence of the start methionine only (when it is not removed by post-translational processes) would ensure proper labeling.

Cool down the 2 ml cell samples immediately to ~0 °C in liquid nitrogen (do NOT freeze), spin down at 10,000 × *g* for 4 min at 4 °C, discard the supernatants and either store the cell pellets in liquid nitrogen or use them immediately for further preparation.

3.2 Cell Lysis and Immunoprecipitation

Resuspend the cell pellets in 50 μl Lysis buffer and incubate the mix at 42 °C for 10 min. Then add 8 μl of SDS (10%), mix well and transfer to a heating block at 95 °C for 5 min. Cells are lysed completely during this step and proteins are completely denatured. Now add 700 μl of KI-buffer to each sample, vortex for 5 s and incubate 15 min on ice prior to spinning down cell debris for 60 min at 14,000 × *g* and 4 °C. After centrifugation transfer the supernatant (protein extract) into a fresh 1.5 ml safe lock Eppendorf tube (avoid disturbance of the pellet) and add 30 μl of primary antibody* (see below: *"Quenching of unspecific antibody interactions"). Incubate the tubes in an overhead shaker at 4 °C overnight.

The next morning vortex the protein A or G coated Dynabeads and immediately transfer an appropriate volume (30 μl per sample, Critical Step: the binding capacity of the beads must exceed the amount of antibody in the sample!) to a fresh 1.5 ml Eppendorf tube. To recover MgsR-antibody conjugates protein G-coated Dynabeads were used. Put the tube in the magnetic stand and separate the beads from the storage buffer for 2 min. Then remove the storage buffer with a pipette while leaving the tube on the magnetic stand. After this, wash the beads with 1 ml of KI-buffer. Therefore, remove the Eppendorf tube from the stand, add the KI-buffer and invert the tube ~15 times. Place the tube on the magnetic stand again and separate the beads from the KI-buffer for 2 min. Discard the supernatant and repeat the washing step one more time. Finally add the same volume of KI-buffer that has been transferred in the first step from the Dynabead stock solution to the bead pellet. Vortex rigorously and transfer 30 μl of bead solution to each sample tube immediately. Incubate the samples for 2 more hours at 4 °C in the overhead shaker (essential to keep the magnetic beads in suspension).

After this period all immune-complexes should be captured by the specific interaction of the protein A or G with the Fcγ region of the antibodies. To precipitate the immune-complexes place the Eppendorf tubes to the magnetic stand again, separate the beads for 2 min and discard the supernatant. Now wash the immune-complexes at least four times. Therefore repeat the inversion (~15 times) in KI-buffer, the separation step on the magnetic stand and the removal of the supernatant. Finally add 20 μl of 1× SDS sample buffer to the separated Dynabead pellet, vortex for 5 s and denature the immune-complexes for 5 min at 95 °C. Spin down shortly and place the samples on the magnetic stand again to separate the beads from your protein samples. Now load the entire protein samples on a SDS-PAGE (*see* below: Subheading 3.4. SDS-PAGE protocol) and run the gels for approximately 45 min at 160 V. Stop the electrophoresis just before the bromophenol blue band runs out of the gel. Immediately transfer the gels to 20 ml of fixation buffer for 10 min (until the bromophenol blue band turns yellow). Then, transfer the gels to 20 ml of distilled water for another 10 min. After this, place the gels on Whatman paper and put them on a preheated (85 °C) vacuum dryer and dry the gels for at least 2 h.

3.3 Optional: Quenching of Unspecific Antibody Interactions

To get rid of unspecific reactions from polyclonal antibody sera a simple trick can be used to block almost all unwanted unspecific reactions. All you need is a deletion mutant in the structural gene encoding your protein of interest (if possible). Here, of course we used a Δ*mgsR*::Erm mutant strain to prepare concentrated protein lysates from both a control culture at early exponential phase and 30 min after exposition to 4% ethanol stress (of course not radioactively labeled). The lysates were prepared in PBS (as the antibody sera) and combined afterward. After this, we added increasing amounts of protein lysate to the antibody sera (10, 20, and 40 μg of protein lysate per microliter of antibody) and incubated the mixtures on an overhead shaker overnight at 4 °C. Thereafter, the antibody sera were used for immune precipitation reactions as described above. Figure 2 shows the results for two different polyclonal MgsR specific antibody sera AB1 and AB2. One can clearly see that the specific MgsR signal (labeled on the right hand side) is not affected by the antibody treatment, whereas the unspecific signals almost completely disappear with increasing amounts of mutant protein lysates.

3.4 SDS-PAGE

SDS-PAGE is carried out according to [9]. For preparing the *resolving gel solution* you have to consider, that different applications will need different percentages of acrylamide gels. As a rough estimation use line 1 in Table 1 below to choose the correct

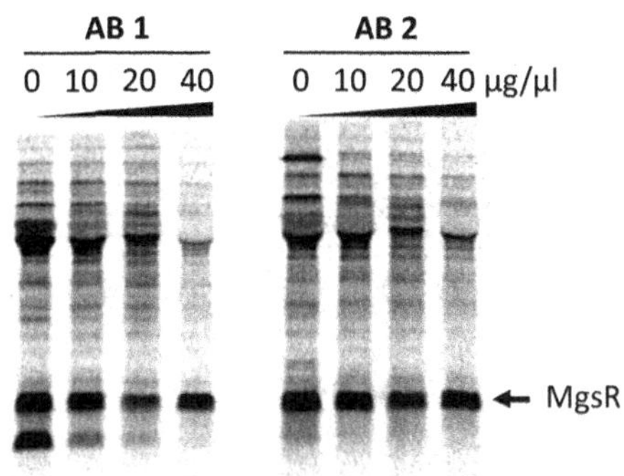

Fig. 2 Blocking of unspecific reactions in the MgsR specific antibody sera AB1and AB2 by incubation with increasing amounts (10, 20, and 40 µg/µl) of protein extract prepared from a Δ*mgsR* deletion mutant strain

Table 1
Pipette scheme for acrylamide resolving gel solutions of different percentages

Molecular weight range kDa	50–500	20–300	10–200	3–100
% Gel	7%	10%	12%	15%
Distilled H_2O	16.6 ml	14.2 ml	12.6 ml	10.2 ml
1.5 M Tris–HCl, pH 9.1	7.5 ml	7.5 ml	7.5 ml	7.5 ml
20% (w/v) SDS	150 µl	150 µl	150 µl	150 µl
Acrylamide–bisacrylamide 37.5:1	5.6 ml	8 ml	9.6 ml	12.0 ml
10% (w/v) APS	150 µl	150 µl	150 µl	150 µl
TEMED	20 µl	20 µl	20 µl	20 µl

percentage of your gel. Then prepare 30 ml of gel solution by selecting one of the following percentages and mix the compounds as shown in Table 1 (*see* **Note 4**).

After complete polymerization of your resolving gel (~45 min) you can prepare 5 ml of *stacking gel solution* (4% acrylamide) as described in Table 2. After pouring the gel do not forget to insert your comb immediately.

3.5 Exposition/ Autoradiography

Samples were subjected to SDS-PAGE (15%) and gels were dried on Whatman paper using a heated vacuum dryer. Now expose the gels to storage phosphor screens (Molecular Dynamics) for a time span ensuring the utilization of the whole dynamic range that will be determined by the strongest signal on the gel. This means, that the strongest signal on your gel must not exceed the maximum intensity level of 100,000 and must not be below a level of 95,000. Therefore, the screens are scanned using a Typhoon 9400 scanner

Table 2
Pipette scheme for acrylamide stacking gel solution

H_2O	3.125 ml
0.5 M Tris–HCl, pH 6.8	1.25 ml
20% (w/v) SDS	25 μl
Acrylamide/Bis-acrylamide 37.5:1	540 μl
10% (w/v) APS	50 μl
TEMED	10 μl

at a resolution of 200 nm and a color depth of 16 bits (65,536 grey levels) and the exposition times must be monitored and adapted to meet the criteria mentioned above. The quantification of the signals can be performed with the ImageQuant TL Software from GE Healthcare Life Sciences or the freely available ImageJ software (Rasband, W.S., ImageJ, US National Institutes of Health, Bethesda, MD, USA).

3.6 Signal Quantification and Half-Life Determination

All experiments were carried out in triplicate. The quantification of the MgsR-specific signals (Fig. 3) was performed using the ImageJ software version 1.43u (Rasband, W.S., ImageJ, US National Institutes of Health, Bethesda, MD). Mean values and standard deviations were calculated and all values were normalized to the initial value of the time-series (t_0 min = 100%). A nonlinear regression model (nonlinear least squares) was used to obtain the best fitting curve function on the logarithmic values (Eq. A, Fig. 4). To determine the half-lifes ($t1/2$) the equation was solved for x (time) in $y = 50$ (50% intensity) (Eq. B, Fig. 4).

3.7 Evaluation of the Results

The degradation pattern observed for the *B. subtilis* wild type shows that the physiological half-life of MgsR is extremely short (5.8 min) (Fig. 3). The three different ATPase components ClpE, ClpC, and ClpX mediate substrate recognition and proteolysis by the ClpP peptidase. MgsR is not stabilized by a deletion of the ClpE ATPase (data not shown). Instead, the deletion of *clpC* causes a slight increase in MgsR stability. Although the half-life of MgsR was not affected ($t1/2 = 5.9$ min), one can observe that the residual background level is increased compared to the wild type (Fig. 3). However, the deletion of *clpX* causes a distinct elevation of MgsR stability resulting in an extended half-life of 23 min (Fig. 3). Interestingly, MgsR stability was strongly enhanced in a double Δ*clpC*/Δ*clpX* mutant ($t1/2 = 120$ min, Fig. 3). Finally, the deletion of *clpP* extended the MgsR half-life beyond the analyzed time window (Fig. 3) clearly indicating that the major fraction of MgsR becomes specifically degraded by the ClpXP and (to a lower extent) by the ClpCP machinery.

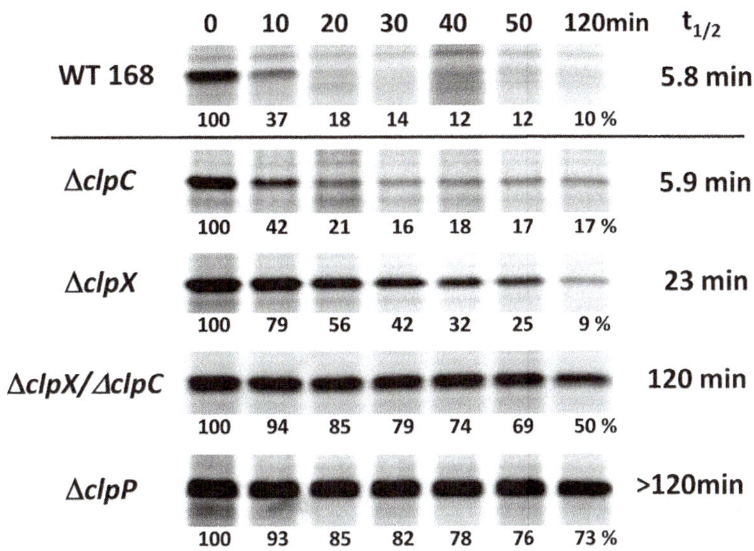

Fig. 3 Detailed kinetics of MgsR stability by pulse-chase labeling and immunoprecipitation obtained from the *B. subtilis* wild-type 168, a Δ*clpC*, Δ*clpX*, and the Δ*clpC*/Δ*clpX* double mutant as well as a Δ*clpP* mutant strain. Samples were taken after 10, 20, 30, 40, 50, and 120 min after exposition to 4% ethanol stress as indicated in the top row. All experiments were carried out in triplicate and values given below each lane represent the median per cent volumes normalized to t_0 (100%). Half-lives ($t1/2$) of MgsR are indicated on the right side of each kinetic. The respective graphs for half-life determination can be found in Fig. 4

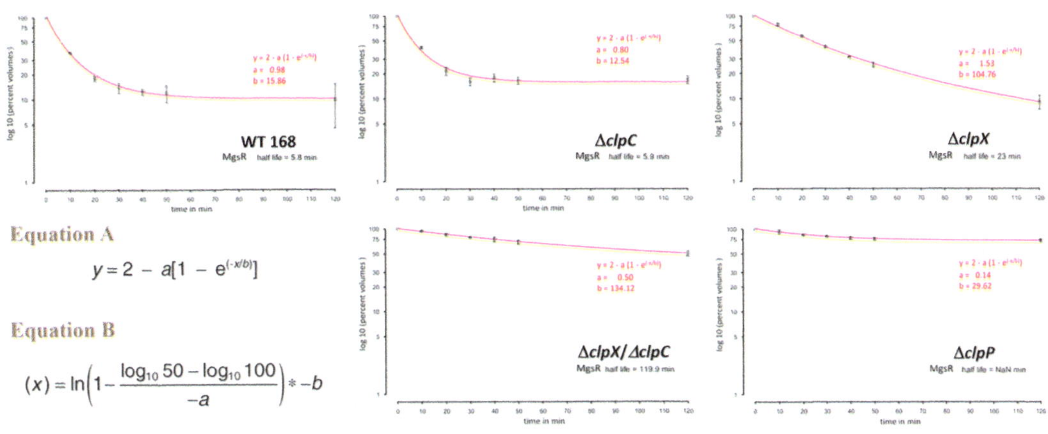

Equation A

$$y = 2 - a[1 - e^{(-x/b)}]$$

Equation B

$$(x) = \ln\left(1 - \frac{\log_{10} 50 - \log_{10} 100}{-a}\right) * -b$$

Fig. 4 Graphical representation and calculation of the MgsR half lifes in the *B. subtilis* WT, a Δ*clpC*, Δ*clpX*, and the Δ*clpC*/Δ*clpX* double mutant as well as a Δ*clpP* mutant strain

4 Notes

1. *Security advice*: The cultivation and labeling of cellular proteins with radioactive ^{35}S-L-methionine as well as all subsequent steps as centrifugation, cell disruption, the SDS–polyacrylamide gel run as well as the drying of the gel must be performed in a certified isotopic lab. Follow the waste disposal regulations when disposing radioactive waste.
2. Calculate the desired amount of activity for your experiment before ordering. So how many samples do you have, how many replicates and strains. Usually 25 μCi/ml final activity in your cell culture is a good starting point.
3. The specificity of protein A or G for the respective primary antibody should be tested in a small scale experiment prior to an extensive analysis).
4. After adding TEMED your gel will polymerize immediately, so make sure you have everything ready to pour the gel.

References

1. Reder A, Höper D, Weinberg C, Gerth U, Fraunholz M, Hecker M (2008) The Spx paralogue MgsR (YqgZ) controls a subregulon within the general stress response of *Bacillus subtilis*. Mol Microbiol 69(5):1104–1120
2. Reder A, Pöther DC, Gerth U, Hecker M (2012) The modulator of the general stress response, MgsR, of *Bacillus subtilis* is subject to multiple and complex control mechanisms. Environ Microbiol 14(10):2838–2850
3. Nakano MM, Lin A, Zuber CS, Newberry KJ, Brennan RG, Zuber P (2010) Promoter recognition by a complex of Spx and the C-terminal domain of the RNA polymerase alpha subunit. PLoS One 5(1):e8664
4. Nakano S, Nakano MM, Zhang Y, Leelakriangsak M, Zuber P (2003) A regulatory protein that interferes with activator-stimulated transcription in bacteria. Proc Natl Acad Sci U S A 100(7):4233–4238
5. Lin AA, Walthers D, Zuber P (2013) Residue substitutions near the redox center of *Bacillus subtilis* Spx affect RNA polymerase interaction, redox control, and Spx-DNA contact at a conserved cis-acting element. J Bacteriol 195 (17):3967–3978
6. Rochat T, Nicolas P, Delumeau O, Rabatinova A, Korelusova J, Leduc A, Bessieres P, Dervyn E, Krasny L, Noirot P (2012) Genome-wide identification of genes directly regulated by the pleiotropic transcription factor Spx in *Bacillus subtilis*. Nucleic Acids Res 40(19):9571–9583
7. Stülke J, Hanschke R, Hecker M (1993) Temporal activation of beta-glucanase synthesis in *Bacillus subtilis* is mediated by the GTP pool. J Gen Microbiol 139(9):2041–2045
8. Gerth U, Kock H, Kusters I, Michalik S, Switzer RL, Hecker M (2008) Clp-dependent proteolysis down-regulates central metabolic pathways in glucose-starved *Bacillus subtilis*. J Bacteriol 190 (1):321–331
9. Laemmli UK (1970) Cleavage of structural proteins during the assembly of the head of bacteriophage T4. Nature 227(5259):680–685

Chapter 11

Absolute Protein Quantification Using AQUA-Calibrated 2D-PAGE

Sandra Maaß

Abstract

Absolute protein quantification for the analysis of proteome dynamics is more and more required by the scientific community. Therefore a number of methods have recently been reported that aim at determining concentrations of single proteins in complex samples, all of them having their advantages and limitations. However, for all of these methods an accurate and protein unspecific determination of the total protein amount in a given sample is urgently needed. Here a ninhydrin-based assay established to reach this goal is described. Moreover, an optimized protocol for protein digestion is an inevitable prerequisite for all mass spectrometry-based approaches aiming at absolute protein quantification. In this chapter, various aspects are described which have to be considered during validation of a suitable digestion method and a detailed protocol is presented that can be applied to the digestion of soluble proteins originated from microbes.

In order to provide an absolute protein quantification workflow applicable for small scale and large scale approaches, a step-by-step guide is provided for the so-called AQUA-strategy (AQUA = *ab*solute *qua*ntification), including selection of suited standard peptides, the development of optimized MS methods and the determination of absolute protein concentration using stable isotope dilution and selected reaction monitoring (SID-SRM). Subsequently, a workflow is introduced that combines targeted mass spectrometry and two-dimensional polyacrylamide gel electrophoresis for the large-scale determination of absolute protein amounts.

Key words Proteomics, Absolute protein quantification, Protein, Digestion, AQUA, Stable isotope dilution, Selected reaction monitoring, 2D-PAGE

1 Introduction

Gelfree and gel-based relative protein quantification is widely used to compare protein amounts between different samples. Thereby protein abundances are represented by number and intensity of detected mass spectra, respectively, or by the intensity of a protein staining after polyacrylamide gel electrophoresis (PAGE). Determining relative protein abundances allows monitoring of changes in protein concentration but does not necessarily reflect the ratios of different proteins within one sample. In contrast, absolute protein quantification gives rise to protein amounts as fraction of total

Dörte Becher (ed.), *Microbial Proteomics: Methods and Protocols*, Methods in Molecular Biology, vol. 1841, https://doi.org/10.1007/978-1-4939-8695-8_11,

protein mass, as concentration or even in molecules per cell, representing the abundance of every protein within the sample. The availability of such data is mandatory for the growing field of systems biology aiming at mathematical modeling of biological contexts and deduction of new regulatory relationships of proteins [1, 2]. Additionally, data on protein abundance allow for calculating stoichiometries in protein complexes, operon structures or between proteins in metabolic or regulatory networks [3–7]. Furthermore, time-resolved determination of protein concentrations allows estimation of protein synthesis rates and protein half-lives without the use of traditional methods normally based on radioactive protein labeling [8, 9].

1.1 Determination of Total Protein Amount for Absolute Protein Quantification

One of the most important prerequisites for absolute protein quantification is a suitable protocol for determination of the total protein concentration of a given sample. Although a variety of colorimetric assays have been developed for this purpose, the accuracy of most of them is affected by the sample composition, as they are based on the interaction of a dye to specific amino acids [10–13]. However, when proteomics aims at absolute protein quantification, in most cases a defined amount of sample is applied for downstream analysis and this defined amount is often used as reference parameter for determined concentration of target proteins. Hence, the determination of total protein amount has to be sensitive and independent of protein composition. Comparison of the traditional coomassie-based Bradford assay [10] and a ninhydrin-based assay [14] has shown that the latter fits the requirements for absolute protein quantification best [15].

The ninhydrin-assay is based on the detection of free primary amino residues by the reagent 2,2-dihydroxyindane-1,3-dione (ninhydrin). Starting from a protein sample free amino acids are generated by acidic hydrolysis. During the detection reaction ninhydrin forms a Schiff base with the amino residue of primary amino acids. After decarboxylation of the amino acid and subsequent release of the amino residue amino-ninhydrin is formed. Dimerization of amino-ninhydrin leads to the generation of the purple compound known as Ruhemann's purple (diketohydrindylidenediketohydrindamine). The change of color can be determined at 575 nm. The usage of a calibration curve allows calculating protein concentration based on the measured absorbance.

1.2 In-Solution Digest for Absolute Protein Quantification

The second very important prerequisite for mass spectrometry (MS)-based absolute protein quantification is the complete and reproducible digestion of the sample into specific peptides. Incomplete digestion will lead to an underestimation of real protein concentration. Indeed, about 3.1% of the technical variability in protein quantification is caused during protein digestion [16]. Sample to sample differences on the digestion yield may occur due to the heterogeneity of the sample, the physicochemical properties of proteins, and the diversity in the number and quantity of the proteins present in the sample [17]. Hence, the selection of the most suitable enzyme or chemical reagent depends on the characteristics of the target proteins as well as pH and composition of the buffer used during digestion [18]. Usually it is worth to optimize the applied proteases and their efficiency by testing different buffers, protein-to-enzyme ratios, detergents that may support protease efficiency (e.g., RapiGest), digestion temperatures, and digestion durations [19] (for an example *see* Fig. 1).

The maximum peptide yield is an important indicator of complete digestion. Another possibility is to monitor missed cleavage products by adding the corresponding transitions into a targeted MS analysis. Moreover, if two or more unique peptides from different sites of a protein are quantified and compared, differences in quantitative amounts obtained by the analysis of these

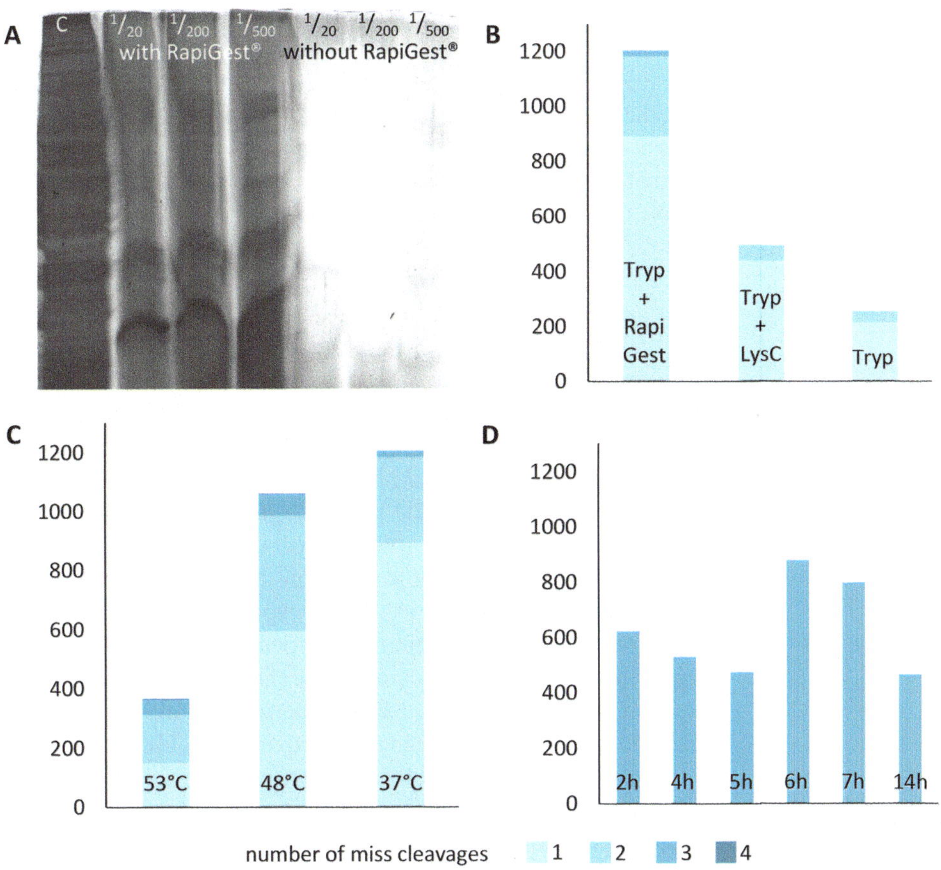

Fig. 1 Optimization of digestion parameters for absolute protein quantification. Selected results are shown for the optimization of digestion buffers and protein-to-enzyme ratios (**A**), the protease applied (**B**), incubation temperature (**C**), and duration of incubation (**D**). *y*-axis indicates the number of identified proteins from a complex protein sample. *C* undigested control, *Tryp* trypsin

peptides would indicate possible problems in digestion efficiency as well [19].

In the presented protocol materials, buffers, and reagents are optimized toward complete digestion of microbial soluble proteins and subsequent purification of the peptides in order to generate an MS-compatible sample, which is suited for absolute protein quantification. Thereby reduction and alkylation of cysteine residues are performed to increase protease susceptibility. The use of an acid-cleavable reagent aims at improving the proteolytic activity of trypsin. Furthermore, the protein-to-enzyme ratio, digestion duration, and digestion temperature are optimized in order to achieve a specific and complete cleavage of all proteins to tryptic peptides [15].

1.3 Method Development for Absolute Protein Quantification Using Stable Isotope Dilution and Selected Reaction Monitoring

One of the best established MS-based strategies for absolute protein quantification is to add stable-isotope-labeled reference standards of known absolute quantity to a sample. This method is also known as stable isotope dilution (SID) approach. Several workflows utilizing such reference peptides and proteins exist, including the AQUA-technique (AQUA = *a*bsolute *qua*ntification). Here, chemically synthesized and stable isotope labeled reference peptides (so-called AQUA-peptides) are spiked into a complex sample in known amounts. As the native and the reference peptide have exactly the same amino acid sequence, they share the same physicochemical properties like chromatographic coelution, ionization efficiency, and relative distribution of fragment ions. However, the native and the reference peptide can be distinguished by the introduced mass shift. Therefore spiked-in AQUA-peptides allow for simultaneous LC-MS analysis and peak area detection of both native and reference peptides. As the analyzed peptides function as surrogates for their corresponding proteins, the amount of native target protein can be subsequently calculated by comparing the respective peak areas [20, 21].

In recent years, the SID principle has been widely successfully applied [22–25]. Thereby this method is most often coupled to an MS technique called selected reaction monitoring (SRM). In contrast to typical LC-MS/MS shotgun experiments, in SRM measurements no complete fragment ion spectra are recorded. Instead, signal intensities for selected predefined SRM transitions are acquired over time. Such SRM experiments are typically carried out with triple quadrupole mass spectrometers. Here, the capability of these instrument types to function as ion filters is used. In the first quadrupole (Q1) the precursor ion of interest is selected and subsequently guided to the second quadrupole (Q2), where the selected ions are fragmented. The generated fragment ions are then led to the third quadrupole (Q3), where the selection of fragment ions of interest occurs. Thus, the detected signal, specific for the transition of the specific precursor ion to the fragment ion of interest, is monitored in a given time window [21, 26]. Hence, each transition is defined by a combination of a Q1 and a Q3 *m*/*z* value resulting in a filtering at the peptide and fragment ion level. Moreover, these transitions can be easily monitored for native and reference peptides in parallel, which allows integrating and comparing peak areas of both molecules for quantification. The selection of AQUA-peptides, the development of optimized MS-methods and the calculation of absolute protein amounts using AQUA-SRM is part of this chapter.

1.4 Absolute Protein Quantification Using AQUA-Calibrated 2D-PAGE

To date, simultaneous absolute protein quantification of hundreds of proteins is a challenge for the scientific community as transferring established methods for absolute quantification of single proteins to large-scale studies is not a trivial task. In this chapter a workflow is presented that makes use of two well-established techniques: absolute protein quantification using stable isotope dilution and selected reaction monitoring and 2D-PAGE. Here, a subset of proteins in a complex sample is absolutely quantified using the AQUA-technique and subsequently used to calibrate the corresponding 2D gels. By comparing spot intensities of the calibration proteins, of which the absolute amount is known after AQUA-SRM, to intensities of other protein spots, the absolute amount of all protein spots on the 2D gel can be determined (*see* Fig. 2). Identification of these protein spots by *matrix-assisted laser desorption/ionization-mass spectrometry* (MALDI-MS) allows calculating abundances of all detected isoforms belonging to a protein.

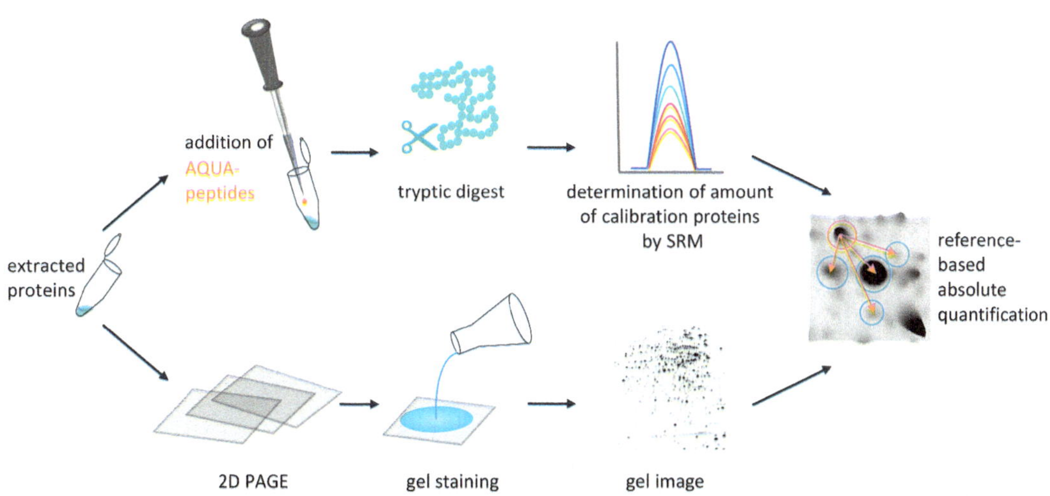

Fig. 2 Workflow described in this chapter. Absolute protein quantification is achieved by combining quantification using stable isotope dilution and selected reaction monitoring and 2D-PAGE

2 Materials

2.1 Determination of Total Protein Amount

Prepare all solutions in aqua bidest if not indicated elsewise. Prepare and store all reagents at room temperature (unless indicated otherwise).

1. BSA (bovine serum albumin) stock solution: 10 mg/ml BSA (*see* **Note 1**). Transfer 150 mg BSA to a plastic reaction tube (*see* **Note 2**). Add 15 ml aqua bidest and mix gently. Store the solution at 4 °C.
2. Hydrochloric acid (HCl), fuming 37% (also traded as "concentrated hydrochloric acid").
3. Protein sample with unknown concentration in suitable buffer (*see* **Note 3**).
4. 4 N sodium acetate buffer: Dissolve 272 g Na $(CH_3COO^-) \times 3H_2O$ in 100 ml pure acetic acid and fill up to 450 ml with aqua bidest (*see* **Note 4**). Adjust pH to 5.5 using 1 M NaOH. Fill up to 500 ml with aqua bidest.
5. Stannous chloride solution: Add 500 mg of $SnCl_2$ to 5 ml of ethylene glycol (*see* **Notes 5** and **6**).
6. Ninhydrin reagent: Solve 1 g ninhydrin in 37.5 ml ethylene glycol and 12.5 ml 4 N sodium acetate buffer (*see* **Note 6**). Mix gently in the dark for at least 1 h thereby taking care that no ninhydrin is left at the bottom of the tube (*see* **Note 7**). It is preferable to prepare this solution at the same day the determination of protein concentration is performed.
7. Ninhydrin working solution: Add 1.25 ml stannous chloride solution to the ninhydrin reagent shortly before adding it to the sample and mix well by inverting the tube several times. The solution should change color from pale yellow to pale red.

2.2 In-Solution Digest

All chemicals used should have at least LC-MS-grade purity. Prepare all solutions in aqua bidest and do not adjust pH of the reagent if not indicated elsewise. Prepare and store all reagents at room temperature (unless indicated otherwise). Use low-binding reaction tubes.

For the digestion of proteins the following materials are needed:

1. 50 mM TEAB (triethylammonium bicarbonate): 1.0 M stock solution is commercially available. Dilute 1 ml TEAB in 19 ml aqua bidest. Store the solution at 4 °C.
2. Sample in a digestion compatible buffer (*see* **Note 8**).
3. 0.5% (w/v) RapiGest SF Surfactant: solve 1 mg RapiGest in 200 μl 50 mM TEAB. Aliquots of the solution can be stored at −20 °C.

4. 500 mM TCEP (tris [2-carboxyethyl]phosphine): dissolve 1 mg TCEP in 7 μl 50 mM TEAB (*see* **Note 9**).
5. 500 mM IAA (iodoacetamide): dissolve 1 mg IAA in 10.8 μl 50 mM TEAB (*see* **Note 9**).
6. 10× trypsin: solve 20 μg in 100 μl 50 mM acetic acid or 100 μl vendor-provided resuspension buffer.
7. Trifluoroacetic acid (*see* **Note 10**).

For the purification of the generated peptide samples before MS analysis the following materials are needed:

8. 100% (v/v) methanol.
9. C18 chromatographic material.
10. Solvent A: 0.1% (v/v) acetic acid.
11. Solvent B: 0.1% (v/v) acetic acid in acetonitrile.
12. Stage Tips, C18 material, GELoader pipette tip (Thermo).

2.3 Method Optimization for Absolute Protein Quantification Using Stable Isotope Dilution and Selected Reaction Monitoring

Please use low-binding reaction tubes.

1. Digested protein sample.
2. AQUA-peptides: suited AQUA-peptides will be selected during method development described in this protocol. They can be purchased by different vendors with different levels of purity and concentration precision (*see* **Note 11**).

2.4 Absolute Protein Quantification Using AQUA-Calibrated 2D-PAGE

1. Sample (*see* **Notes 12** and **13**) in a digestion compatible buffer (*see* **Note 14**).
2. Images of high quality 2D gels of exactly the same sample in three to five technical replicates (*see* **Note 15**). It is important, that the quantitative information of protein spots in the 2D gel is kept during image capturing. Staining of the gels should have been performed with a sensitive and protein unspecific dye exhibiting a linear response to the protein amount over a wide dynamic range (*see* **Note 16**).

3 Methods

3.1 Determination of Total Protein Amount

1. Dilute the BSA stock solution to the following concentrations: 0.25, 0.5, 1.0, 1.5, 2.0, 4.0, 6.0, 8.0, and 10.0 μg/μl.
2. Prepare 20 μl aliquots of the different BSA concentrations for the calibration curve.
3. Prepare a 20 μl aliquot of the protein sample buffer used (*see* **Note 3**), which will be used as a blank later on.

4. Prepare a single 20 μl aliquot for each protein sample to be analyzed.
5. Add 20 μl HCl to each aliquot of the BSA dilution series, the samples, and the blank.
6. Close the reaction tubes thoroughly (*see* **Note 17**) to avoid opening of the tubes during the following boiling process.
7. For protein hydrolysis incubate all tubes for 24 h at 100 °C in a heating block (*see* **Notes 18** and **19**).
8. Carefully remove samples from the heating block and let them cool to room temperature (*see* **Note 19**) or let the samples cool down in the deactivated heating block.
9. Spin down every sample (including protein sample, BSA samples, and blank) to avoid residual liquid in the lid of the reaction tube.
10. Prepare a 1:100 dilution in two steps for every sample with aqua bidest: I—Dilute the sample 1:10 using 10 μl hydrolyzed sample and 90 μl aqua bidest (*see* **Note 20**), II—Dilute an additional time in the ratio 1:10 using 50 μl of the first dilution (**step I**) and 450 μl aqua bidest.
11. For every technical replicate (*see* **Note 21**) transfer 200 μl of the 1:100-dilution to a dark (brown) reaction tube (*see* **Note** 7).
12. Add 200 μl aqua bidest and 600 μl of freshly prepared ninhydrin working solution (*see* **Note 6**) and mix well by inverting the tube several times.
13. Incubate for 10 min at 100 °C in a heating block.
14. Carefully remove samples from the heating block and cool them shortly on ice.
15. Transfer the samples in microcuvettes (*see* **Notes 6** and **22**).
16. Measure absorbance of protein and BSA samples at 575 nm using the prepared blank value in comparison.
17. Generate a calibration curve using the absorbance values of the BSA samples. Determine protein concentrations according to this calibration curve (*see* **Notes 23** and **24**).

3.2 In-Solution Digest and Sample Purification

1. Aliquot 100 μg of sample in a reaction tube (*see* **Note 25**).
2. Add 20 μl 0.5% (w/v) RapiGest SF Surfactant.
3. Fill up to 97 μl using 50 mM TEAB.
4. For reduction of disulfide bonds add 1 μl 500 mM TCEP and incubate at 60 °C for 45 min.
5. For alkylation of reduced cysteine residues add 2 μl 500 mM IAA and incubate in the dark at room temperature for 15 min.

In parallel, activate 10× trypsin by incubating at 37 °C for 15 min.

6. Add 2.5 μl of 10× activated trypsin. If necessary, at this point the internal standard has to be added as well (*see* **Note 26**). Incubate at 37 °C for 5 h while shaking gently.
7. Add 1 μl trifluoroacetic acid (*see* **Note 10**) und incubate at 37 °C for 45 min without shaking to hydrolyze and precipitate the RapiGest.
8. Spin down the precipitate at least three times at a table top centrifuge at maximum speed and 4 °C for 10 min each. Transfer the supernatant in a new reaction tube after every centrifugation.
9. Store samples at −20 °C up to 1 week or start purification of the peptides samples (next steps).
10. Prepare appropriate holders for the Stage Tip by cutting a hole [Ø 3 mm] in the lid of a 2 ml-reaction tube.
11. Cut the tip of the Stage Tips about 5 mm below the frit and mount the Stage Tip in the holder.
12. Prepare slurry of C18 chromatographic material and methanol. Add as much volume on the top of the original column as needed to obtain a packing of additional 5 mm C18 material (*see* **Notes 27** and **28**). For loading of the C18 slurry, centrifuge Stage Tips in the holder with a table top centrifuge at 10,000 × *g* and room temperature. The column is now capable of binding up to 1 mg proteins or peptides, respectively.
13. For initial washing rinse the column three times with 100 μl solvent A: for every washing load 100 μl solvent A on top of the column material (*see* **Note 29**) and centrifuge Stage Tips in the holders with a table top centrifuge at 9,500 × *g* and room temperature.
14. For further washing rinse the column two times with 100 μl solvent B.
15. Equilibrate the column with 100 μl solvent A. From here on it is important to avoid the resin from drying (*see* **Note 30**).
16. Load the column by adding the digested sample on top of the column material and spinning the Stage Tips in the holders with a table top centrifuge at 9,500 × *g* and room temperature.
17. Wash the bound sample two times with 100 μl solvent A.
18. Weigh a sample vial for every sample using a special accuracy weighing machine.
19. Elute the sample in a weighted sample vial using 30 μl solvent B (*see* **Notes 31** and **32**).

20. Add 20 μl of ultrapure water and dry down the sample below a volume of 10 μl using a vacuum centrifuge to remove acetonitrile originating from solvent B.
21. Weigh the vial to get the exact volume (=mass) of the sample.
22. Fill up to 100 μl with ultrapure water. In case predigested standard peptides should serve as standard for quantification, the peptides have to be spiked into the sample at this point (*see* **Note 26**).
23. Sample volume not needed during MS analysis can be stored at −70 °C.

3.3 Selection of AQUA-Peptides

1. Select 5–10 calibration proteins on the 2D gel image regarding the following criteria: I—presence on 2D gels during all conditions examined, II—arrangement in good quality spots, which can be reliably detected by custom software packages (*see* **Note 33**), and III—the sum of calibration spots should cover the molecular mass and isoelectric point range examined on the gel as good as possible (*see* **Note 34**).
2. Deduce 3–5 AQUA-peptides for every selected calibration protein according to the protocol presented in **steps 5–9** in Subheading 3.3.
3. Order or synthesize the requested peptides for further method development and absolute quantification of calibration proteins (*see* **Note 35**) following Subheading 3.5–3.7.
4. Generate a reference data set for AQUA-peptide selection. This, for example, can be done by extensive protein mapping in the sample by tandem MS. For some proteomes already existing data sets can be downloaded from public resources [27–29]. Another possibility is the in silico digestion of the proteome to be analyzed.
5. Filter for proteotypic peptides. The targeted peptides should uniquely identify the protein of interest in the background proteome (*see* **Note 36**). In cases where no experimental data is available, it has been proposed that proteotypic peptides may be predicted based on physicochemical properties [8, 30, 31].
6. Filter for peptides with missed cleavage sites. Suitable AQUA-peptides should be efficiently cleaved off the native protein during protease treatment resulting in no missed cleavage site (*see* **Note 37**).
7. Filter for peptides in a mass range of 800–2400 Da as peptides of intermediate length (6–13 residues) will give easily interpretable CID (*c*ollision *i*nduced *d*issociation) spectra on Triple Quadrupole and Ion Trap instruments with a handful of predominant product ion species. Hence, a better limit of

detection can be achieved by SRM compared to the analysis of longer peptides [32].

8. Filter for reactive and labile residues. Unwanted modifications and missing peptide stability will reduce the actual amount of targeted internal standard peptide and leading to erroneous quantification values. Therefore, it is recommended to obviate peptides containing cysteine, tryptophan, or methionine, if possible, as these amino acids may undergo oxidation. Tryptophan is prone to undergo alkylation during synthesis. Peptides containing asparagines or aspartate are prone to undergo modification (deamidation or racemization, respectively), especially when followed by glycine in the peptide sequence. The peptide bond between aspartate and proline is labile to formic acid. Therefore, peptides containing this amino acid combination should be obviated.
9. Filter for known posttranslational modifications. If the experiment aims at quantifying the protein, it is desirable to choose peptides that do not contain putative motifs for modifications such as phosphorylation and glycosylation. If the intent is to quantify the amount of modified peptide species, these peptides should be of course included (*see* **Note 36**) [20, 32].
10. Select labeling strategy. AQUA-peptides can be synthesized by different commercial vendors or expressed in vivo or in vitro in your own lab. Therefore various strategies are available differing in labeling and purification strategies [33–36]. One may choose between either complete stable isotope labeling of single elements in the peptide (e.g., every carbon or every nitrogen) or labeling of single amino acid residues instead. Thereby it is preferable to use ^{13}C- and/or ^{15}N-labeled amino acids as they do not show chromatographic retention shift as deuterium-labeled molecules do [21, 37]. As tryptic peptides give predominant y- fragment ion series, it is desirable to have the isotope-labeled residue toward the C-terminus such that y-ions from the native and the labeled reference peptide will have the mass difference. Furthermore, it is preferable to have an addition of at least 6 amu by isotope labeling, since usually multiply charged peptides are chosen for SRM and overlapping m/z isolation windows for the native and reference peptides should be avoided [21]. Whatever labeling strategy you may choose for your intended experiment, the resulting peptide should be very pure and its concentration should be, either directly or indirectly, accurately determined by amino acid analysis.

3.4 Optimization of MS Parameters for SRM Experiments

1. Select Transitions. Calculate transitions with precursor charges 2+ and 3+ and y-ions with m/z larger than the m/z of the precursor for every peptide. SRM chromatograms using a product ion with m/z less than that of the parent ion are

likely to be noisy due to the product ions arising from bulk solvent (chemical noise), thus reducing the overall sensitivity.

2. Analyze the selected transitions. Measure your sample with a shortened LC-gradient in unscheduled SRM mode to characterize the selected transitions (*see* **Note 38**). Choose up to six of the best performing transitions (represented by highest peak areas) for every peptide to be included in subsequent collision energy optimization (*see* **Note 39**).
3. Select collision energies (CEs). Use CEs calculated by optimized formulas published for different MS instruments (like for example in Lange et al. [26]) or optimize the CEs by following the next step.
4. Optimize CEs. Ramp CEs for all monitored transitions in 5–7 steps around the predicted CE (*see* **Note 40**).
5. Analyze the optimized CEs. Measure your sample with a shortened LC-gradient in unscheduled SRM mode to characterize the selected CEs for every transition (*see* **Note 38**). Choose the three best performing transitions (represented by highest peak areas and best signal/noise ratio) for every peptide to be included in the MS method.
6. Select transitions for standard peptides. Calculate the transition of the labeled AQUA-peptide, which correspond to those selected for the native peptide in the previous step. This should result in six transitions per peptide sequence.
7. Check the generated instrument method. Inspect specificity of SRM experiments by checking the obtained data for overlaying retention time profiles for native and reference peptide. All monitored transition belonging to the same peptide sequence should show comparable retention times and similar peak profiles.
8. Validate the generated instrument method. Confirm the selected transitions by MS2 spectrum acquisition to avoid false identification and quantification due to unspecific signals derived from more abundant peptides in the same *m/z* but different retention time window. The combined information of SRM traces and MS/MS spectra with overlaid theoretical fragment ion spectra and associated probabilities will help to validate the peptide identifications.
9. Scheduled SRM. The mass spectrometric information collected during method development and method validation will enable you to set up an instrument method in scheduled SRM mode (*see* **Notes 41** and **42**).

3.5 Mass Spectrometry-Based Absolute Protein Quantification

1. Calculate the peptide ratio of native and heavy species based on three transitions. It is suggested to weigh the peptide ratios according to their signal to noise (S/N) ratios, as explained in the following steps (*see* **Note 43**).
2. Multiply the peak area ratios for every transition with the lowest respective S/N ratio of either the native and or the heavy peptide, leading to three values for "area times lowest S/N" (*see* Fig. 3).
3. Sum up all "area times lowest S/N" for a given peptide sequence (*see* Fig. 3).
4. Sum up the lowest S/N ratios used for calculation in **step 2** in Subheading 3.5 (*see* Fig. 3).
5. Divide the sum obtained in **step 3** in Subheading 3.5 by the sum obtained in **step 4** in Subheading 3.5 to get the weighted peptide ratio (*see* Fig. 3).
6. Average at least three (S/N weighted) peptide ratios to get the final protein ratio (*see* **Notes 37** and **44**).
7. Calculate the absolute quantity of target anchor proteins based on the added amount of AQUA-peptides.
8. Calculate the coefficient of variation (CV) between at least three technical replicates of SRM measurements as additional tool for method evaluation. Accurate protein quantification can be achieved if CV is lower than 15% [15].

3.6 Analysis of Digital Gel Images and Selection of Calibration Proteins for Absolute Protein Quantification Using AQUA-Calibrated 2D-PAGE

1. Prepare 2D Gels of exactly the same samples used during Subheadings 3.1, 3.2, and 3.5. Use protocols published elsewhere [38] or *see* Chapter 7.
2. Load the acquired gel images (*see* **step 2** in Subheading 2.4) into suitable analysis software (*see* **Note 33**).
3. Align all gel images to correct for differences during gel electrophoresis.

	Peptide species	Transition	Area Ratio	Signal / Noise	Area ratio times lowest S/N
PeptideSequence1	AQUA-peptide	heavy precusor to y5	3.8	4000	(3.8*4000)=15200
PeptideSequence1	native peptide	native precursor to y5		15000	
PeptideSequence1	AQUA-peptide	heavy precusor to y6	3.3	900	(3.3*900)=2970
PeptideSequence1	native peptide	native precursor to y6		3700	
PeptideSequence1	AQUA-peptide	heavy precusor to y7	3.7	68000	(3.7*36000)=133200
PeptideSequence1	native peptide	native precursor to y7		36000	

Step 3: Sum (area ratio times lowest S/N) (15200+2970+13320)=151370

Step 4: Sum lowest S/N (4000+900+36000)=40900

Step 5: **Weighted peptide ratio** **(151370/40900)=3.7**

Fig. 3 Calculation of weighted peptide ratios. Example for the calculation of weighted peptide ratios like explained in **steps 2–5**, Subheading 3.5. Subtotals are colored differently

4. Create a fusion image that contains all protein spots in the experiment.
5. Create a consensus spot pattern on the fusion gel.
6. Propagate the consensus spot pattern to all gels of the experiment.
7. Extract quantitative data within the consensus spots by integration of grey levels on every gel image. Use normalized spot volumes (*see* **Note 45**) as basis for protein quantification.

3.7 Absolute Quantification of Calibration Proteins, Calibration of 2D-Gel Images, and Absolute Protein Quantification in 2D Gels

1. Digest your sample with spiked-in AQUA-peptides of calibration proteins selected according to Subheading 3.3. Use a highly reproducible protocol optimized toward completeness of specific protein cleavage. A suitable protocol for microbial soluble proteins is described in Subheading 3.2.
2. Determine the absolute amount of calibration proteins in the original sample following the protocol described in Subheading 3.5.
3. Check quality of quantification by 2D-PAGE for calibration proteins. Calibration proteins must show a relative standard deviation of less than 35% among technical replicates of 2D gels. If this is not the case, exclude the affected protein from the list of calibration proteins.
4. Check quality of quantification by AQUA-SRM for calibration proteins. Calibration proteins must show a relative standard deviation of less than 15% among technical replicates of MS measurements. If this is not the case, exclude the affected protein from the list of calibration proteins.
5. Calculate the absolute amount of all spots detected on the 2D gels according to Eq. 1. Here, V represents the normalized spot volume of any spot, m represents the absolute amount of any spot, Spot_X represents any spot, Spot_Cal represents the spot(s) of the calibration protein(s).

$$m_{\text{Spot}_x} = \frac{V_{\text{Spot}_x} \times m_{\text{Spot}_{\text{Cal}}}}{V_{\text{Spot}_{\text{Cal}}}}. \qquad (1)$$

6. If necessary, derive protein amounts out of spot amounts. To do this, sum up the absolute amounts for all spots of the same protein. If a protein spot is comprised by different proteins it is feasible to multiply the spot by the number of protein species within and use the whole spot amount for either of the proteins (*see* **Note 46**).
7. The standard deviation between technical replicates for proteins comprised out of multiple spots can be calculated by error propagation as shown in Eq. 2, where n represents the different spots belonging to the same protein.

$$\mathrm{SD} = \sqrt{(\mathrm{SD}_1)^2 + (\mathrm{SD}_2)^2 + \ldots (\mathrm{SD}_n)^2}. \tag{2}$$

8. The final standard deviation of the calculated absolute protein amounts results from the combination of technical and biological replicates. Calculation of the final standard deviation can be performed following a random effect model described in Eq. 3. Here, n is the number of technical replicates and p is the number of biological replicates. Furthermore, it applies that $m = 1 \ldots n$ and $k = 1 \ldots p$.

$$\mathrm{SD}^2 = \left(\sum_{k=1}^{p} (\bar{x} - \bar{x}_k)^2 \times \frac{1}{p-1} \right) - \left(\left(\sum_{k=1}^{p} \left(\sum_{m=1}^{n} (\bar{x}_k - x_{km})^2 \right) \right) \times \frac{1}{(p \times n \times (n-1))} \right). \tag{3}$$

To be valid:

$$\bar{x} = \frac{1}{p} \sum_{i=1}^{p} \left(\frac{1}{n} \sum_{j=1}^{n} x_{ij} \right) \text{ with } \mathrm{i} = 1 \ldots p \text{ and } j = 1 \ldots n$$

$$\bar{x}_k = \frac{1}{n} \sum_{l=1}^{n} x_{kl} \text{ with } k = 1 \ldots p \text{ and } \mathrm{l} = 1 \ldots n.$$

4 Notes

1. Some of the commercially available BSA contains large amounts of contaminant salts. For the use in calibration curves it is recommended to purchase at least "sequencing grade".
2. Storage of the stock solution at 4 °C is possible for several months. In order to guarantee maximum accuracy of the assay the usage of low adsorption polypropylene tubes is preferable.
3. Do not use buffers containing amino groups (e.g., urea-buffers or TEAB [triethylammonium bicarbonate]). These amino groups will react with the ninhydrin reagent and therefore bias the results. Recommended buffers are TE-buffer (10 mM Tris, 1 mM EDTA, pH 7.5), PBS-buffer (13.7 mM NaCl, 2.7 mM KCl, 10 mM Na_2HPO_4, 2 mM KH_2PO_4), or comparable buffers.
4. Dissolving the sodium acetate can be very difficult. Gentle warming of the solution can help to accelerate this process.
5. Mix well before pipetting as this solution is a fine suspension.

6. Due to the ethylene glycol this solution is very viscous. Therefore, it is recommended to pipette slowly up and down before taking the determined amount of solution. All pipetting steps should be carried out slowly.
7. Ninhydrin is light sensitive. Therefore, the reagent should be stored in dark containers and every experimental step should be carried out as light protected as possible.
8. Buffers at pH of 7–9 containing *up to* 50 mM ammonium bicarbonate, 50 mM Tris, 50 mM TEAB, 1 M urea, or 10% acetonitrile can be used as they allow for highest tryptic activity.
9. Prepare the reagent freshly before use as they are unstable either in neutral solutions (TCEP) or are light sensitive (IAA).
10. Alternatively, another strong acid can be used. However, the pH of the sample after adding of the acid should be below 2. The added volume may differ from that sufficient for trifluoroacetic acid and should be considered when calculating absolute protein amounts in the sample.
11. For protein quantification it is mandatory to use pure peptides whose amount has been accurately determined.
12. For optimal results sample preparation conditions should be carefully controlled. It is essential to prevent or to exactly determine sample losses which can be caused by fractionation, enrichment, or cleaning. Therefore, the most accurate results are achieved if sample preparation is kept to a minimum number of steps.
13. If quantitative results are required in amounts per cell, it is necessary to determine the number of cells the sample originates from, e.g., by cell counting. In addition, all proteins of a cell, tissue, or compartment have to be extracted for the exact determination. Therefore, cell disruption and protein extraction have to be rechecked and optimized for the biological sample of interest if required.
14. Buffers at pH of 7–9 containing *up to* 50 mM ammonium bicarbonate, 50 mM Tris, 50 mM TEAB, 1 M urea, or 10% acetonitrile can be used.
15. Technical replicates of 2D gels will be necessary to evaluate quality of quantification by 2D-PAGE. However, biological replicates are also important in order to statistically secure the physiological results of the study. At least three biological replicates are necessary.
16. Suitable protein dyes have been evaluated for absolute quantification. Due to their protein sequence unspecificity, sensitivity, and wide dynamic range fluorescent dyes (e.g., Krypton or Flamingo) met the requirements to be used for this protocol.

More detailed information can be found elsewhere [15, 39, 40].

17. Microcentrifuge tubes can be closed securely by lid locks. Another possibility is the usage of PCR-reaction tubes. However, used tubes have to fit the heating block used during cooking in the following steps.
18. In order to avoid opening of the reaction tubes during incubation it can be helpful to cover the tubes with a glass plate, which may also be weighted with a heat-resistible heavy material (like for instance a glass beaker filled with water).
19. Ensure lab safety as this step requires prolonged cooking of concentrated acids.
20. Before pipetting the dilutions ensure a proper mixing of the sample. Especially in highly concentrated protein samples (including the higher concentrations of the BSA standard) ashed amino acids tend to precipitate. Use only the clear supernatant in the sample for determination of protein concentration.
21. It is recommended to repeat the measurement at least two times. The protocol described in this chapter provides enough material to measure two technical replicates from every hydrolyzed sample.
22. The Ruhemann's purple ninhydrin complex is unstable. Therefore, measurement of absorbance should be performed immediately after the boiling procedure.
23. The calibration curve has to be prepared for every analysis batch as the response factor is strictly depending on the exact composition and the age of the ninhydrin working solution.
24. Use only absorbance values that are within the calibration curve. If this is not the case the hydrolyzed samples can be further diluted with aqua bidest, mixed with the ninhydrin working solution and cooked (starting from **step 10** in Subheading 3.1). Do not dilute the already boiled sample, this will give erroneous results!
25. The sample size is scalable; here a 100 μl digestion volume will be described. The protein concentration during digestion should be around 1 μg/μl. The final volume of the reaction mixture should be considered when calculating absolute protein amount in the sample.
26. Internal standards for absolute protein quantification using the isotope-dilution principle may be added at this time point. This is the case for AQUA-peptides for absolute protein quantification described in Subheadings 3.3, 3.4, and 3.5. Please be aware of the fact that absolute quantification results obtained this way cannot be corrected for quantification errors during

sample preparation before this step as standards are not added earlier. To compensate for as many quantification errors as possible the earliest spike-in of standards should be considered [41, 42].

27. The volume needed depends on the concentration of the slurry. It is recommended to start with about 5 μl of the slurry and check the amount of added C18 material volume afterward.
28. In order to add a comparable amount of C18 material to the single Stage Tips it is recommended to vortex the slurry thoroughly before every pipetting step.
29. It is preferable to use GELoader Tips for loading as they allow to apply the solvent directly to the column without extended air bubbles.
30. Ensure that there is always solvent remaining on top of the column. To do so, the durations for centrifuging have to be adapted. If necessary the speed of centrifuging can be reduced.
31. Elution works best when using an appropriate syringe (e.g., Eppendorf Combitips Advanced 2.5 ml) and pressing the liquid through the column instead of centrifuging.
32. For sample fractionation different concentrations of solvent B (60–100%) may be applicable.
33. Suitable software and the advantages and limitations of different analysis workflows have been extensively discussed elsewhere [39, 43]. The workflow presented in Subheading 3.6 is supported by Delta2D (Decodon).
34. Furthermore, it is beneficial, not inevitable, if selected calibration proteins would exhibit only one protein spot per gel. Quantification will get more accurate if selected calibration proteins do not share their 2D-gel spots with other protein species.
35. It is highly important to properly store these AQUA-peptides, as any loss of standard will result in an underestimation of the real amount of target protein in the sample. Therefore it is recommended to store the standards at −20 or −80 °C in low-adsorption polypropylene tubes or glass vials [44, 45]. Moreover, freeze–thaw cycles should be avoided by using aliquots of the standards.
36. Depending on the experiment's intention, AQUA-peptides can be chosen either for quantifying a specific isoform/polymorphism or for quantifying the total amount of all the isoforms/polymorphisms.
37. For accurate absolute protein quantification of the total amount of targeted proteins it is desirable to analyze at least

three peptides per protein (*see* **step 6** in Subheading 3.5). This is even more important since there may be still unknown modification occurring on single peptides that may lead to erroneous quantification results. Moreover, a higher number of quantified peptides per protein will introduce an additional level of result validation. In cases where less than three peptides remain after filtering of AQUA-peptide candidates (*see* **steps 5–9** in Subheading 3.3) the acceptance of peptides with missed cleavage sites may be applicable. However, these isoforms should be investigated with care. In any case it is preferable if the miss cleavage occurs reproducibly.

38. In order to gain enough measurements over a peak do not analyze more than 600 transitions within one method.
39. The number of transitions selected at this step depends on the sighted number of transitions monitored in the final instrument method. It is recommended to select double the number of planned final transitions per peptide for subsequent method optimization.
40. CE-prediction is frequently included in SRM software tools from different vendors and in the open-source software Skyline (MacCoss Lab).
41. Scheduled SRM [46] will enable you to monitor more than 600 transition per MS run. However, the number of analyzed transitions is limited by the cycle time of the instrument at a specific retention time. Keep transitions down to a number that ensures at least ten measurements of every precursor along its elution from the LC column.
42. Successful scheduled SRM strictly depends on reproducible LC runs. Therefore it is recommended to choose a retention time window of at least five times bottom peak width and checking for retention time shifts frequently.
43. Weighting of peptide ratios functions as normalization strategy against outliers due to unspecific measurements. However, any other published strategy, e.g., that introduced by Loziuk et al. [47], may be applicable.
44. The accordance of peptide ratios assigned to the same protein can be used as additional level of method validation.
45. Normalized spot volumes represent the grey values in the spot as fraction of all grey values in all detected spots. Therefore, the sum of assigned normalized spot volumes is "1".
46. The method described in Subheading 3.7 has been published elsewhere [15, 48]. Another option is to divide the spot amount by the number of protein species within the spot and calculate equal fractions of spot amount for every protein species. Be aware that both options can only be approximations to the real value.

Acknowledgments

This work was supported by the German Research Foundation Grant SFB/TR34.

References

1. Bennett EJ, Rush J, Gygi SP, Harper JW (2010) Dynamics of cullin-RING ubiquitin ligase network revealed by systematic quantitative proteomics. Cell 143(6):951–965
2. Kuepfer L, Peter M, Sauer U, Stelling J (2007) Ensemble modeling for analysis of cell signaling dynamics. Nat Biotechnol 25 (9):1001–1006
3. Holzmann J, Pichler P, Madalinski M, Kurzbauer R, Mechtler K (2009) Stoichiometry determination of the MP1-p14 complex using a novel and cost-efficient method to produce an equimolar mixture of standard peptides. Anal Chem 81(24):10254–10261
4. Muntel J, Fromion V, Goelzer A, Maaβ S, Mäder U, Büttner K, Hecker M, Becher D (2014) Comprehensive absolute quantification of the cytosolic proteome of *Bacillus subtilis* by data independent, parallel fragmentation in liquid chromatography/mass spectrometry (LC/MS^E). Mol Cell Proteomics 13 (4):1008–1019
5. Nanavati D, Gucek M, Milne JLS, Subramaniam S, Markey SP (2008) Stoichiometry and absolute quantification of proteins with mass spectrometry using fluorescent and isotope-labeled concatenated peptide standards. Mol Cell Proteomics 7(2):442–447
6. Schmidt C, Lenz C, Grote M, Lührmann R, Urlaub H (2010) Determination of protein stoichiometry within protein complexes using absolute quantification and multiple reaction monitoring. Anal Chem 82(7):2784–2796
7. Wepf A, Glatter T, Schmidt A, Aebersold R, Gstaiger M (2009) Quantitative interaction proteomics using mass spectrometry. Nat Methods 6(3):203–205
8. Lu P, Vogel C, Wang R, Yao X, Marcotte EM (2007) Absolute protein expression profiling estimates the relative contributions of transcriptional and translational regulation. Nat Biotechnol 25(1):117–124
9. Schwanhäusser B, Busse D, Li N, Dittmar G, Schuchhardt J, Wolf J, Chen W, Selbach M (2011) Global quantification of mammalian gene expression control. Nature 473 (7347):337–342
10. Bradford MM (1976) A rapid and sensitive method for the quantitation of microgram quantities of protein utilizing the principle of protein-dye binding. Anal Biochem 72:248–254
11. Compton SJ, Jones CG (1985) Mechanism of dye response and interference in the Bradford protein assay. Anal Biochem 151(2):369–374
12. Lowry OH, Rosebrough NJ, Farr AL, Randall RJ (1951) Protein measurement with the Folin phenol reagent. J Biol Chem 193(1):265–275
13. Smith PK, Krohn RI, Hermanson GT, Mallia AK, Gartner FH, Provenzano MD, Fujimoto EK, Goeke NM, Olson BJ, Klenk DC (1985) Measurement of protein using bicinchoninic acid. Anal Biochem 150(1):76–85
14. Starcher B (2001) A ninhydrin-based assay to quantitate the total protein content of tissue samples. Anal Biochem 292(1):125–129
15. Maass S, Sievers S, Zühlke D, Kuzinski J, Sappa PK, Muntel J, Hessling B, Bernhardt J, Sietmann R, Völker U, Hecker M, Becher D (2011) Efficient, global-scale quantification of absolute protein amounts by integration of targeted mass spectrometry and two-dimensional gel-based proteomics. Anal Chem 83 (7):2677–2684
16. Piehowski PD, Petyuk VA, Orton DJ, Xie F, Moore RJ, Ramirez-Restrepo M, Engel A, Lieberman AP, Albin RL, Camp DG, Smith RD, Myers AJ (2013) Sources of technical variability in quantitative LC-MS proteomics: human brain tissue sample analysis. J Proteome Res 12 (5):2128–2137
17. Cañas B, Piñeiro C, Calvo E, López-Ferrer D, Gallardo JM (2007) Trends in sample preparation for classical and second generation proteomics. J Chromatogr A 1153(1-2):235–258
18. Switzar L, Giera M, Niessen WMA (2013) Protein digestion: an overview of the available techniques and recent developments. J Proteome Res 12(3):1067–1077
19. Norrgran J, Williams TL, Woolfitt AR, Solano MI, Pirkle JL, Barr JR (2009) Optimization of digestion parameters for protein quantification. Anal Biochem 393(1):48–55
20. Gerber SA, Rush J, Stemman O, Kirschner MW, Gygi SP (2003) Absolute quantification of proteins and phosphoproteins from cell

lysates by tandem MS. Proc Natl Acad Sci U S A 100(12):6940

21. Kirkpatrick DS, Gerber SA, Gygi SP (2005) The absolute quantification strategy: a general procedure for the quantification of proteins and post-translational modifications. Methods 35(3):265–273
22. Anderson L, Hunter CL (2006) Quantitative mass spectrometric multiple reaction monitoring assays for major plasma proteins. Mol Cell Proteomics 5(4):573–588
23. Keshishian H, Addona T, Burgess M, Kuhn E, Carr SA (2007) Quantitative, multiplexed assays for low abundance proteins in plasma by targeted mass spectrometry and stable isotope dilution. Mol Cell Proteomics 6 (12):2212–2229
24. Kuhn E, Wu J, Karl J, Liao H, Zolg W, Guild B (2004) Quantification of C-reactive protein in the serum of patients with rheumatoid arthritis using multiple reaction monitoring mass spectrometry and 13C-labeled peptide standards. Proteomics 4(4):1175–1186
25. Picotti P, Bodenmiller B, Mueller LN, Domon B, Aebersold R (2009) Full dynamic range proteome analysis of *S. cerevisiae* by targeted proteomics. Cell 138(4):795–806
26. Lange V, Malmstrom JA, Didion J, King NL, Johansson BP, Schafer J, Rameseder J, Wong CH, Deutsch EW, Brusniak MY (2008) Targeted quantitative analysis of *Streptococcus pyogenes* virulence factors by multiple reaction monitoring. Mol Cell Proteomics 7(8):1489
27. Desiere F, Deutsch EW, Nesvizhskii AI, Mallick P, King NL, Eng JK, Aderem A, Boyle R, Brunner E, Donohoe S, Fausto N, Hafen E, Hood L, Katze MG, Kennedy KA, Kregenow F, Lee H, Lin B, Martin D, Ranish JA, Rawlings DJ, Samelson LE, Shiio Y, Watts JD, Wollscheid B, Wright ME, Yan W, Yang L, Yi EC, Zhang H, Aebersold R (2005) Integration with the human genome of peptide sequences obtained by high-throughput mass spectrometry. Genome Biol 6(1):R9
28. Deutsch EW, Lam H, Aebersold R (2008) PeptideAtlas: a resource for target selection for emerging targeted proteomics workflows. EMBO Rep 9(5):429–434
29. Prince JT, Carlson MW, Wang R, Lu P, Marcotte EM (2004) The need for a public proteomics repository. Nat Biotechnol 22 (4):471–472
30. Mallick P, Schirle M, Chen SS, Flory MR, Lee H, Martin D, Ranish J, Raught B, Schmitt R, Werner T, Kuster B, Aebersold R (2007) Computational prediction of proteotypic peptides for quantitative proteomics. Nat Biotechnol 25(1):125–131
31. Tang H, Arnold RJ, Alves P, Xun Z, Clemmer DE, Novotny MV, Reilly JP, Radivojac P (2006) A computational approach toward label-free protein quantification using predicted peptide detectability. Bioinform Oxf Engl 22(14):e481–e488
32. Mayya V, Rezual K, Wu L, Fong MB, Han DK (2006) Absolute quantification of multisite phosphorylation by selective reaction monitoring mass spectrometry: determination of inhibitory phosphorylation status of cyclin-dependent kinases. Mol Cell Proteomics 5 (6):1146–1157
33. Brun V, Dupuis A, Adrait A, Marcellin M, Thomas D, Court M, Vandenesch F, Garin J (2007) Isotope-labeled protein standards: toward absolute quantitative proteomics. Mol Cell Proteomics 6(12):2139–2149
34. Hanke S, Besir H, Oesterhelt D, Mann M (2008) Absolute SILAC for accurate quantitation of proteins in complex mixtures down to the attomole level. J Proteome Res 7 (3):1118–1130
35. Singh S, Springer M, Steen J, Kirschner MW, Steen H (2009) FLEXIQuant: a novel tool for the absolute quantification of proteins, and the simultaneous identification and quantification of potentially modified peptides. J Proteome Res 8(5):2201–2210
36. Zeiler M, Straube WL, Lundberg E, Uhlen M, Mann M (2012) A Protein Epitope Signature Tag (PrEST) library allows SILAC-based absolute quantification and multiplexed determination of protein copy numbers in cell lines. Mol Cell Proteomics 11(3):O111.009613
37. Ong S-E, Foster LJ, Mann M (2003) Mass spectrometric-based approaches in quantitative proteomics. Methods San Diego CA 29 (2):124–130
38. Büttner K, Bernhardt J, Scharf C, Schmid R, Mäder U, Eymann C, Antelmann H, Völker A, Völker U, Hecker M (2001) A comprehensive two-dimensional map of cytosolic proteins of *Bacillus subtilis*. Electrophoresis 22 (14):2908–2935
39. Berth M, Moser FM, Kolbe M, Bernhardt J (2007) The state of the art in the analysis of two-dimensional gel electrophoresis images. Appl Microbiol Biotechnol 76(6):1223–1243
40. Kuntumalla S, Braisted JC, Huang S-T, Parmar PP, Clark DJ, Alami H, Zhang Q, Donohue-Rolfe A, Tzipori S, Fleischmann RD, Peterson SN, Pieper R (2009) Comparison of two label-free global quantitation methods, APEX and

2D gel electrophoresis, applied to the *Leptospira interrogans* proteome. Proteome Sci 7:22

41. Bantscheff M, Lemeer S, Savitski MM, Kuster B (2012) Quantitative mass spectrometry in proteomics: critical review update from 2007 to the present. Anal Bioanal Chem 404 (4):939–965
42. Maaß S, Becher D (2016) Methods and applications of absolute protein quantification in microbial systems. J Proteome 136:222–233
43. Bandow JE, Baker JD, Berth M, Painter C, Sepulveda OJ, Clark KA, Kilty I, VanBogelen RA (2008) Improved image analysis workflow for 2-D gels enables large-scale 2-D gel-based proteomics studies—COPD biomarker discovery study. Proteomics 8(15):3030–3041
44. Kraut A, Marcellin M, Adrait A, Kuhn L, Louwagie M, Kieffer-Jaquinod S, Lebert D, Masselon CD, Dupuis A, Bruley C, Jaquinod M, Garin J, Gallagher-Gambarelli M (2009) Peptide storage: are you getting the best return on your investment? Defining optimal storage conditions for proteomics samples. J Proteome Res 8(7):3778–3785
45. van Midwoud PM, Rieux L, Bischoff R, Verpoorte E, Niederländer HAG (2007) Improvement of recovery and repeatability in liquid chromatography-mass spectrometry analysis of peptides. J Proteome Res 6 (2):781–791
46. Stahl-Zeng J, Lange V, Ossola R, Eckhardt K, Krek W, Aebersold R, Domon B (2007) High sensitivity detection of plasma proteins by multiple reaction monitoring of N-glycosites. Mol Cell Proteomics 6(10):1809–1817
47. Loziuk PL, Sederoff RR, Chiang VL, Muddiman DC (2014) Establishing ion ratio thresholds based on absolute peak area for absolute protein quantification using protein cleavage isotope dilution mass spectrometry. Analyst 139(21):5439–5450
48. Maaß S, Wachlin G, Bernhardt J, Eymann C, Fromion V, Riedel K, Becher D, Hecker M (2014) Highly precise quantification of protein molecules per cell during stress and starvation responses in *Bacillus subtilis*. Mol Cell Proteomics 13(9):2260–2276

Chapter 12

Sulfur-^{34}S and ^{36}S Stable Isotope Labeling of Amino Acids for Quantification (SULAQ34/36) of Proteome Analyses

Florian-Alexander Herbst, Nico Jehmlich, Martin von Bergen, and Frank Schmidt

Abstract

Quantitative proteome profiling of microorganisms by isotopic labeling of amino acids is still a challenge, because only microorganisms with auxotrophic character are able to embed amino acids into their biomass in a quantitatively correct manner. Here, we describe an isotopic labeling technique (sulfur stable isotope labeling of amino acids for quantification, SULAQ) for the sulfur-containing amino acids cysteine and methionine in a broad range of organisms. The metabolic labeling approach is suitable for gel-based and gel-free protein analysis.

Key words Sulfur stable isotope labeling of amino acids for quantification, SULAQ, Sulfur-34S, Sulfur-36S, Quantitative proteomics

1 Introduction

One of the key issues in systems biology is the assessment of quantitative dynamic changes of the proteome [1]. For achieving this goal, the development of nanoflow liquid chromatography coupled with high sensitive and high resolution mass spectrometer has elevated the analysis into a new era. One challenge in nano liquid chromatography based quantitative proteomics was overcome by introducing metabolic labeling procedures such as stable isotope labeled amino acids in cell culture (SILAC) for eukaryotic systems [2]. The usage of SILAC for bacteria is often hampered by the versatile metabolism of microorganisms. In fact, many bacteria tend to metabolize the provided labeled amino acids and dilute the isotope label throughout their biomass.

However, the dynamic proteome requires reliable protein abundance values to address systems biology research questions. Compared to gel-based techniques that lack higher throughput and higher identification rates, LC-based techniques have clear

Dörte Becher (ed.), *Microbial Proteomics: Methods and Protocols*, Methods in Molecular Biology, vol. 1841, https://doi.org/10.1007/978-1-4939-8695-8_12, © Springer Science+Business Media, LLC, part of Springer Nature 2018

advantages where several biological replicates, time points, or treatments are measured with an increase in reproducibility and reliability of the adjacent conclusion. In order to obtain high-quality proteomics data a set of requirements has to be fulfilled: (1) the experimental design should enable accurate protein quantification, (2) the introduction of isotopic labeling and the unification of sample/control should occur as early as possible within the experimental procedure to minimize technical variability, (3) the selected technique should be suitable for high-throughput, and (4) the selected technique should allow for a wide coverage of the proteome.

Metabolic amino acid labeling is well suited compared to other approaches such as ICAT and iTRAQ experiments because the unification of sample and control occurs earlier [3]. A physiological side effect of the label [4] can be visualized by switching the light/heavy label between the conditions. SILAC has been applied to samples from eukaryotic cell culture or tissues from completely labeled animals of different phyla [5]. However, there are two very important taxa where SILAC cannot be used without applying specific efforts, namely plants and bacteria. In plants the labeling efficiency is normally too low for a proper SILAC [6] and in bacteria the labeled amino acids can be catabolized and the label is widely distributed within the biomass [7]. It became clear that SILAC could not be applied to almost all microorganism-related approaches except for example *Staphylococcus aureus* or *Helicobacter pylori* [8, 9]. Therefore metabolic labeling techniques with substrates containing stable isotopes of carbon (^{13}C) and nitrogen (^{15}N) were developed [10]. Using these approaches also helped to assess the lifetime of proteins in bacteria and eukaryotes [11]. Within such a labeling experiment, cultures are grown in different media containing either isotopic labeled substrate as sole carbon source or isotopically labeled nitrogen as major source of nitrogen [12]. The metabolization leads to the incorporation of stable isotopes into the biomass of the cell, including proteins, and to an altered isotopic pattern of a given peptide or protein. This approach can be used for quantification of samples grown under different conditions but has several disadvantages. The most prominent challenge in ^{13}C and ^{15}N labeling is the dynamic incorporation, meaning that peptides containing different amounts of exchanged atoms (carbon or nitrogen) are showing a dynamic mass difference. This fact makes the option of automatic detection and comparison of incorporation from mass spectra extremely difficult, leading often to manual analysis of mass spectra and thereby disabling high-speed data stream analysis. Furthermore, the identification of the labeled peptides by MS/MS often suffers from the changed distribution of the fragments within the spectra using conventional search engines.

In previous studies we introduced a new technique of direct labeling combining the advantages of early metabolic labeling, high-throughput capacity and accurate quantification. Instead of using carbon or nitrogen the stable isotope ^{36}S was used and proved to be useful (Fig. 1).

In proof of principle studies, *Pseudomonas putida* was grown with ^{34}S-sulfate or ^{36}S-sulfate as sole sulfur source [13, 14] and the incorporation of heavy sulfur into the amino acids cysteine and methionine was detected by mass spectrometry (Fig. 2).

The incorporation into defined amino acids has four major advantages: (1) the constant mass shift from control to sample displays, comparable to SILAC, of 2 or 4 Da for each sulfur containing amino acid per peptide, (2) the incorporation in exclusively cysteine and methionine still allows the identification of proteins on the basis of PMF or tandem-MS measurements while simultaneously the abundance can be quantified on the basis of the sulfur containing peptides, (3) the difference of 4 Da has been proven to be sufficient for unambiguous picking of the corresponding

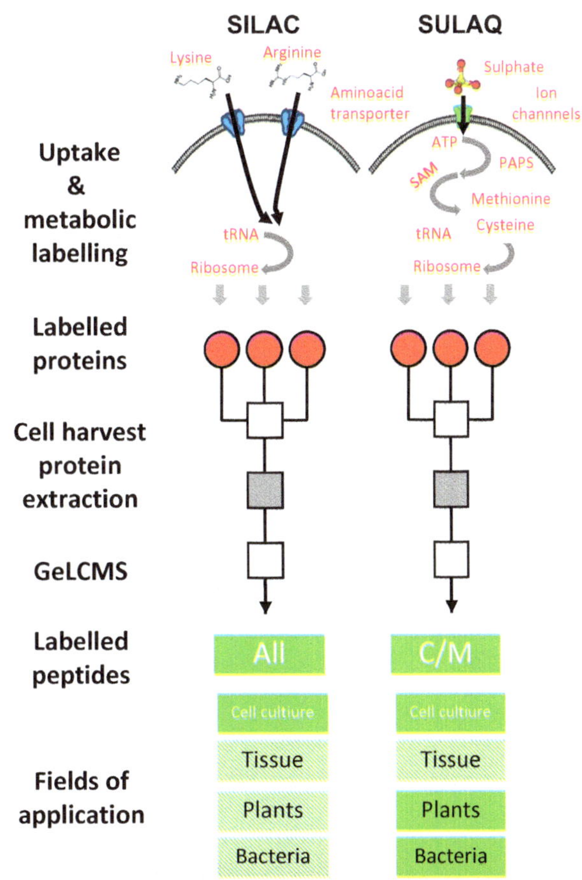

Fig. 1 Comparison of the SILAC and SULAQ labeling strategies and there fields of application

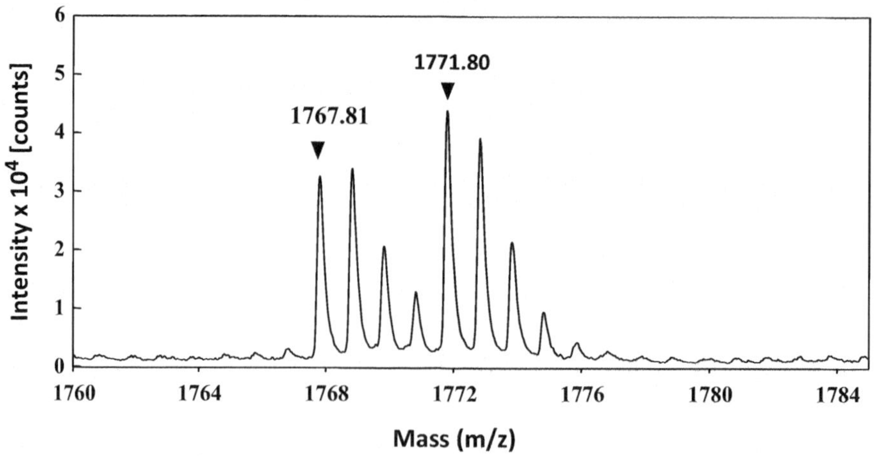

Fig. 2 Sulfur containing peptide CDVLTSDKSQLFAQR (modification: carbamido-methyl) with the *m*/*z* of 1767.81 from the amino acid ABC transporter protein obtained from the culture of *P. putida* grown in the presence of ^{32}S and ^{36}S respectively. The samples were mixed after growth and simultaneously processed

peptides for quantification, (4) cysteine containing peptides can be separated by sulfide-affinity chromatography and therefore the complexity of samples can be efficiently reduced.

In the following sections, the step-by-step procedures of the SULAQ ^{34}S and ^{36}S approaches are described. The utility of amino acid specific labeling (^{34}S and ^{36}S) has been adapted for universal use in a wide range of organisms, especially in microorganisms.

2 Materials

2.1 Equipment

1. Incubators for cultivations and enzymatic digestions.
2. Heating block.
3. SDS-PAGE equipment and gels.
4. Sterile materials (cylinders and ten sterile flasks (250 ml)).
5. C_{18} microcolumns (Gel loader tips loaded with C_{18} resins or a commercial equivalent).

2.2 General Reagent Setup

1. Sulfate solution: ^{32}S sodium sulfate and either ^{34}S or ^{36}S sodium sulfate as stock solution (1 M in H_2O).
2. Mineral medium: 1441 mg/l disodium succinate, 760 mg/l NH_4Cl, 680 mg/l KH_2PO_4, 871 mg/l K_2HPO_4, 5.5 mg/l $CaCl_2 \cdot 6H_2O$, 200 mg/l $MgCl_2 \cdot 6H_2O$, 0.25 mg/l $Na_2MoO_4 \cdot 4H_2O$, pH 7.0, trace element solution SL-10 (1 ml/l) according to Table 1.
3. Washing buffer: 50 mM Phosphate buffer (100 ml).

Table 1
Composition of trace element solution

Ingredient	1 l
Milli-Q H_2O	800 ml
KH_2PO_4	0.42 g
K_2HPO_4	0.22 g
NH_4Cl	0.2 g
KCl	0.38 g
NaCl	0.36 g
$CaCl_2{\cdot}2H_2O$	0.04 g
$MgCl_2{\cdot}6H_2O$	0.10 g
Selenate/tungstate solution (Widdel)	1 ml
Trace mineral mix no NTA (Widdel)	1 ml
0.1% Resazurin	1 ml
Complete volume with Milli-Q H_2O to	1000 ml

4. SDS-PAGE sample buffer (10 ml): 60.55 mg Tris (0.1 M), 0.2 g SDS, 2.92 mg EDTA (2 mM), 1 ml glycerol 10% (v/v), a small pinch of bromophenol blue, fill up with deionized water to 10 ml.
5. SDS-PAGE running buffer (1 l): 3.21 g Tris (250 mM), 1 g SDS, 14.44 g glycine (1.92 M), fill up with deionized water to 1 l.
6. SDS-PAGE denaturing solution: acetic acid (10%v/v), ethanol (40%v/v).

2.3 Solutions for Trypsin Digestion

1. Acetonitrile.
2. 0.1 M NH_4HCO_3.
3. 10 mM dithiothreitol (DTT) in 0.1 M NH_4HCO_3.
4. 55 mM iodoacetamide in 0.1 M NH_4HCO_3.
5. 0.1 M NH_4HCO_3 containing 12.5 ng/μL of trypsin.

2.4 Solutions for Peptide Desalting and Reconstitution

1. Formic acid (5% v/v).
2. Acetonitrile (66% v/v) containing trifluoroacetic acid (0.1% v/v).
3. LC sample buffer: 2% Acetonitrile–0.1% TFA.

2.5 Data Handling and Bioinformatics

SULAQ34 analysis: MsConvert (Proteowizard Suite) [15], PeakExtractor and 34Scalc Excel script [14], Krönik [16], Microsoft Excel, Java Runtime Environment SULAQ36 analysis: Rosetta Elucidator, Microsoft Excel.

3 Methods

In the protocol described below, the comparison of the log-phase proteome to the stationary phase proteome is the anticipated scenario, but this might be adjusted for other experiments (e.g., different carbon sources).

3.1 Precultivation of Bacteria using 34S or 36S

1. Fill ten flasks with 100 ml mineral medium each (*see* **Note 1**).
2. Add 0.2 ml sulfate solution ($Na^{32}SO_4$) to five of the flasks.
3. Add 0.2 ml sulfate solution ($Na^{34}SO_4$) or sulfate solution ($Na^{36}SO_4$) to the remaining five flasks.
4. Start two precultures by inoculating one ^{32}S-flask and one $^{34/36}S$-flask with *P. putida*.
5. Cultivate at 28 °C and 200 rpm overnight.

3.2 Main Cultivation Experiment

1. Inoculate all remaining flasks from the respective ^{32}S or $^{34/36}S$ preculture (1:100) and measure the optical density (OD_{600}) at regular intervals to assess the bacterial growth (*see* **Note 2**).
2. Harvest ^{32}S and $^{34/36}S$ cultures each at mid-logarithmic growth and stationary phases by centrifugation (8000 × *g*, 4 °C, and 10 min).

3.3 Sample Preparation for Mass Spectrometric Measurements

1. Reconstitute the cell pellets in phosphate buffer and mix the mid-logarithmic and stationary samples with the ratio 1:1 according to OD measurements (always ^{32}S and $^{34/36}S$ samples) (*see* **Note 3**).
2. Remove the supernatant after centrifugation (8000 × *g*, 4 °C, and 10 min).
3. Dissolve the cell pellet in 20 μl SDS sample buffer and heat to 95 °C for 15 min (*see* **Note 4**).
4. Remove any cell debris by centrifugation (8000 × *g*, 5 min) and perform SDS-PAGE with the supernatant.
5. Stop the electrophoresis after the leading band moved 5 cm into the separation gel.
6. Add denaturing solution and cut the gel into ten pieces and distribute them to 1.5 ml tubes (*see* **Notes 5** and **6**).
7. To **perform in gel digestion** add 500 μl 0.1 M NH_4HCO_3 to the gel particles and incubate for 5 min (*see* **Note** 7).
8. Add 500 μl acetonitrile and incubate for 5 min.
9. Replace the complete supernatant with 500 μl acetonitrile and incubate for 5 min.
10. Remove the acetonitrile.

11. Repeat **steps 7–10** of this section until gel pieces do not release any color in case of for example Coomassie staining.
12. Dry gel particles (*see* **Note 8**).
13. Cover gel particles with 10 mM dithiothreitol (DTT)–0.1 M NH_4HCO_3 for 45 min at 56 °C to reduce the protein.
14. Chill tubes to room temperature.
15. Remove any liquid and cover gel particles with 55 mM iodoacetamide–0.1 M NH_4HCO_3.
16. Incubate for 30 min at room temperature in the dark.
17. Remove any liquid and add 500 μl 0.1 M NH_4HCO_3 for 5 min.
18. Remove any liquid and add 500 μl 0.1 M NH_4HCO_3/acetonitrile for 5 min.
19. Remove all liquid and cover gel particles with acetonitrile for 5 min.
20. Remove acetonitrile and dry gel particles.
21. Cover gel particles with trypsin solution (0.1 M NH_4HCO_3 and 12.5 ng/μl of trypsin).
22. Incubate at 37 °C overnight (<16 h) (*see* **Note 9**).
23. Recover peptides by covering gel pieces in 5% formic acid for 10 min.
24. Add an equal amount of acetonitrile and incubate for 10 min.
25. Recover the supernatant (*see* **Note 10**).
26. Repeat **steps 23–25** this section and pool the extracts.
27. Dry the extracts in a vacuum centrifuge.
28. **For desalting and clean-up of peptides** reconstitute peptides in 20 μl 5% formic acid (*see* **Notes 11** and **12**).
29. Wash C18 microcolumn with 20 μl 5% formic acid.
30. Load peptides onto the microcolumn.
31. Wash microcolumn with 20 μl 5% formic acid.
32. Elute peptides into a fresh tube with 20 μl acetonitrile (66%) containing trifluoroacetic acid (0.1%).
33. Dry the eluent in a vacuum centrifuge.

3.4 nanoLC-MS/MS Analysis and Database Search

1. Reconstitute peptides in 20 μl LC sample buffer .
2. Perform nanoLC-MS/MS measurements (a suggestion for settings on an Orbitrap XL can be found in Table 2).
3. Peptide identification from nanoLC-MS/MS data for the SULAQ34 approach can be performed by various search engines as long as charge, m/z, and ScanNumber or retention time (RT) are available as output. Add replacement of ^{32}S by ^{34}S as variable modification and perform the search.

Table 2
Settings for the measurement on an Orbitrap XL mass spectrometer (Thermo Scientific)

Step	Setting
nLC-separation	
Buffer system	Binary buffer system consisting of 0.1% acetic acid, 5% Acetonitrile (LC-buffer A) and 100% Acetonitrile in 0.1% acetic acid (LC-buffer B)
Trap column	75 μm inner diameter, packed with 3 μm C18 particles (Acclaim PepMap100, Thermo Scientific)
Analytical column	25 cm analytical column packed with 2 μm C18 particles (Acclaim PepMap RSLC, Thermo Scientific)
Gradient	120 min. With a total run time of 145 min. The percentage for LC buffer A decreased during the run and the percentage for LC buffer B increased steadily: 2% (v/v) for 10 min, 2% to 25% (v/v) for 120 min, 25% to 40% (v/v) for 5 min, 40% to 90% (v/v) for 2 min, and 90% (v/v) for 5 min.
Analytical column temperature	40 °C
Flow rate	300 nl/min during elution
Mass spectrometry	
Instrument	Orbitrap XL mass spectrometer (Thermo Scientific)
Electrospray	nano ESI source
Acquisition mode	Data-dependent, profile mode with positive polarity
Full scan MS	
Full scan mass range	300–1650 *m/z*
Resolution on MS level	70,000
AGC target	3×10^6
Maximum ion injection time	120 ms
MS/MS acquisition	
Resolution on MS/MS level	17,500
AGC target	2×10^5
Maximum ion injection time	120 ms
MS/MS scan acquisition	8 most intense ions with charge ≥2 from the survey scan
Fragmentation mode	Higher-energy collisional dissociation (HCD)
Intensity threshold	8.3×10^4

(continued)

Table 2
(continued)

Step	Setting
Normalized collision energy	27.5%
Monoisotopic precursor selection	Rejecting +1 and ≥ +7 charged ions
Dynamic exclusion	30 s

3.5 SULAQ34 Analysis

1. Convert all raw files to peak picked mzML files. In the case of Thermo's RAW files: msconvert *.RAW—filter "peakPicking true 1- "(*see* **Note 13**).
2. Create a list containing the sequence information (methionine oxidation as small m and carbamidomethylation of cysteines as small c, everything else in capital letters), a unique ID number, name of the MS file (everything before the dot), Modifications, Charge, *m/z*, and ScanNumber or RT (*see* **Note 14**).
3. Use *PeakExtractor* (edit the PeakExtractor.ini accordingly): java -jar PeakExtractor.jar PeakExtractor.ini.
4. Use *Krönik* to process the *PeakExtractor* output: Kronik.exe -d 5 ReplaceWithFilename.hkf ReplaceWithFilename.txt.
5. Load your peptide list (name it XXX_List) and the *Krönik* output (name it XXX) into the excel script. Start the quantify macro (View → Macros) from the XXX_List worksheet.
6. For all quantifiable peptides in your list you will get heavy/light ratios and the R^2 as a measure of the fit. You can now further process your data, e.g., define a minimum R^2 and mean/median your peptide ratios per protein to receive protein abundance changes (*see* **Note 15**).

3.6 SULAQ36 Analysis

1. Data analysis can be done using the software Rosetta Elucidator (Rosetta Biosoftware v3.3.0.SP3.19, MA, USA) with the following settings: 1000 count intensity threshold, 10 ppm mass accuracy, 10 min alignment search distance applied for the binning process (*see* **Note 16**).
2. All data searches were performed with SEQUEST/Sorcerer (Sorcerer v3.5, Sage-N Research, CA, USA) using a *P. putida* FASTA sequence (downloaded from NCBI repository, 2009) and identified peptides/proteins were accepted if they exceeded a PeptideTeller score of 0.8 and a ProteinTeller score of 0.95.
3. For quantification only labeled pairs were considered if at least one of the isotope groups in the labeled pair was annotated by peptide identification, reaching a labeled pair status of "good" [13].

3.7 Data Handling and Statistics

1. In most cases normalization/transformation of your quantitative data will be the first step. Divide each protein H/L ratio by the median of all ratios (*see* **Notes 17** and **18**).
2. Perform a log2 transformation of all ratios to receive normal distributed protein ratios.
3. Identify proteins that changed in abundance. Although fold-change and correlation of variance thresholds may give usable results, testing for significance by statistics is recommended (*see* **Note 19**).

4 Notes

1. This can easily be scaled down to lower amounts.
2. Make sure to assess the specific growth curve beforehand.
3. Dry weight can also be used instead of OD. Please, aliquot your samples for different enrichments, e.g., membrane or surface proteins [17].
4. It might be recommended to use more sophisticated lysis methods and to assess the protein concentration.
5. To waive Coomassie staining can simplify successive cleaning steps, but proteins should be fixated by application of a denaturing solution to prevent diffusion.
6. Adjust this prefractionation step according to the performance of the available MS instruments. State-of-the-art equipment will need less fractionation.
7. Further information about the in gel digestion can be found in the original publication [18].
8. Air drying is possible, but ~5 min vacuum centrifugation is recommended.
9. Digestion times of ~8 h are preferable. Long digestion times can lead to unspecific cleavage and abiotic modifications of peptides.
10. The supernatant contains the peptides. Do not discard.
11. Further information about the microcolumns can be found in the literature [19].
12. To enhance solubilization 5 min ultrasonic bath is recommended.
13. For more information and examples for all the steps see the SULAQ34_HowTo.pdf [14].
14. This has to be done for each of the replicates. Fractions will be united in this step.

15. If you add protein identifiers in column "K", you can use the QuantifyPostProcessing macro, which was used in the original SULAQ34 study.
16. For quantitative analysis, the data pairs were built using a binning tolerance of 5 ppm, a RT location tolerance of 0.2 min, and a mass label shift of 3.995 Da. It is important that for the peptide identification the following data-dependent settings were indicated: mass shift of methionine and cysteine by 3.995 Da.
17. Division by the median should remove systematic influences such as inaccurate sample mixing.
18. Starting from the peptide ratios, QPPC enables fast permutation based assessment of significances online [20] and takes care of log transformation as well as normalization.
19. Perseus (www.perseus-framework.org) offers the possibilities do perform significance A and B testing [21] among much other useful functionality (such as median division or log2 transformation). Depending on the organism, functional enrichment analysis with for example DAVID [22] might help.

References

1. Bachi A, Bonaldi T (2008) Quantitative proteomics as a new piece of the systems biology puzzle. J Proteome 71(3):357–367
2. Ong SE, Mann M (2005) Mass spectrometry-based proteomics turns quantitative. Nat Chem Biol 1(5):252–262
3. Chen X, Sun L, Yu Y, Xue Y, Yang P (2007) Amino acid-coded tagging approaches in quantitative proteomics. Expert Rev Proteomics 4 (1):25–37
4. Filiou MD, Varadarajulu J, Teplytska L, Reckow S, Maccarrone G, Turck CW (2012) The ^{15}N isotope effect in *Escherichia coli*: a neutron can make the difference. Proteomics 12(21):3121–3128
5. Hoedt E, Zhang G, Neubert TA (2014) Stable isotope labeling by amino acids in cell culture (SILAC) for quantitative proteomics. Adv Exp Med Biol 806:93–106
6. Matthes A, Kohl K, Schulze WX (2014) SILAC and alternatives in studying cellular proteomes of plants. Methods Mol Biol 1188:65–83
7. Petriz BA, Franco OL (2014) Application of cutting-edge proteomics technologies for elucidating host-bacteria interactions. Adv Protein Chem Struct Biol 95:1–24
8. Schmidt F, Volker U (2011) Proteome analysis of host-pathogen interactions: Investigation of pathogen responses to the host cell environment. Proteomics 11(15):3203–3211
9. Muller SA, Pernitzsch SR, Haange SB, Uetz P, von Bergen M, Sharma CM, Kalkhof S (2015) Stable isotope labeling by amino acids in cell culture based proteomics reveals differences in protein abundances between spiral and coccoid forms of the gastric pathogen *Helicobacter pylori*. J Proteome 126:34–45
10. Oda Y, Huang K, Cross FR, Cowburn D, Chait BT (1999) Accurate quantitation of protein expression and site-specific phosphorylation. Proc Natl Acad Sci U S A 96(12):6591–6596
11. Snijders AP, de Koning B, Wright PC (2005) Perturbation and interpretation of nitrogen isotope distribution patterns in proteomics. J Proteome Res 4(6):2185–2191
12. Jehmlich N, Schmidt F, Hartwich M, von Bergen M, Richnow HH, Vogt C (2008) Incorporation of carbon and nitrogen atoms into proteins measured by protein-based stable isotope probing (Protein-SIP). Rapid Commun Mass Spectrom 22(18):2889–2897
13. Jehmlich N, Kopinke FD, Lenhard S, Vogt C, Herbst FA, Seifert J, Lissner U, Volker U, Schmidt F, von Bergen M (2012) Sulfur-^{36}S stable isotope labeling of amino acids for quantification (SULAQ). Proteomics 12(1):37–42

14. Herbst FA, Taubert M, Jehmlich N, Behr T, Schmidt F, von Bergen M, Seifert J (2013) Sulfur-^{34}S stable isotope labeling of amino acids for quantification (SULAQ34) of proteomic changes in *Pseudomonas fluorescens* during naphthalene degradation. Mol Cell Proteomics 12(8):2060–2069

15. Kessner D, Chambers M, Burke R, Agus D, Mallick P (2008) ProteoWizard: open source software for rapid proteomics tools development. Bioinformatics 24(21):2534–2536

16. Hoopmann MR, MacCoss MJ, Moritz RL (2012) Identification of peptide features in precursor spectra using Hardklor and Kronik. Curr Protoc Bioinformatics Chapter 13: Unit13 18

17. Otto A, Bernhardt J, Meyer H, Schaffer M, Herbst FA, Siebourg J, Mader U, Lalk M, Hecker M, Becher D (2010) Systems-wide temporal proteomic profiling in glucose-starved *Bacillus subtilis*. Nat Commun 1:137

18. Shevchenko A, Tomas H, Havlis J, Olsen JV, Mann M (2006) In-gel digestion for mass spectrometric characterization of proteins and proteomes. Nat Protoc 1(6):2856–2860

19. Rappsilber J, Mann M, Ishihama Y (2007) Protocol for micro-purification, enrichment, pre-fractionation and storage of peptides for proteomics using StageTips. Nat Protoc 2 (8):1896–1906

20. Chen D, Shah A, Nguyen H, Loo D, Inder KL, Hill MM (2014) Online quantitative proteomics p-value calculator for permutation-based statistical testing of peptide ratios. J Proteome Res 13(9):4184–4191

21. Tyanova S, Temu T, Sinitcyn P, Carlson A, Hein MY, Geiger T, Mann M, Cox J (2016) The Perseus computational platform for comprehensive analysis of (prote)omics data. Nat Methods 13(9):731–740

22. Huang da W, Sherman BT, Lempicki RA (2009) Systematic and integrative analysis of large gene lists using DAVID bioinformatics resources. Nat Protoc 4(1):44–57

Chapter 13

Metabolic Labeling of Microorganisms with Stable Heavy Nitrogen Isotopes (^{15}N)

Andreas Otto

Abstract

Quantitative Proteomics based on stable isotope labeling by ^{15}N is a versatile method in determination of changes in microbial protein expression profiles. Once set up, the workflow relies on defined chemical or commercially available complex media leading to universal heavy labeling by the stable isotope ^{15}N. This chapter provides a detailed protocol for labeling of different bacteria and the strategy for evaluating the incorporation rate of the heavy stable isotopes in the sample is discussed.

Key words Microorganisms, Gel-free proteomics, Quantitative proteomics, ^{15}N

1 Introduction

Proteomics is regarded as a versatile technique for analyzing complex biological samples in modern systems biology studies [1]. Whilst classical proteomics was mainly relying on differential 2D SDS PAGE for determination of quantitative changes, today mostly LC-MS/MS based proteomics techniques are used. Here, advances are driven by the analytical depth and versatility of mass spectrometry based shot-gun proteomics [2]. In order to allow for differential analyses in LC-MS/MS based proteomics, two main pillars of quantitative LC-MS based proteomics techniques exist. The first is label free quantitative proteomics that compares basically different LC-MS runs either by signal intensities of resolved peptides or by the number of spectra measured in the mass spectrometer. The second quantitative proteomics technique is based on stable heavy isotopes. Stable isotope labeling was originally introduced as chemical labeling of proteins or peptides and has since been further developed and refined (e.g., iCAT, iTRAQ, TMT). Another possibility of introducing the heavy label for relative quantification is the "metabolic labeling" during cell growth. Metabolic labeling is regarded as gold standard in quantitative

Dörte Becher (ed.), *Microbial Proteomics: Methods and Protocols*, Methods in Molecular Biology, vol. 1841,
https://doi.org/10.1007/978-1-4939-8695-8_13,

proteomics: both the robustness of the labeling method against possible biases introduced in the sample preparation process due to the early introduction of the label and the accuracy of the quantification results lead to this notion [3, 4].

In stable isotope based quantitative proteomics, samples can be compared with the help of different *m/z* signals in the mass spectrum. Eluting masses leading back to the same peptide primary sequence differ by distinct mass changes caused by isotope labeling. Changes in the masses of labeled peptides leave the remaining physicochemical properties largely unaffected. Mass traces encoding the light and the heavy version of the peptide can be compared by the intensity of the signals in the same mass spectrum or by the extracted ion chromatograms of the two masses over time. Metabolic labeling can be performed either based on the introduction of isotope labeled amino acids that are fed to microorganisms, higher organisms in cell culture or even to small animals (SILAC, Stable isotope labeling of amino acids in cell culture) or by growing cells in chemically defined media composed of e.g., nitrogen sources that are isotopically labeled with ^{15}N [5].

The latter has proven to be very successful in microbial proteomics: not in need of any auxotrophy of the organism under investigation, cells can be labeled by using heavy nitrogen containing ammonia salts or even using complex growth media provided by commercial vendors mostly consisting of algae based formulae [6].

In this chapter a detailed protocol for metabolic labeling using two different approaches will be provided. First, an example for defined cell growth media exemplified for the gram positive *Bacillus subtilis* shows the necessities in exchange for available nitrogen sources in defined media. Second, the use of complex, commercially available media is exemplified for the gram positive pathogen *Staphylococcus aureus.* Both approaches have proven to be successful in current proteomics research [7–9]. Further, a workflow for evaluation of incorporation of the heavy isotopes in the bacterial cultures is proposed.

2 Materials

It is mandatory to use LC-MS grade chemicals along with LC-MS grade deionized water (ASTM type I) for sample preparation purposes. Store all solutions at 4 °C unless otherwise indicated. Provide liquid handling pipettes including sterile pipette tips of appropriate sizes. For the determination of the incorporation rate it is necessary to get access to high resolution mass spectrometry equipment coupled to liquid chromatography (*see* Subheading 2.6).

2.1 Metabolic Labeling of *Bacillus subtilis* in Minimal Media

1. Glycerol stock of *Bacillus subtilis* 168 (DSMZ Braunschweig, DSM No.: 402).
2. BMM light (2× concentrated): For 1 l, dissolve 4 g $(NH_4)_2SO_4$, 4 g $MgSO_4 \cdot 7H_2O$, 4 g KCl, 4 g Na_3Citrate$\cdot 2H_2O$, and 12.12 g Tris–HCl in 700 ml of water. Adjust the pH to 7.5 using HCl and adjust to the final volume of 1 l. Sterilize by autoclaving.
3. BMM heavy (2× concentrated): For 1 l, dissolve 4 g heavy $(^{15}NH_4)_2SO_4$, 4 g $MgSO_4 \cdot 7H_2O$, 4 g KCl, 4 g Na_3Citrate$\cdot 2H_2O$, and 12.12 g Tris–HCl in 700 ml of water. Adjust the pH to 7.5 using HCl and adjust to the final volume of 1 l. Sterilize by autoclaving.
4. 0.2 M KH_2PO_4 (Stock solution): For 50 ml, dissolve 1.35 g KH_2PO_4 in 50 ml of water. Sterilize by autoclaving.
5. 1 M $CaCl_2$ (Stock solution): For 50 ml, dissolve 7.35 g $CaCl_2 \cdot H_2O$ in 50 ml of water. Sterilize by autoclaving.
6. 0.5 μM $FeSO_4$: For 50 ml, dissolve 6.95 mg $FeSO_4 \cdot 7H_2O$ in 50 ml of water. Sterilize by filtration. (*see* **Note 1**).
7. 25 μM $MnSO_4$: For 50 ml, dissolve 275 mg $MnSO_4 \cdot H_2O$ in 50 ml of water. Sterilize by filtration.
8. 0.5 M glutamic acid: For 100 ml, dissolve 7.36 g glutamic acid in 100 ml of water. Sterilize by filtration.
9. 0.5 M heavy ^{15}N-glutamic acid: For 100 ml, dissolve 7.41 g ^{15}N-glutamic acid in 100 ml of water. Sterilize by filtration.
10. 39 μM tryptophan: For 10 ml, dissolve 80 mg tryptophan in 10 ml of water. Sterilize by filtration.
11. 39 μM heavy ^{15}N-tryptophan: For 10 ml, dissolve 80 mg ^{15}N-tryptophan in 10 ml of water. Sterilize by filtration.
12. 20% (w/v) D-glucose: For 200 ml, dissolve 40 g D-glucose in 200 ml of water. Sterilize by filtration.
13. Deionized water. Sterilize by autoclaving.
14. Sterile **light** Belitsky minimal medium (BMM) (ready to use solution): For preparation of light, fully supplemented BMM work in a safety cabinet type II for sterile conditions. For 100 ml of light BMM, use 50 ml of BMM light (2× concentrated), add 0.5 ml of KH_2PO_4 stock solution, 0.2 ml of tryptophan stock solution, 0.9 ml of glutamic acid stock solution, 0.2 ml of $FeSO_4$ stock solution, 0.04 ml of $MnSO_4$ stock solution to 70 ml of water. Mix thoroughly. Then add 1 ml of glucose stock solution. Mix thoroughly. Finally add 0.2 ml of $CaCl_2$ stock solution and adjust to 100 ml with sterile deionized water (*see* **Note 2**).
15. Sterile **heavy** Belitsky minimal medium (BMM) (ready to use solution): For preparation of heavy, fully supplemented BMM

work in a safety cabinet type II for sterile conditions. For 100 ml of heavy BMM, use 50 ml of BMM light (2× concentrated), add 0.5 ml of KH_2PO_4 stock solution, 0.2 ml of heavy tryptophan stock solution, 0.9 ml of heavy glutamic acid stock solution, 0.2 ml of $FeSO_4$ stock solution, 0.04 ml of $MnSO_4$ stock solution to 70 ml of water. Mix thoroughly. Then add 1 ml of glucose stock solution. Mix thoroughly. Finally add 0.2 ml of $CaCl_2$ stock solution and adjust to 100 ml with sterile deionized water (*see* **Note 2**).

16. Shaking incubator with temperature control (e.g., 37 °C for *Bacillus subtilis*) for bacterial growth in batch cultures.
17. Sterile Erlenmeyer flasks of 500 ml and 100 ml size (sealed e.g., sterile cotton wool plugs).
18. Cuvettes for determination of optical density of 600 nm.
19. Spectrophotometer capable of measuring optical densities at a range of 500–600 nm.
20. NaCl solution for dilution (0.9% (w/v)): To prepare 100 ml, dissolve 0.9 g of NaCl in 100 ml of water. Sterilize by autoclaving.
21. Washing buffer: For 100 ml of 100 mM Tris–HCl, 5 mM $MgCl_2$ add 1.576 g Tris–HCl and 262 mg of $MgCl_2$ to 90 ml of water. Adjust to a final pH of 7.5 using HCl. Adjust to the final volume of 100 ml. Sterilize by autoclaving.
22. High performance centrifuge suitable for 50 ml conical sample tube at a speed of 10,000 × *g*.
23. Conical sample tubes, 50 ml.

2.2 Metabolic Labeling of *Staphylococcus aureus* in Complex Media

1. Glycerol stock of *Staphylococcus aureus* COL (Culture Collection of Switzerland, CCOS 461).
2. Complex growth medium unlabeled: Bioexpress cell growth media (10× concentrate) CGM-1000-U (Cambridge Isotopes).
3. Complex growth medium ^{15}N labeled: Bioexpress cell growth media (U-^{15}N, 98%) (10× concentrate) CGM-1000-U (Cambridge Isotopes).
   (a) 20% (w/v) D-glucose: For 200 ml, dissolve 40 g D-glucose in 200 ml of water. Sterilize by filtration.
4. Shaking incubator with temperature control (37 °C for *Staphylococcus aureus*) for bacterial growth in batch cultures.
5. Sterile Erlenmeyer flasks of 500 ml and 100 ml size (sealed e.g., sterile cotton wool plugs).
6. Cuvettes for determination of optical density of 500 nm.

7. Spectrophotometer capable of measuring optical densities at a range of 500–600 nm.
8. High performance centrifuge suitable for 50 ml conical sample tube at a speed of 10,000 × *g*.
9. Conical sample tubes, 50 ml.
10. Washing buffer: For 100 ml of 100 mM Tris–HCl, 5 mM $MgCl_2$ add 1.576 g Tris–HCl and 262 mg of $MgCl_2$ to 90 ml of water. Adjust to a final pH of 7.5. Adjust to the final volume of 100 ml. Sterilize by autoclaving.

2.3 Cell Disruption

1. 2 ml polypropylene tubes (caps with O-ring).
2. 1.5 ml polypropylene reaction tubes.
3. Lysis buffer: For 100 ml of 10 mM Tris–HCl, 1 mM EDTA add 0.158 g Tris–HCl and 29.24 mg of EDTA to 90 ml of water. Adjust to a final pH of 7.5 using HCl. Adjust to the final volume of 100 ml. Sterilize by autoclaving.
4. Glass disruptor beads, 0.1–0.11 mm diameter (e.g., Sartorius or USA Scientific).
5. High-throughput tissue homogenizer (e.g., Precellys 24, Bertin Technologies).
6. High performance centrifuge suitable for 2 ml polypropylene reaction tubes.
7. Insulated beaker/container with crushed ice.

2.4 Determination of Protein Concentration

1. Roti-Nanoquant reagent (Carl Roth, Germany).
2. BSA stock solution (80 mg/ml) (e.g., Gibco or other high quality BSA in solution) (*see* **Note 3**).

2.5 In-Solution Digest

1. Triethylammonium bicarbonate buffer (TEAB) 1 M TEAB, pH 8.5 (commercially available).
2. TEAB 50 mM: dilute the TEAB stock solution to 50 mM (1:20) with A. bidest.
3. Tris(2-carboxyethyl)phosphine hydrochloride (TCEP): dissolve 1 mg TCEP in 7 μl 50 mM TEAB.
4. Iodoacetamide (IAA): dissolve 1 mg IAA in 10.8 μl 50 mM TEAB. Keep in the dark.
5. 10× trypsin: dissolve 20 μg MS-grade (for mass spectrometry) Trypsin in 100 μl A. bidest.
6. Trifluoroacetic acid, ≥99% analytical grade.
7. Shaking incubator with temperature control (capable of maintaining 37 °C) for protein digests suitable for small reaction tubes of 1.5 ml (*see* **Note 4**).

2.6 Evaluation of Incorporation Rates

For the evaluation of incorporation rates it is advisable to consult the contact person of the central mass spectrometry facility/operator of the LC-MS system. Regardless of the equipment/service that will actually be provided, the LC-MS data needs to be processed by database searching and the search results should be used as the basis for the following evaluation. Depending on the guidelines of the respective facility, additional measures for sample preparation/clean-up will be necessary.

1. A minimum requirement for evaluation of incorporation rates is the acquisition of an LC-MS/MS run on a state-of-the-art LC-MS/MS system that will yield data both for overview scans (MS1) at resolutions of $R = 30{,}000$ together with fragment ion scans (MS2) automatically triggered by data-dependent acquisition of a comparable or lower resolution.
2. According to the organism that is used for the determination of incorporation rates (e.g., *Bacillus subtilis* or *Staphylococcus aureus* in the examples given), an appropriate fasta database containing the amino acid sequences of all predicted proteins will be needed that can be used to perform database searches with the LC-MS/MS data for peptide/protein identification. A resource for fasta-databases is the "National Center for Biotechnology" (http://www.ncbi.nlm.nih.gov/).
3. For database searching of heavy labeled samples, the searches must take into account that all amino acids have an increased mass according to the number of nitrogen atoms in the molecule. This may either be considered automatically by checkboxes (e.g., MASCOT, Matrix Science) or by defining fixed modifications for the proteinogenic amino acids (e.g., COMET [10]).
4. For assessment of incorporation rates based on specific spectra, it is useful to rely on tools simulating spectra of predetermined composition. A freeware tool that is useful for this task is for example the "IDCalc—Isotope Distribution Calculator" provided by the MacCoss lab at University of Washington (https://proteome.gs.washington.edu/software/IDCalc/).

3 Methods

Metabolic labeling has proven to be manageable for most bacteria that are growing in classical defined minimal media or in commercially available complex media. Choosing an appropriate formulation of growth media should be based on the following questions:

1. Are the strains of interest growing in defined media based on ammonia containing salts (e.g., NH_4Cl, $(NH_4)_2SO_4$)? Defined media will exist for a wide range of bacteria that fulfill this

criterion. While checking media formulae, it is only necessary to consider all biologically available nitrogen sources to be supplied as heavy labeled compounds: buffer substances like Tris salts may be neglected as these are not biologically available. Cofactors that are needed for certain strains like biotin may also be neglected if only needed in minute amounts.

2. Is there a commercially available complex medium that allows for cell growth of the strains of interest and therefore for a complete metabolic labeling? Meanwhile there is a substantial choice of commercially available complex media. Complex heavy labeled media include for example Bioexpress (e.g., Cambridge Isotopes), Rich growth medium (e.g., Silantes), or Isogro (e.g., Sigma).
3. Is there a possibility of supplementing defined media with smaller amounts of commercially available complex media replacing classical complex supplements like casein hydrolysate? Exchange of recipe formulations can be cumbersome in some cases. Composition trials should be carried out with the light (unlabeled) complex media that are in most cases not as expensive as the heavy (labeled) complex media. If changes do not lead to satisfactory results (cell growth/ incorporation rate), addition of very low concentrations of unlabeled complex media constituents like casein hydrolysate may help to recover growth as trace amounts of undefined biologically important molecules are provided. In any case, care should be taken if adding low concentrations of unlabeled nitrogen sources as this may cause insufficient labeling efficiencies.

Based on a growth media composition with interchangeable nitrogen salts, cell can be grown under the same conditions both in the heavy and the light formulations of the growth media.

3.1 Metabolic Labeling of *Bacillus subtilis* in Minimal Media

For the metabolic labeling of *Bacillus subtilis*, it has been shown that a defined medium like Belitzky minimal medium (BMM) is amenable for metabolic labeling [9]. This type of growth medium is representative of other defined media like M9 medium or DAVIS medium, often used for *Escherichia coli* [11]. In the following, preparation of the heavy labeled media is described. For preparation of the light media use the light media components.

1. Prepare sterile heavy fully supplemented Belitsky minimal medium (BMM).
2. Use 30 ml of the fully supplemented heavy BMM for three sterile Erlenmeyer flask each and inoculate the first with the glycerol stock (first dilution). Mix the Erlenmeyer flask thoroughly and transfer 500 µl of the first dilution to the second Erlenmeyer flask (second dilution). Proceed as with flask 1 and also inoculate the third flask with 500 µl of the second dilution

(third dilution). Work in a safety cabinet type II to ensure sterile solutions.

3. Incubate the Erlenmeyer flasks overnight at 180 rpm, 37 °C in a temperature controlled shaking incubator.
4. Determine the optical density at a wavelength of 600 nm. Dilute if needed with NaCl solution to keep the absorbance below 0.8.
5. Prewarm an Erlenmeyer flask of 500 ml filled with 100 ml of fully supplemented heavy BMM in a water bath at 37 °C (main culture flask).
6. Use the overnight culture being closest to an $OD_{600\ nm}$ of 1 for inoculation of the main culture flask to reach a starting $OD_{600\ nm}$ of 0.05.
7. Incubate the Erlenmeyer flasks at 180 rpm, 37 °C and determine the $OD_{600\ nm}$ every 30 min. Dilute samples for determination of optical density with 0.9% (w/v) NaCl solution to keep the absorbance below 0.8 (*see* **Note 5**).
8. After six or seven generations harvest the cells by centrifugation in 2 × 50 ml conical flask with 25 ml each.
9. Wash with 40 ml of ice-cold washing buffer.
10. Store cell pellets at −20 °C or immediately proceed to cell lysis.

3.2 Metabolic Labeling of *Staphylococcus aureus* in Complex Media

For the metabolic labeling of *Staphylococcus aureus* it has been shown that a commercially available complex medium based on algae lysates is amenable for metabolic labeling. To be fully useable, addition of glucose leads to markedly higher optical densities avoiding glucose limitations and therefore stress and starvation during growth. In the following preparation of the heavy labeled media is described. For preparation of the light media use the light media components. In any case, a standard for light “labeled” cells should also be generated.

1. Prepare 300 ml of 1× Bioexpress cell growth medium (heavy ^{15}N labeled, Cambridge Isotopes): Dilute the 10× concentrate according to manufacturer’s instructions and prior to adding water to the final volume, add 2.22 ml of the glucose stock solution (*see* **Note 6**).
2. Use 30 ml of the readily prepared heavy Bioexpress medium for three sterile Erlenmeyer flask each and inoculate the first with the glycerol stock (first dilution). Mix the Erlenmeyer flask thoroughly and transfer 500 μl of the first dilution to the second Erlenmeyer flask (second dilution). Proceed as with flask 1 and also inoculate the third flask with 500 μl of the second dilution (third dilution). Work in a safety cabinet type II to ensure sterile conditions.

3. Incubate the Erlenmeyer flasks overnight at 120 rpm, 37 °C in a temperature controlled Shaking incubator.
4. Determine the optical density at a wavelength of 540 nm. Dilute if needed with sterile 1× Bioexpress cell growth medium to keep the absorbance below 0.8.
5. Prewarm an Erlenmeyer flask of 500 ml filled with 100 ml of sterile 1× Bioexpress cell growth medium in a water bath at 37 °C (main culture flask).
6. Use the overnight culture being closest to an $OD_{540\ nm}$ of 1 for inoculation of the main culture flask to reach a starting $OD_{540\ nm}$ of 0.05.
7. Incubate the Erlenmeyer flasks at 180 rpm, 37 °C and determine the $OD_{540\ nm}$ every 30 min. Dilute samples for determination of optical density with Bioexpress medium to keep the absorbance below 0.8.
8. After six or seven generations harvest the cells centrifugation in 2 × 50 ml conical flask with 25 ml each.
9. Wash with 40 ml of ice-cold washing buffer.
10. Store cell pellets at −20 °C or immediately proceed to cell lysis.

3.3 Cell Disruption

1. Resuspend cell pellet in 1 ml of lysis buffer.
2. Add glass beads to the 2 ml polypropylene tubes correlating to a volume of approximately 500 μl.
3. Add resuspended cell pellet to the polypropylene tube with glass beads and close thoroughly (*see* **Note 7**).
4. Perform cell disruption e.g., in the Precellys 24 at 6.5 m/s, 1 min for three times. It is mandatory to cool down the samples for 1 min on ice between the cell disruption cycles.
5. Centrifuge after the last cell disruption at 5000 × *g* at 4 °C for 5 min.
6. Transfer the supernatant into a new, 1.5 ml reaction tube.
7. Repeat centrifugation (at 20,000 × *g* and 4 °C for 10 min).
8. Transfer the supernatant into a new, 1.5 ml reaction tube. With the performed repeated centrifugation the cell debris is completely discarded with the pellets obtained.
9. Freeze sample at −20 °C (*see* **Note 8**).

3.4 Determination of Protein Concentration

1. Determine the protein concentration of the cell extract(s) by Bradford assay according the manufacturer's instruction (*see* **Note 9**).

3.5 In-Solution Digest

1. Pipette a volume of the sample equaling 50 μg of the protein as determined by the Bradford assay into a 1.5 ml reaction tube.
2. Fill with 50 mM TEAB up to 50 μl of total volume.

3. Add 0.5 μl TCEP and incubate for 45 min at 60 °C.
4. Add 1 μl IAA and incubate in the dark for 15 min at RT.
5. Add 1.25 μl of 10x trypsin and incubate overnight at 37 °C.
6. Add 1 μl TFA to stop the digest.
7. Store at −20 °C.

3.6 LC-MS/MS Analysis of the Digested Sample

For the evaluation of the incorporation rate it is important to analyze an average population of peptides from the organism under investigation. With this, it is possible to assess the rate of incorporation based on the isotope pattern of the spectra leading to identification of the light/heavy peptides during the database search. To allow for a convenient evaluation of both the light and the heavy peptides in the same spectrum, it is advisable to generate cell extracts of light and heavy labeled cells. These extracts should be mixed in equal amounts based on protein concentrations and prepared according to the protocol provided. MS-based proteomics in general is strongly dependent on the mass spectrometric instrumentation used and therefore only the general strategy for acquisition of the test dataset needed and the general workflow for assessing the incorporation rate is described.

1. For LC MS/MS experiments, use digested peptides relating to 500 ng to 1 μg of protein sample used for in-solution digests (either light or heavy cell extracts or a mixture of both).
2. LC-MS/MS runs should be performed according to standard conditions. Please refer to the mass spectrometer's operator. The complex sample should be first trapped/ desalted and then be resolved by reversed phase C18 chromatography. The mass spectrometer should determine first an overview scan at a resolution of at least $R = 30{,}000$. In this experiment, the information of the isotope patterns of all precursors eluting at a specific time is acquired. To allow for automatic identification of peptides based on fragmentation experiments, MS/MS spectra should be acquired for, for example, 5–50 precursors per duty cycle in the mass spectrometer.
3. Following LC-MS/MS acquisition, the data should be subjected to database searches. Here, the searches should either consider that the light or the heavy labeled cells are analyzed to choose the correct settings for a regular database search or a search considering the mass changes according to the increase in mass based on ^{15}N.
4. The search result will yield a table of proteins and the respective peptides. The experimentalist should then choose a number of peptides (10–20) of different proteins and the scan numbers of the spectra leading to positive identifications have to be determined.
5. Open the raw mass spectrometry files and display the chosen spectra.

6. Open the "IDCalc—Isotope Distribution Calculator" (*see* Subheading 2.6). Insert the peptide primary amino acid sequence (one letter code) based on database searches in the field denoted with "Amino acid composition" (*see* Fig. 1). Adjust the charge (e.g., 2). Press "calculate." The software displays the unlabeled, calculated spectrum on the upper right panel as calculated centroid spectrum. This should match the spectrum of the chosen peptide from the light sample. On the lower right you will find the calculated isotope peaks of the respective spectrum of the nominal masses and the m/z values according to the charge chosen. Next, Please chose the radio button "Elemental composition." The software calculates the elemental composition. Click on "^{15}N" and add the number of nitrogen atoms as calculated and displayed under "Nitrogen." Declick the radio button "Nitrogen." Press "calculate" again. On the lower right you will find the calculated isotope peaks of the respective spectrum of the nominal masses and the m/z values according to the charge chosen for the enrichment of 99% ^{15}N. Upon changing this percentage, the calculated spectrum will change its relative distributions. Start displaying the spectrum with the value "100" "Atom%". On lowering this percentage and by comparison of the height of the isotopic peak that appears left to the monoisotopic peak ($z = 2$: monoisotopic mass 0.5 m/z; $z = 3$: monoisotopic mass 0.33 m/z) with the monoisotopic peak and then to the spectrum that appears in the actual heavy labeled sample, it is possible to judge on the incorporation rate (The comparison must be done between the theoretical spectrum appearing e.g., in Fig. 1b and the actual spectrum as in Fig. 2). It should be noted that already at incorporation of only 99.2%, the peak appearing with incorporated ^{14}N may represent 10% of the monoisotopic peak of fully labeled peptides. This ratio should be accepted as threshold for successful labeling if highly enriched chemicals were used for the growth media.

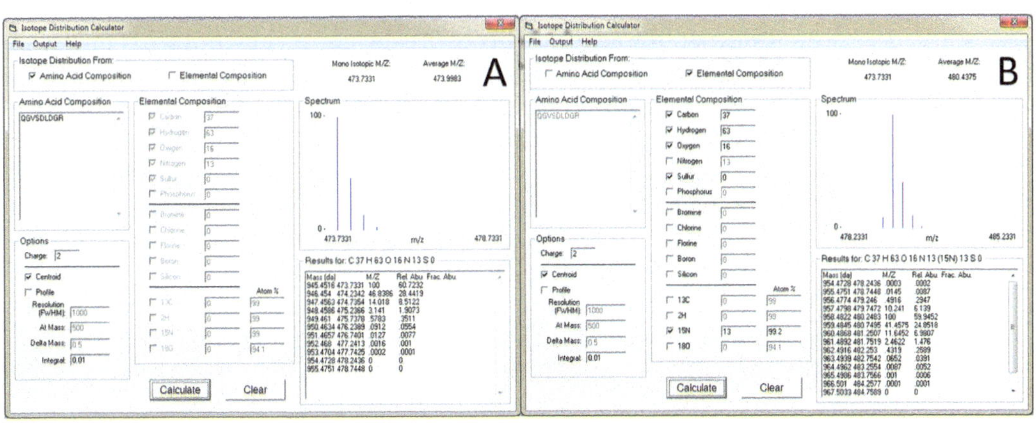

Fig. 1 (**a**) Theoretical spectrum of the light peptide "QGVSDLDGR." (**b**) Theoretical spectrum of the heavy peptide "QGVSDLDGR" with an incorporation rate of 99.2% ^{15}N. Capture taken from the Isotope pattern calculator

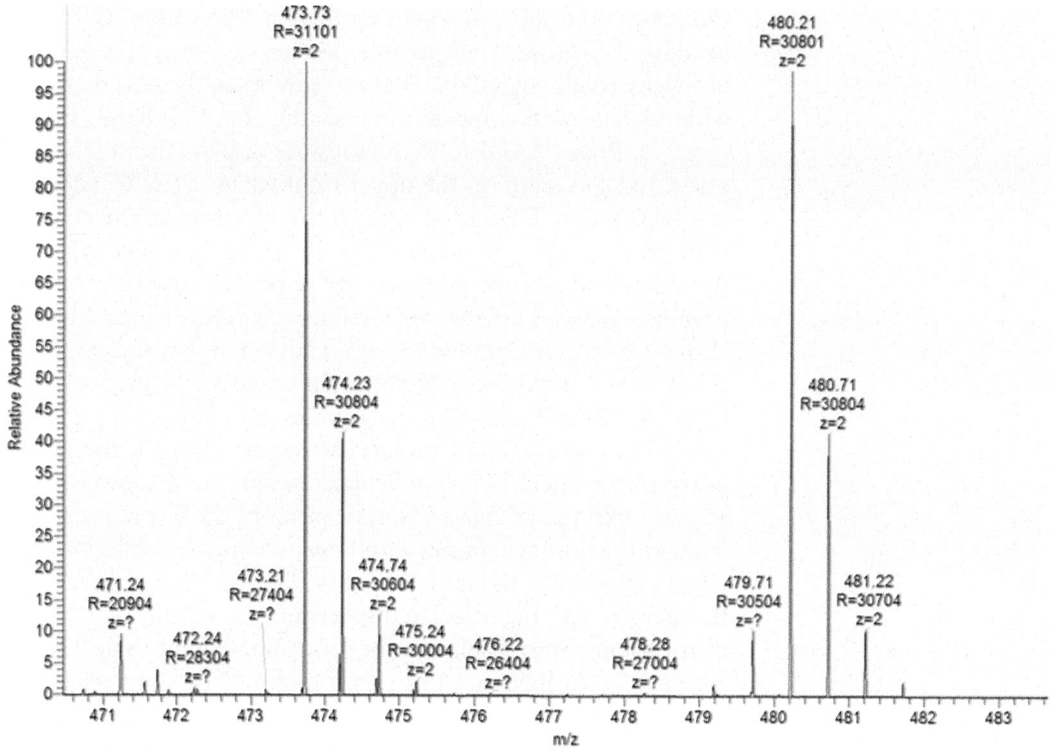

Fig. 2 Spectrum acquired of a mixed sample with light (473.73 *m/z*) and heavy (480.21 *m/z*) peptides "QGVSDLDGR." The ratio between the monoisotopic peak (480.21 *m/z*) and the peak with ^{14}N incorporation (479.71 *m/z*) correlates to an incorporation rate of 99.2% ^{15}N

3.7 Practical Considerations for Quantitative Studies Based on ^{15}N Metabolic Labeling

In principle there are two ways of quantitative studies based on ^{15}N metabolic labeling: the first is often referred to as "label switching" and the second is named the "pooled reference approach" (for the latter *see* Fig. 3). The first approach, the "label switching" considers both the light and the heavy media as exchangeable in terms of resource for cell growth and cell physiology. With experimental evidence for adverse effects of labeling mainly both for microbes [12] and for mammals [13], the "pooled reference approach" is nowadays mostly used [14]. Here, cultures are grown in heavy labeled media to represent the "average" of the conditions under investigation. The cell extracts are then mixed to yield a pooled heavy reference. Then, the proper samples/replicates are grown in the regular, "light" media and then mixed according to the protein content with the pooled reference. With this strategy, samples of larger cohorts or larger numbers of replicates/time course samples can be compared as all samples share the same pooled reference.

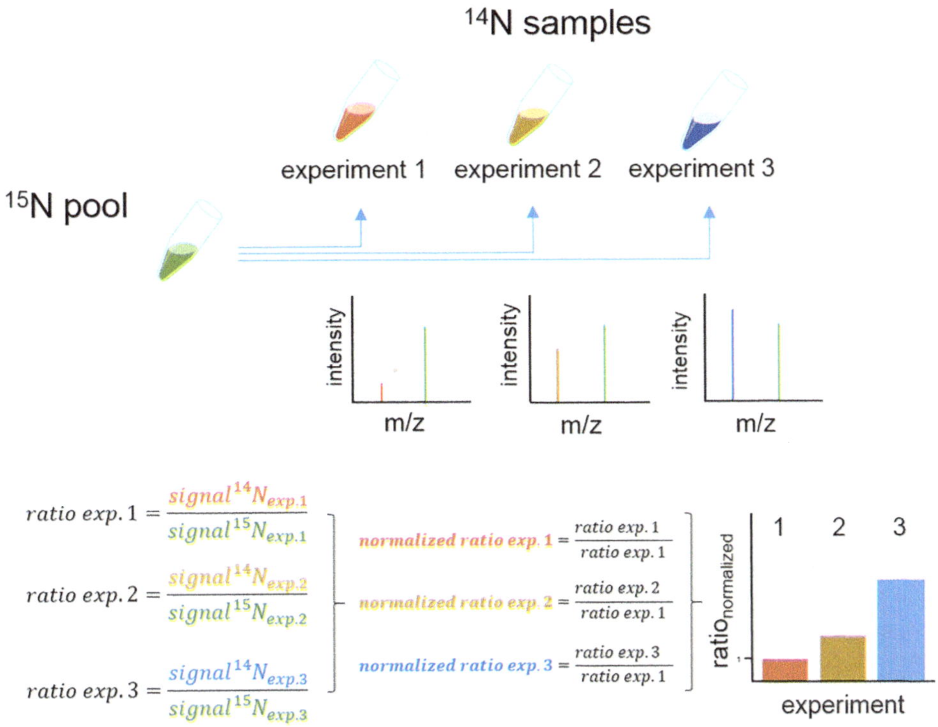

Fig. 3 Pooled reference approach. The samples are grown in light medium (^{14}N) and the pooled reference is made up of samples grown in heavy medium (^{15}N). Samples and the pooled reference are mixed and prepared for LC-MS/MS measurement. Signals of the light and the heavy peptides are compared to yield ratios of light to heavy intensities. Comparison of samples is done by normalizing the ratios, in this example to experiment 1. Figure according to Geiger et al. [15]

4 Notes

1. Solution must be freshly prepared before use.
2. The components of the Belitsky minimal medium (BMM) must be added in the order indicated. Most attention is needed on adding the $CaCl_2$ which has to be added slowly at the end.
3. The BSA stock solution must be vigorously agitated prior to use. The stock can be stored frozen as aliquots.
4. Check the maximum horizontal oscillation. Best mixing for volumes of up to 2 ml in reaction tubes up to 2 ml is seen with oscillations of 3 mm minimum.
5. Dilutions should be made beginning from densities of 0.3 to avoid technical errors based on the absorbance measurements. Larger ranges may lead to nonlinearity of calibration curves.
6. According to the manufacturer, precipitates may be seen. Try to get the complete concentrate in solution by gentle warming up to 40 °C or by shaking.

7. Ensure proper labeling both on the sides of the tubes as well as on top of the tube as the processing will lead to damage on the screw lids.
8. A single freeze–thaw cycle is important to yield a reproducible protein concentration determination. The duration of freezing is not crucial.
9. Protein samples that are used for calibration curves must be prepared always fresh for proper determination.

Acknowledgments

This work was supported by the German Research Foundation Grant SFB/TR34.

References

1. Sabido E, Selevsek N, Aebersold R (2012) Mass spectrometry-based proteomics for systems biology. Curr Opin Biotechnol 23 (4):591–597
2. Cravatt BF, Simon GM, Yates JR 3rd (2007) The biological impact of mass-spectrometry-based proteomics. Nature 450 (7172):991–1000
3. Bantscheff M, Lemeer S, Savitski MM, Küster B (2012) Quantitative mass spectrometry in proteomics: critical review update from 2007 to the present. Anal Bioanal Chem 404 (4):939–965
4. Bantscheff M, Schirle M, Sweetman G, Rick J, Küster B (2007) Quantitative mass spectrometry in proteomics: a critical review. Anal Bioanal Chem 389(4):1017–1031
5. Gouw JW, Krijgsveld J, Heck AJ (2010) Quantitative proteomics by metabolic labeling of model organisms. Mol Cell Proteomics 9 (1):11–24
6. Otto A, Becher D, Schmidt F (2014) Quantitative proteomics in the field of microbiology. Proteomics 14(4–5):547–565
7. Otto A, Bernhardt J, Meyer H, Schaffer M, Herbst FA, Siebourg J, Mäder U, Lalk M, Hecker M, Becher D (2010) Systems-wide temporal proteomic profiling in glucose-starved *Bacillus subtilis*. Nat Commun 1:137
8. Becher D, Hempel K, Sievers S, Zühlke D, Pané-Farré J, Otto A, Fuchs S, Albrecht D, Bernhardt J, Engelmann S, Völker U, van Dijl JM, Hecker M (2009) A proteomic view of an important human pathogen–towards the quantification of the entire *Staphylococcus aureus* proteome. PLoS One 4(12):e8176
9. Dreisbach A, Otto A, Becher D, Hammer E, Teumer A, Gouw JW, Hecker M, Völker U (2008) Monitoring of changes in the membrane proteome during stationary phase adaptation of *Bacillus subtilis* using in vivo labeling techniques. Proteomics 8(10):2062–2076
10. Eng JK, Jahan TA, Hoopmann MR (2013) Comet: an open-source MS/MS sequence database search tool. Proteomics 13(1):22–24
11. Price CE, Otto A, Fusetti F, Becher D, Hecker M, Driessen AJ (2010) Differential effect of YidC depletion on the membrane proteome of *Escherichia coli* under aerobic and anaerobic growth conditions. Proteomics 10 (18):3235–3247
12. Filiou MD, Varadarajulu J, Teplytska L, Reckow S, Maccarrone G, Turck CW (2012) The ^{15}N isotope effect in *Escherichia coli*: a neutron can make the difference. Proteomics 12(21):3121–3128
13. Frank E, Kessler MS, Filiou MD, Zhang Y, Maccarrone G, Reckow S, Bunck M, Heumann H, Turck CW, Landgraf R, Hambsch B (2009) Stable isotope metabolic labeling with a novel N-enriched bacteria diet for improved proteomic analyses of mouse models for psychopathologies. PLoS One 4(11):e7821
14. MacCoss MJ, Wu CC, Liu H, Sadygov R, Yates JR 3rd (2003) A correlation algorithm for the automated quantitative analysis of shotgun proteomics data. Anal Chem 75(24):6912–6921
15. Geiger T, Wisniewski JR, Cox J, Zanivan S, Kruger M, Ishihama Y, Mann M (2011) Use of stable isotope labeling by amino acids in cell culture as a spike-in standard in quantitative proteomics. Nat Protoc 6(2):147–157

Chapter 14

Next-Generation Trapping of Protease Substrates by Label-Free Proteomics

Claudia Lindemann, Nikolas Thomanek, Katja Kuhlmann, Helmut E. Meyer, Katrin Marcus, and Franz Narberhaus

Abstract

AAA^+ proteases (ATPases associated with various cellular activities) shape the cellular protein pool in response to environmental conditions. A prerequisite for understanding the underlying recognition and degradation principles is the identification of as many protease substrates as possible. Most previous studies made use of inactive protease variants to trap substrates, which were identified by 2D-gel based proteomics. Since this method is known for limitations in the identification of low-abundant proteins or proteins with many transmembrane domains, we established a trapping approach that overcomes these limitations. We used a proteolytically inactive FtsH variant ($FtsH^{trap}$) of *Escherichia coli* (*E. coli*) that is still able to bind and translocate substrates into the proteolytic chamber but no longer able to degrade proteins. Proteins associated with $FtsH^{trap}$ or $FtsH^{wt}$ (proteolytically active FtsH) were purified, concentrated by an 1D-short gel, and identified by LC-coupled mass spectrometry (LC-MS) followed by label-free quantification. The identification of four known FtsH substrates validated this approach and suggests that it is generally applicable to AAA^+ proteases.

Key words Label-free proteomics, Mass spectrometry, Trapping, protease substrate, FtsH, Regulated proteolysis, AAA^+ protease

1 Introduction

Regulated proteolysis allows for the rapid adaptation of the bacterial proteome to changing environmental conditions. As proteolysis is an energy-dependent and irreversible process, it has to be tightly controlled and permitted only under certain conditions (e.g., temperature, growth rate) [1–3]. Correct delivery of substrates to their appropriate protease often needs specific degradation signals (degrons) that mark them for degradation [4] and/or requires specific interaction partners (adaptor proteins) [2, 5].

The key players of regulated proteolysis in bacteria are AAA^+ proteases (ATPases associated with various cellular activities). They serve two functions: first, protein quality control by degradation of

Dörte Becher (ed.), *Microbial Proteomics: Methods and Protocols*, Methods in Molecular Biology, vol. 1841,
https://doi.org/10.1007/978-1-4939-8695-8_14,

misfolded or aberrant proteins and second, regulated proteolysis of intact transcription factors and enzymes under certain conditions. Various cellular processes are regulated by proteases [6, 7]. The *E. coli* AAA$^+$ protease FtsH for example is involved in quality control of membrane proteins, degradation of SsrA-tagged proteins, regulation of heat shock gene expression and control of lipopolysaccharide (LPS) biosynthesis [8]. The latter is the essential function of *E. coli* FtsH. Conditional proteolysis of LpxC, which catalyzes the first committed step of the LPS biosynthesis, prevents a toxic LpxC accumulation [2, 9, 10].

In general, the knowledge about substrates of AAA$^+$ proteases is limited. Recently developed substrate trapping approaches exploit the potential of ATPase-active but protease-deficient protease variants to capture substrate proteins. Client proteins of the ClpP proteases of *E. coli*, *Staphylococcus aureus* and *Caulobacter crescentus* have been identified by this approach [11–13]. It is also applicable to the FtsH protease. A proteolytically inactive FtsH variant (FtsH^{trap}) carries a H417Y amino acid exchange, which abrogates Zn^{2+}-binding in the catalytic center [14]. Substrate binding and translocation into the proteolytic chamber are not affected. Four new FtsH substrates were enriched with FtsH^{trap} in *E. coli* [14]. However, it was not possible to identify all known FtsH substrates, for example the low-abundant heat shock transcription factor RpoH (Sigma32).

Most of the documented substrate trapping experiments include a 2D-gel protein separation, which limits the separation and/or the identification of low-abundant, hydrophobic, acidic, very large and small proteins. To uncover the entire substrate repertoire more advanced and 2D-gel free technologies are needed. Knowledge about the full substrate repertoire of proteases, like the essential FtsH protease, will help to understand their physiological role and might offer therapeutic targets for the defense of bacterial infections by controlled blockade or activation of proteases [15]. Here, we report on a straightforward substrate identification strategy using FtsH^{trap} and a 1D-short gel MS-based technique coupled to label-free data analysis. In our workflow, an *E. coli* Δ*ftsH* deletion strain was complemented either with plasmid-encoded FtsH^{wt} or FtsH^{trap} (Fig. 1). Cell cultivation, cell harvest and FtsH purification were followed by label-free LC/MS-based analysis of both purified FtsH complexes (FtsH^{wt} and FtsH^{trap}). Proteins with a higher normalized intensity copurified with FtsH^{trap} in comparison with proteins copurified with FtsH^{wt} were considered as putative FtsH substrates (shown for LpxC in Fig. 2). Unspecifically bound proteins showed similar normalized intensities in both samples (shown for TufA in Fig. 2). By this label-free trapping technique, we identified 60 putative FtsH substrates including proteins with one or more transmembrane domains, low-abundant proteins and four already known substrates

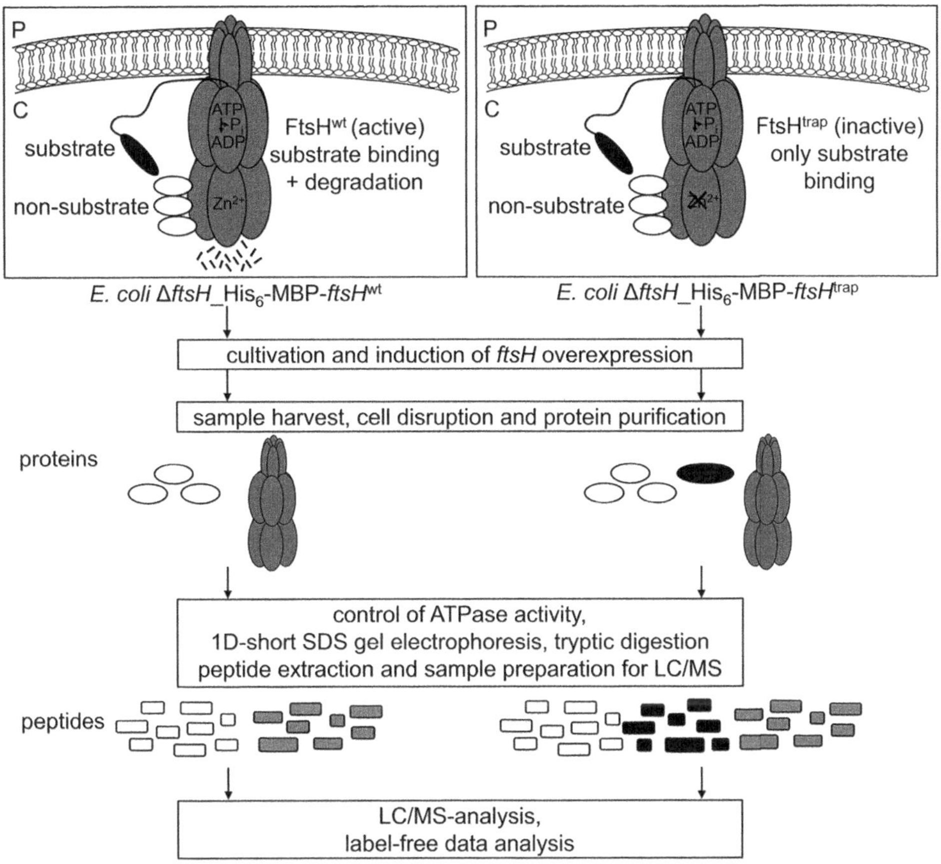

Fig. 1 Label-free trapping approach to identify FtsH substrates. The *E. coli* Δ*ftsH* strain carrying the His_6-MBP-$FtsH^{wt}$ or His_6-MBP-$FtsH^{trap}$ plasmids are used. $FtsH^{wt}$ and $FtsH^{trap}$ are anchored in the inner cell membrane by an N-terminal transmembrane domain (C = cytoplasm, P = periplasm). FtsH substrates are unfolded and translocated by the ATPase domain under ATP hydrolysis and degraded in the protease domain in a Zn^{2+}-dependent manner. $FtsH^{trap}$ carries a H417Y amino acid exchange in the Zn^{2+}-binding motif resulting in the loss of Zn^{2+}-binding and protease activity. An FtsH substrate (black circle) can interact with both FtsH variants ($FtsH^{wt}$ and $FtsH^{trap}$). However, only $FtsH^{wt}$ is able to degrade the substrate. *E. coli* strains are cultivated and *ftsH* overexpression is induced in the exponential growth phase. After cell harvest, cell disruption and purification of the FtsH interactome, the resulting protein mixture consists of putative FtsH substrates copurified with $FtsH^{trap}$ and accessory binding partners (e.g., interaction partners, but nonsubstrates; white circle), which are equally abundant in the $FtsH^{wt}$ and $FtsH^{trap}$ samples. In contrast, the amount of putative FtsH substrates copurified with $FtsH^{wt}$ is lower compared to proteins copurified with $FtsH^{trap}$ as substrates are translocated into the ATPase domain but not degraded by $FtsH^{trap}$. After 1D-short gel, protein bands are tryptically digested and peptides extracted from the gel are analyzed by LC/MS

(LpxC, RpoH, SecY, and YfgM) [9, 14, 16, 17]. The substantial increase in the number of putative FtsH substrates in *E. coli* suggests that this 1D-short gel label-free trapping approach might be applicable to other bacteria and proteases like ClpXP and Lon.

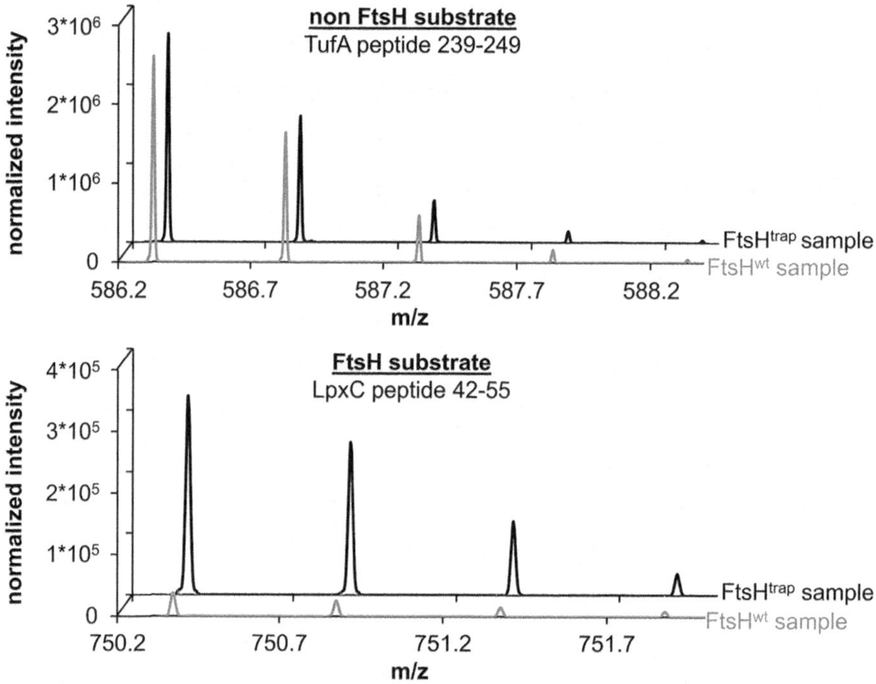

Fig. 2 The known FtsH substrate LpxC was identified by label-free proteomics. FTMS-spectra of LpxC peptide 42–55 detected with a higher normalized intensity in the $FtsH^{trap}$ sample (black line) compared to the $FtsH^{wt}$ sample (grey line) due to the absence of protease activity of $FtsH^{trap}$. Peptides with similar normalized intensities in the $FtsH^{wt}$ or $FtsH^{trap}$ samples represent non-FtsH substrates, illustrated by the FTMS-spectra of the elongation factor TufA peptide 239–249

2 Materials

The *E. coli* Δ*ftsH* deletion strain carrying the His_6-MBP-$FtsH^{wt}$ or His_6-MBP-$FtsH^{trap}$ plasmid is described in [14]. Prepare all solutions using ultrapure water and analytic-grade reagents. Prepare and store all reagents at room temperature unless indicated otherwise. Carefully follow all waste disposal regulations. Autoclave all materials from Subheading 2.1 and filtrate heat-sensitive materials (e.g., antibiotics) sterile.

2.1 Plasmids and Strains

1. Plasmids: His_6-MBP-$FtsH^{wt}$ (vector: pMAL-C, 1149 bp *ftsH* gene fragment (lacking the second transmembrane domain)) [18], His_6-MBP-$FtsH^{trap}$ (vector: pMAL-C, 1149 bp *ftsH* gene fragment (mutation in *ftsH* leads to H417Y in the Zn^{2+}-binding domain of $FtsH^{trap}$, lacking the second transmembrane domain)) [14].
2. Bacterial strain: *E. coli* Δ*ftsH* (F-, IN (*rrnD-rrnE*)1, *zad220*::Tn10, *sfhC21*, Δ*ftsH3*::kan) [3], single colonies containing the His_6-MBP-$FtsH^{wt}$ or His_6-MBP-$FtsH^{trap}$ plasmid on an

lysogeny broth (LB) agar plate with ampicillin (Amp, 100 μg/mL) and kanamycin (Kan, 50 μg/mL).

3. LB-medium: 1% (w/v) tryptone, 0.5% (w/v) sodium chloride (NaCl), 0.5% (w/v) yeast extract, 100 μg/mL Amp, and 50 μg/mL Kan in distilled water.
4. Isopropyl-β-D-thiogalactopyranoside (IPTG): Prepare 100 mM stock solution in distilled water and store at −20 °C until further use.

2.2 Cell Harvest, Washing, and Disruption

1. 1× Phosphate buffered saline (PBS), pH 7.3: Prepare 1× PBS by 1:10 dilution of 10× PBS stock solution. 10× PBS: 1400 mM NaCl, 27 mM potassium chloride (KCl), 100 mM disodium hydrogen phosphate (Na_2HPO_4), 18 mM potassium dihydrogen phosphate (KH_2PO_4). Autoclave the buffer and store at room temperature.
2. RNaseI, DNaseI, and the complete, mini, EDTA-free protease inhibitor cocktail.
3. Tris–EDTA buffer (TE buffer): 10 mM Tris–HCl (pH 8.0), 1 mM EDTA in distilled water.

2.3 Protein Purification

1. Washing (W) and elution (E) buffers (pH 7.9): 500 mM KCl, 20 mM Tris(hydroxymethyl)-aminomethane (Tris) in distilled water. Use buffers with different (0–1000 mM) imidazole concentrations for protein purification (W1-3 and E1-3). W1 contains 0 mM, W2 25 mM and W3 50 mM imidazole. E1 contains 100 mM, E2 250 mM, and E3 1000 mM imidazole. Store the different buffers at 4 °C.
2. Ni-Nitrilotriacetic acid (NTA) agarose.
3. Empty purification columns.
4. Storage buffer: Prepare 1× storage buffer by 1:10 dilution of 10× storage buffer stock solution. 10× storage buffer contains 2000 mM NaCl, 5% (v/v) Octylphenoxypolyethoxyethanol (Nonidet P-40), 200 mM Monoethanolamine (MEA) in distilled water. Add 1 mM dithiothreitol (DTT) to the 1× storage buffer before use (*see* **Note 1**).

2.4 Bradford Assay

1. 5× Roti®-Quant reagent.
2. Protein standard: 1 mg/mL bovine serum albumin (BSA) in distilled water.

2.5 ATPase Assay

Solve all chemicals or buffers in distilled water unless otherwise specified.

1. ATPase buffer (*see* **Note 2**): 1 mM adenosine triphosphate (ATP), 50 mM Tris–hydrochloric acid pH 7.5 (Tris–HCl), 5 mM magnesium chloride ($MgCl_2$), 25 μM zinc acetate

($Zn(O_2CCH_3)_2$), 2.5 mM potassium acetate ($C_2H_3KO_2$), 1 mM DTT, 0.1% (v/v) Nonidet P-40.

2. Detection solution: 0.027% (w/v) malachite green, 0.1% (v/v) Triton X-100, 0.95% (w/v) ammonium molybdate ($(NH_4)_2MoO_4$) solved in 4 M HCl.
3. Citrate solution: 34% trisodium citrate ($C_6H_5Na_3O_7$).
4. Use 2-200 μM monopotassium phosphate (KH_2PO_4) for the calibration curve.

2.6 SDS Gel Electrophoresis

1. 4–12% Bis–Tris SDS gel (1.0 mm × 10 well).
2. Protein ladder (e.g., 10–170 kDa).
3. SDS lysis buffer (5×): 250 mM Tris–HCl (pH 6.8), 10% (w/v) SDS, 5% (v/v) β-mercaptoethanol, 0.5% (w/v) bromophenol blue (BPB), 50% (v/v) glycerol.
4. SDS running buffer (1×): Prepare 1× SDS running buffer by 1:20 dilution of 20× MES SDS running buffer (1 M MES, 1 M Tris-Base, 69.3 mM SDS, 20.5 mM EDTA in distilled water).

2.7 Colloidal Coomassie Brilliant Blue Staining

1. Fixing solution: 2% (v/v) phosphoric acid, 50% (v/v) methanol in distilled water.
2. Incubation solution: 2% (v/v) phosphoric acid, 34% (v/v) methanol, 17% (w/v) ammonium sulfate in distilled water.
3. Coomassie brilliant blue G-250.

2.8 Peptide Production and Extraction

1. Buffer A: 10 mM ammonium bicarbonate (NH_4HCO_3) in distilled water.
2. Buffer B: Mix 10 mM NH_4HCO_3 with 100% acetonitrile (ACN) (v/v = 1:1).
3. Sequencing grade modified trypsin (100 μg).
4. 10 mM HCl in distilled water.
5. 100 mM NH_4HCO_3 in distilled water.
6. Mix 0.1% trifluoroacetic acid (TFA) with 100% ACN (v/v = 1:1).
7. Vacuum concentrator.

2.9 HPLC-Mass Spectrometry

1. HPLC-glass inlet.
2. HPLC (high performance liquid chromatography) instrument such as HP Ultimate 3000 system (Thermo Fisher Scientific, Waltham, MA).
3. Precolumn for enrichment and desalting of samples: 75 μm × 20 mm Acclaim PepMap100 C18 with 3 μm particle size (Thermo Fisher Scientific, Waltham, MA).

4. Mobile phase (precolumn): solvent A (0.1% TFA (v/v)) and solvent B (0.1% TFA, 50% ACN (v/v)).
5. Analytical column: 75 μm × 50 cm Acclaim PepMap100 RSLC C18 with 2 μm particle size (Thermo Fisher Scientific, Waltham, MA).
6. Mobile phase (analytical column): solvent A (0.1% formic acid (v/v)) and solvent B (0.1% formic acid, 84% ACN (v/v)).
7. Mass spectrometer: Thermo Orbitrap Elite™ hybrid ion trap-Orbitrap mass spectrometer with electrospray ion source (Thermo Fisher Scientific, Waltham, MA).

2.10 Label-Free Data Analysis

1. MaxQuant version 1.5.0.0 [19].
2. Microsoft Office program: Microsoft Excel (e.g., version 2010).
3. RawMeat version 2.1. (Vast Scientific).

3 Methods

The trapping technique was applied to identify FtsH substrates in the exponential growth phase of *E. coli* from at least three biological replicates. In general the approach should work at various conditions (e.g., different growth phases). If you plan to use the technique for other proteases, construct a protease-inactive and protease-active variant as documented for Clp proteases [11]. Carry out all procedures at room temperature, unless otherwise specified.

3.1 Cell Culture and Protein Overproduction

Steps 1–3 will be done at the first day and **steps 4** and **5** will be done at the second day of the experiment.

1. Inoculate an overnight culture from a single colony of *E. coli* Δ*ftsH* with His_6-MBP-$FtsH^{wt}$ or His_6-MBP-$FtsH^{trap}$ plasmid in 20 mL LB media with 100 μg/mL Amp and 50 μg/mL Kan in a sterile 250 mL Erlenmeyer shake flask. Incubate at 30 °C with shaking at 180 rpm (*see* **Note 3**).
2. Prepare sterile 5000 mL Erlenmeyer flasks with 1000 mL LB-medium, 100 μg/mL Amp and 50 μg/mL Kan. Prewarm the flasks at 30 °C for the next day.
3. Prepare clean 250 mL and 50 mL centrifuge tubes for cell harvest.
4. Inoculate 1000 mL LB media in a 5000 mL Erlenmeyer flask with the *E. coli* Δ*ftsH* carrying His_6-MBP-$FtsH^{wt}$ or His_6-MBP-$FtsH^{trap}$ plasmid preculture to an $OD_{580\ nm}$ of 0.05. This main culture is grown at 30 °C with shaking at 180 rpm until an $OD_{580\ nm}$ of 0.5 is reached. Take a 1 mL cell culture aliquot before adding IPTG.

5. Induce the expression of *ftsH*^{wt} (His$_6$-MBP-FtsH^{wt} plasmid) or *ftsH*^{trap} (His$_6$-MBP-FtsH^{trap} plasmid) at an $OD_{580\ nm}$ of 0.5 (exponential growth phase) by adding 1 mM IPTG to the culture following incubation overnight at room temperature with shaking at 180 rpm (*see* **Note 4**). Take a 1 mL cell culture aliquot at the next day.

3.2 Cell Harvest, Washing, and Disruption

Make sure that the following steps are carried out on ice.

1. Harvest the cell culture in 250 mL centrifuge tubes (prepared in Subheading 3.1, **step 5**) 16–18 h after adding IPTG. Centrifuge cell culture at 2704 × *g*-force (rotor type: SLA-3000) and 4 °C for 20 min and discard the supernatant.
2. Before proceeding with the next step centrifuge samples taken before and after protein overproduction at 16,100 × *g*-force at room temperature. Discard the supernatant and resuspend the cells in TE buffer according to their $OD_{580\ nm}$ (at an $OD_{580\ nm}$ of 0.5 add 50 μL TE buffer) and add 5× SDS lysis buffer to a final concentration of 1× SDS lysis buffer (*see* **Note 5**). Perform SDS gel electrophoresis and a colloidal Coomassie brilliant blue staining (compare Subheadings 3.6 and 3.7).
3. Wash the cell pellet (prepared in Subheading 3.2, **step 1**) with ice-cold 40 mL 1× PBS buffer and transfer the cells to a 50 mL centrifuge tubes. Centrifuge cells at 1912 × *g*-force (rotor type: SS-34) and 4 °C for 20 min and discard the supernatant.
4. Repeat **step 2** twice each with 40 mL ice-cold 1× PBS buffer.
5. Resuspend cells in 12 mL ice-cold 1× PBS buffer and add a small amount (tip of a spatula) of RNaseI and DNaseI and half a pill of the protease inhibitor cocktail before cell disruption.
6. Disrupt the cells using a French press cell disrupter (three runs with 900 psi).
7. Centrifuge the sample at 16,100 × *g*-force and 4 °C for at least 45 min until a pellet of cell debris is formed. Use the supernatant for protein purification.

3.3 Protein Purification

Perform all purification steps in a cooling chamber at 4 °C.

1. Mix the Ni-NTA agarose by inverting it 8 to 10 times. Pipet 2 mL of the agarose (final column bed volume = 1 mL) into an empty purification column, equilibrate the column with 5 mL distilled water and 5 mL washing buffer (W1) (*see* **Note 6**).
2. Transfer the supernatant (from Subheading 3.2, **step 7**) on the column and collect the flow-through (FT). Mix 40 μL of the flow-through with 10 μL 5× SDS lysis buffer.
3. Wash the column with 1.5 mL buffer W1, 4.5 mL buffer W2, and 1.5 mL buffer W3. Collect 40 μL of the flow-through from

every washing fraction (W1, W2, and W3) and mix it with 10 μL 5× SDS lysis buffer.

4. Elute the proteins by sequentially pipetting 1 mL of buffer E1, E2, and E3 to the column. Collect each elution fraction (E1, E2, and E3) separately and save 5 μL from every fraction for determination of the protein concentration by Bradford assay (compare Subheading 3.4).
5. Mix 800 μL of each elution fraction with 200 μL 5× SDS lysis buffer for SDS gel electrophoresis.
6. Mix the remaining 195 μL of each elution-fraction with 195 μL 1× storage buffer and 20% (v/v) glycerin for ATPase activity assay. Store the samples at −80 °C.

3.4 Bradford Assay

Determine the elution fraction (E1, E2, and E3) with the highest protein amount by Bradford assay. Use this elution fraction for the ATPase assay and 1D-short gel (compare Subheading 3.6). Perform a modified Bradford assay according to the manufacturer's protocol.

1. Fill five 1 mL cuvettes for the calibration curve and three cuvettes for the elution fraction (E1, E2 and E3; compare Subheading 3.3) with 795 μL distilled water.
2. Add 0, 1, 3, 5, or 10 μL, respectively, of the 1 mg/mL BSA-solution to five cuvettes for the calibration line. Add 5 μL of every elution fraction (E1, E2, or E3) to three different cuvettes.
3. Start the reaction by adding 200 μL of the 5× Roti®-Quant reagent. Close the cuvette immediately with a small piece of Parafilm M®, invert the cuvette 5 to 6 times and incubate for 10 min at room temperature.
4. Set blank value of the photometer by measuring the 0 μL BSA sample at $OD_{595\ nm}$ and measure the samples in the same order and time as you prepared them.
5. Create a calibration line of the BSA samples and set a line of best fit. Use the formula of this line to determine the protein concentrations of the elution fraction (E1, E2 or E3).

3.5 ATPase Assay

Measure the ATPase activity of $FtsH^{wt}$ or $FtsH^{trap}$ to ensure the correct translocation of protease substrates. Use the sample of each elution fraction (E1, E2 or E3) stored in storage buffer containing glycerin (compare Subheading 3.3, **step 6**). Perform the assay as described in [14].

1. Mix 1 μg of purified $FtsH^{wt}$ or $FtsH^{trap}$ with ATPase buffer in a total volume of 500 μL and incubate the solution at 37 °C without shaking.

2. After 0, 30, 60, and 90 min mix 120 μL of the reaction mixture with 480 μL of detection solution and 40 μL of 34% citrate solution. The malachite green and molybdate in the detection solution reacts with free orthophosphate which results of ATP hydrolysis by ATPase domain of FtsH.
3. After 40 min incubation at room temperature, measure the absorption at 660 nm.
4. Perform the ATPase assay without FtsH in absence and present of 1 mM ATP as controls.
5. Perform the ATPase assay with increasing concentrations of KH_2PO_4 to calculate the amount of free phosphate. Create a calibration curve of the KH_2PO_4 samples and set a line of best fit. The formula of this line can be used to determine the concentration of free phosphate in the FtsH samples.

3.6 SDS Gel Electrophoresis

Perform SDS gel electrophoresis to verify the success of gene expression or protein purification.

1. Heat the protein samples (FT, W1, W2, W3, E1, E2, and E3) at 95 °C for 10 min and centrifuge at 16,100 × *g*-force for 1 min.
2. Add 10 μL of each protein sample and 4 μL of the prestained protein ladder on a prefabricated gel.
3. Use an appropriate gel chamber and fill it with 1× SDS running buffer.
4. Connect the gel chamber with a power supply and use running conditions according to the manufacturer's protocol.

   To preprocess samples for LC-MS use the following protocol:
5. Heat the protein sample (elution fraction with highest protein amount, compare Subheading 3.4) at 95 °C for 10 min and centrifuge at 16,100 × *g*-force for 1 min.
6. Load 3 μg protein of the elution fraction on an SDS gel (compare Subheading 3.3).
7. Use an appropriate gel chamber and fill it with 1× SDS running buffer.
8. Connect the gel chamber with a power supply and use running conditions according to the manufacturer's protocol.
9. Run the gel until the protein mixture has completely entered the gel. A separation of proteins is not necessary.

3.7 Colloidal Coomassie Brilliant Blue Staining

Protein bands are visualized on the SDS gel by colloidal Coomassie brilliant blue staining (*see* **Note 7**). Perform all steps at room temperature.

1. Fill a small plastic bowl with 50 mL fixing solution and add the SDS gel (compare Subheading 3.6) to the solution.
2. Incubate the gel for 30 min on a shaker at 40 to 60 rpm.

3. Remove the fixing solution and wash the gel 3 times with 50 mL distilled water for 10 min on a shaker at 40–60 rpm.
4. Remove the distilled water and incubate the gel in 50 mL incubation solution for 30 min on a shaker at 40–60 rpm.
5. Add Coomassie brilliant blue G 250 (tip of a spatula) to the incubation solution and incubate the gel overnight on a shaker at 40 to 60 rpm.
6. Remove the incubation solution and wash the gel with 50 mL distilled water on a shaker at 40 to 60 rpm until the protein bands are visible and the background is nearly decolored.

3.8 Gel Slicing and Washing

1. Excise the protein band with a surgical disposable scalpel and transfer the gel piece to a microcentrifuge tube. The size of one gel pieces should not exceed 1–3 × 5–6 × 1 mm^3 (*see* **Note 8**).
2. Add 40 μL of buffer A to the sample and incubate for 10 min at room temperature without shaking.
3. Remove buffer A and add 40 μL buffer B to the sample and incubate for 10 min at room temperature without shaking and remove buffer B.
4. Repeat **steps 2** and **3** twice. After the last washing step remove remaining liquid by evaporate the sample to dryness in a vacuum concentrator (*see* **Note 9**). Store the dry gel piece at −80 °C or directly use it for tryptic digestion (compare Subheading 3.9).

3.9 Trypsin Digestion ("In Gel-Digestion")

1. Solve 100 μg (1 vial) trypsin in 50 μL 10 mM HCl ($c = 2.0$ μg/μL trypsin), aliquot in 5 μL samples and store at −20 °C for further use.
2. Add 55 μL 100 mM NH_4HCO_3 to one 5 μL trypsin aliquot ($c = 0.17$ μg/μL trypsin).
3. Add 0.17 μg/μL trypsin solution to the protein sample (final protease–protein ratio: 1:100 to 1:20 (w/w)) and incubate overnight at 37 °C without shaking.

3.10 Peptide Extraction and Sample Preparation

1. Add 20 μL 0.1% TFA–100% ACN (v/v = 1:1) to each tryptically digested sample and incubate 10 min in an ultrasonic bath.
2. Centrifuge the sample to bring all solution to the bottom and transfer the supernatant into an HPLC-glass inlet.
3. Repeat **steps 1** and **2** and transfer each supernatant in the corresponding HPLC-glass inlet.
4. Remove ACN by evaporating the sample to dryness for approximately 2 min in the vacuum concentrator.
5. Refill the sample volume to 15 μL with 0.1% TFA and measure the sample by LC/MS (*see* **Note 10**).

3.11 Reversed-Phase HPLC Coupled with High-Resolution Mass Spectrometry

Label-free analysis requires high technical reproducibility. Therefore it is important to measure all samples in one batch with the same LC-MS system and method (*see* **Note 11**).

3.11.1 HPLC Analysis

1. Inject 15 μL sample volume on a precolumn for enrichment and desalting of the complex peptide sample. Perform the loading and washing step with a mixture of 95% solvent A (0.1% TFA) and 5% solvent B (0.1% TFA, 50% ACN) at a flow rate of 30 μL/min for 7 min.
2. Separation of peptides is carried out on an analytical column connected to the precolumn. Peptides are loaded on the analytical column with a mixture of 95% solvent A (0.1% FA) and 5% solvent B (0.1% FA, 84% ACN) at a flow rate of 400 nL/min and eluted with a linear gradient of 5% B to 40% B over 120 min with the same flow rate.

3.11.2 Mass Spectrometry Analysis

Mass spectra are obtained by an Orbitrap instrument online coupled to the HPLC system (or any equivalent system). Data-dependent acquisition in a Top N method is performed on the Orbitrap instrument. In the Top N method 20 of the most intense peaks in each MS spectrum are selected for MS/MS-fragmentation.

1. Perform a Fourier transform mass spectrometry (FTMS) scan in a mass/charge range of 300 to 2000 with a resolution for the first scan of 60,000.
2. Select the following conditions for MS/MS-fragmentation: Select 20 of the most intense peptides in the FTMS-scan with minimal signal intensity of 1500 and a charge range $\geq$ +2 in the collision-induced dissociation mode (CID-mode). Set the exclusion list size to 500 and the exclusion duration time to 35 s.

3.12 Data Analysis Using MaxQuant

Protein identification and label-free quantification can be performed by different software tools [20]. We have made good practical experiences with the freely available MaxQuant software. The software can be used for the analysis of high-resolution MS data from large-scale proteomic raw data sets. The identification and quantification of peptides and proteins is carried out automatically and user can exert influence on different parameters. General information, like the software installation and system requirements, is addressed in the ref. 19 and at the MaxQuant website (http://www.maxquant.org). Check the home page for the recent version and features of MaxQuant. Here, we describe a guide for label-free quantification using this software. Use default parameters, if not addressed in "**step 1**".

3.12.1 Label-Free Quantification (LFQ) by MaxQuant (Version 1.5.0.0)

Create a selected folder in which you save the "raw" files and the "FASTA" database text file (*see* **Note 12**). Save the folder on the same computer where MaxQuant is installed.

1. Open MaxQuant.exe and select the "raw" files for data analysis under parameter "Raw files". All loaded "raw" files will automatically be attached to the same group number (e.g., "Group0"). Click the button "No fractions" and all the "raw" files will automatically get a different experimental name. MaxQuant uses the last characters of the "raw" file name to create the corresponding experimental name. If you want to give the "raw" file a customized name click the button "Set experiment" and type in an experimental name of your choice.
2. Select following parameters:
   (a) "Andromeda Configuration": Use "Andromeda configuration" to add your "FASTA" text file into the default database list of MaxQuant.

   Initially, create a new line in the default database list by clicking the button "Sequence database." Add your "FASTA" text file (saved in your selected folder) by clicking the button "Add," "Fasta file," and "Select." Insert your database source by clicking "Source" (e.g., Ecoprot, Uniprot, NCBI) and add the taxonomy of your organism by clicking "Taxonomy" (e.g., *E. coli*). Choose the parse rules which are specific for your database source. Save changes by clicking the button "Save changes" to include your database to the list.

   (b) "Group-specific parameters." Please go to the following subitems:
   - "General": Select "Oxidation (M)" for variable modification and use no fixed modification. Define "Trypsin" as enzyme (use digestion mode "Specific") with maximum two missed cleavages (use match type "Match from and to"). For label-free search set "Multiplicity" to 1.
   - "Instrument": Select your MS instrument and use the default or your own mass tolerances.
   - "Label-free quantification": Check the box "LFQ", select "Fast LFQ" and set the "LFQ min. ratio count" to 2.
   - "Advanced": Select "Cut peaks" and "Max. number of modification per peptide" to 5.

   (c) "Global parameters": Please go to the following subitems:

- "General": Add your "FASTA" protein database by clicking "Add file." Remove any fixed modifications and check the box "Re-quantify."
- "Sequences": Select "Randomize" as "Decoy mode," use the "Special AAs" "KR" and select "Include contaminants" (*see* **Note 13**).
- "Identification": Use the default settings and change "Min. unique peptides" to 1. Check the box "Second peptides" (*see* **Note 14**) and uncheck "Filter labeled amino acids" for label-free data analysis.
- "Protein quantification": Set "Min. ratio count" to 2. Set "Peptides for quantification" to "Unique + razor". Check "Use only unmodified peptides and..." and add "Oxidation (M)" for "Modifications used in protein quantification". Uncheck the box "Discard unmodified counterpart peptides" (*see* **Note 15**).
- "Label free quantification": Select "Stabilize large LFQ ratios," "Require MS/MS for LFQ comparisons," "iBAQ," and "Advanced site intensities."
- "MS/MS-FTMS" and "MS/MS-ITMS": Select your MS-instrument and use the default or your own mass tolerances.
- Use default parameters for "Isobaric label quantification," "AIF," "MS/MS-TOF," "MS/MS-unknown," and "Advanced."

Save the parameters in your selected folder by clicking "Save parameters".

3. Start the data analysis by clicking the "Start" button (*see* **Note 16**). The results will be saved in your selected folder. Use the file "proteinGroups.txt" in the subfolder: "\combined\txt\" for further data interpretation (compare "**step 2**").

3.12.2 Data Interpretation

The file "proteinGroups.txt" includes all information you need to identify putative protease substrates. In general, a putative substrate (copurified with $\text{FtsH}^{\text{trap}}$) has a higher normalized intensity (LFQ intensity) in comparison with the LFQ intensity of the corresponding protein copurified with FtsH^{wt}.

The following steps help you to simplify data interpretation.

1. Open file "proteinGroups.txt" in Microsoft Office Excel.
2. Activate the function "Filter" and reject positive "Reverse" and "Contaminants" results.

3. Choose the following criteria to identify significantly enriched proteins copurified with FtsH^{trap} in comparison to copurified proteins with FtsH^{wt}.

   (a) The protein was identified in at least three biological replicates in the FtsH^{trap} samples (use function: "count if" (segment, ">0") for columns "LFQ intensity"; check only 3) with 3 unique peptides in at least one biological replicate (use function: "count if" (segment, ">2") for columns "Unique peptides", uncheck only 0). The criteria "1" ensure a good identification and a reliable quantification of proteins in FtsH^{trap} samples.

   Additionally to criteria (a), choose one of the following criteria (b–d) to ensure a significant enrichment of proteins copurified with FtsH^{trap} compared to copurified proteins with FtsH^{wt}.

   (b) The protein was identified in all three biological replicates in the FtsH^{wt} samples and has a ratio (average LFQ-intensity (FtsH^{trap}) /average LFQ-intensity (FtsH^{wt})) ≥ 2 with a p-value ≤ 0.05.

   (c) The protein was identified in one or two biological replicates in the FtsH^{wt} samples and has a ratio (average LFQ-intensity (FtsH^{trap})/average LFQ-intensity (FtsH^{wt})) ≥ 5.

   (d) The protein was not identified in any FtsH^{wt} replicate and was only identified in the FtsH^{trap} samples.

Using these criteria, we were able to identify 60 significantly enriched FtsH substrates including four known substrates (LpxC, RpoH, SecY, and YfgM).

4 Notes

1. Always prepare the storage buffer fresh because DTT is not stable in solution.
2. Always prepare the ATPase buffer fresh because ATP hydrolysis at room temperature and ATPase buffer contains DTT (*see* **Note 1**).
3. The *E. coli* Δ*ftsH* strain is temperature sensitive. Make sure that the cells never grow at a temperature higher than 30 °C. For plasmid transformation perform the heat shock for 1.5 min at 37 °C instead of 42 °C.
4. It is possible to perform the overexpression of *ftsH*^{wt} or *ftsH*^{trap} under other conditions (e.g., different growth media).

5. Perform SDS gel electrophoresis and a Coomassie brilliant blue staining to check for successful protein overproduction before proceeding with the next step.
6. Do not centrifuge the column. The column should not run dry at any time.
7. Colloidal Coomassie brilliant blue staining is compatible with LC/MS-analysis.
8. The size of one gel piece should not exceed the described dimension to make sure that the added trypsin solution covers the gel piece completely to ensure an effective digestion.
9. A third repetition of both washing steps can be performed if the Coomassie staining of the gel pieces is still visible.
10. Measure the exact peptide concentration of each sample (e.g., colorimetric assays (ninhydrin, Lowry, BCA)) to calculate the right peptide amount that is optimal for your LC/MS-system (e.g., 200 ng of peptides for the described LC/MS-system).
11. A prerequisite for label-free studies is a minimization of the biological variances. The evaluation of biological variance can be carried out by *t*-test or by calculation of the stability index (R^2) by comparing the normalized intensities (calculated by MaxQuant) within two biological replicates. Thereby proteins identified in all biological replicates were considered. Additionally, technical variance of LC/MS-analysis should be controlled by the correlation of total-ion-current (TIC) chromatograms within biological replicates using RawMeat.
12. Your database ("FASTA" file) should include the unmodified and modified protease sequences (in this case His_6-MBP-$FtsH^{wt}$ and His_6-MBP-$FtsH^{trap}$). Furthermore, do not choose a decoy database because MaxQuant creates its own.
13. MaxQuant includes a list of contaminants which can be found in the text file "contaminants.fasta" located in the "conf" subfolder. Contaminants are defined as proteins which impurify your sample (e.g., serum albumin or keratin). Checking the subitems "Randomize" leads to creation of a randomized decoy database and checking the box "Special AAs" "KR" interchanges lysine to arginine to avoid false-positive hits. The decoy database enables the calculation of the false discovery rate.
14. Check "Second peptides" to let MaxQuant search for mixed MS/MS-spectra to improve peptide identification.
15. Uncheck the button "Discard unmodified counterpart peptides" to select unmodified and modified peptides for label-free quantification. If you leave the box checked, only modified peptides are used for quantification.

16. The number of threads depends on your computer system. One thread will need 2 gigabytes random access memory (RAM).

Acknowledgments

The authors would like to thank Alexandra Müller, Sina Langklotz, and Thilo Lerari for carefully reading the manuscript and for many helpful comments. The work was supported by grants from the German Research Foundation (DFG, SFB642: GTP- and ATP-dependent membrane processes) and P.U.R.E. (Protein Unit for Research in Europe, funded by the German federal state North Rhine-Westphalia).

References

1. Langklotz S, Narberhaus F (2011) The *Escherichia coli* replication inhibitor CspD is subject to growth-regulated degradation by the Lon protease. Mol Microbiol 80(5):1313–1325
2. Schäkermann M, Langklotz S, Narberhaus F (2013) FtsH-mediated coordination of lipopolysaccharide biosynthesis in *Escherichia coli* correlates with the growth rate and the alarmone (p)ppGpp. J Bacteriol 195(9):1912–1919
3. Tatsuta T, Tomoyasu T, Bukau B, Kitagawa M, Mori H, Karata K, Ogura T (1998) Heat shock regulation in the ftsH null mutant of *Escherichia coli*: dissection of stability and activity control mechanisms of sigma32 in vivo. Mol Microbiol 30(3):583–593
4. Sauer RT, Baker TA (2011) AAA+ proteases: ATP-fueled machines of protein destruction. Annu Rev Biochem 80:587–612
5. Battesti A, Gottesman S (2013) Roles of adaptor proteins in regulation of bacterial proteolysis. Curr Opin Microbiol 16(2):140–147
6. Gur E, Biran D, Ron EZ (2011) Regulated proteolysis in Gram-negative bacteria--how and when? Nat Rev Microbiol 9(12):839–848
7. Konovalova A, Søgaard-Andersen L, Kroos L (2014) Regulated proteolysis in bacterial development. FEMS Microbiol Rev 38 (3):493–522
8. Langklotz S, Baumann U, Narberhaus F (2012) Structure and function of the bacterial AAA protease FtsH. Biochim Biophys Acta 1823(1):40–48
9. Ogura T, Inoue K, Tatsuta T, Suzaki T, Karata K, Young K, Su LH, Fierke CA, Jackman JE, Raetz CR, Coleman J, Tomoyasu T, Matsuzawa H (1999) Balanced biosynthesis of major membrane components through regulated degradation of the committed enzyme of lipid A biosynthesis by the AAA protease FtsH (HflB) in *Escherichia coli*. Mol Microbiol 31(3):833–844
10. Führer F, Langklotz S, Narberhaus F (2006) The C-terminal end of LpxC is required for degradation by the FtsH protease. Mol Microbiol 59(3):1025–1036
11. Flynn JM, Neher SB, Kim YI, Sauer RT, Baker TA (2003) Proteomic discovery of cellular substrates of the ClpXP protease reveals five classes of ClpX-recognition signals. Mol Cell 11 (3):671–683
12. Feng J, Michalik S, Varming AN, Andersen JH, Albrecht D, Jelsbak L, Krieger S, Ohlsen K, Hecker M, Gerth U, Ingmer H, Frees D (2013) Trapping and proteomic identification of cellular substrates of the ClpP protease in *Staphylococcus aureus*. J Proteome Res 12 (2):547–558
13. Bhat NH, Vass RH, Stoddard PR, Shin DK, Chien P (2013) Identification of ClpP substrates in *Caulobacter crescentus* reveals a role for regulated proteolysis in bacterial development. Mol Microbiol 88(6):1083–1092
14. Westphal K, Langklotz S, Thomanek N, Narberhaus F (2012) A trapping approach reveals novel substrates and physiological functions of the essential protease FtsH in *Escherichia coli*. J Biol Chem 287(51):42962–42971
15. Kirstein J, Hoffmann A, Lilie H, Schmidt R, Rübsamen-Waigmann H, Brötz-Oesterhelt H, Mogk A, Turgay K (2009) The antibiotic ADEP reprogrammes ClpP, switching it from a regulated to an uncontrolled protease. EMBO Mol Med 1(1):37–49

16. Kihara A, Akiyama Y, Ito K (1995) FtsH is required for proteolytic elimination of uncomplexed forms of SecY, an essential protein translocase subunit. Proc Natl Acad Sci U S A 92 (10):4532–4536
17. Bertani D, Oppenheim AB, Narberhaus F (2001) An internal region of the RpoH heat shock transcription factor is critical for rapid degradation by the FtsH protease. FEBS Lett 493(1):17–20
18. Katz C, Ron EZ (2008) Dual role of FtsH in regulating lipopolysaccharide biosynthesis in *Escherichia coli*. J Bacteriol 190 (21):7117–7122
19. Cox J, Mann M (2008) MaxQuant enables high peptide identification rates, individualized p.p.b.-range mass accuracies and proteome-wide protein quantification. Nat Biotechnol 26(12):1367–1372
20. Megger DA, Bracht T, Meyer HE, Sitek B (2013) Label-free quantification in clinical proteomics. Biochim Biophys Acta 1834 (8):1581–1590

Chapter 15

In vivo Proteomics Approaches for the Analysis of Bacterial Adaptation Reactions in Host–Pathogen Settings

Henrike Pförtner, Maren Depke, Kristin Surmann, Frank Schmidt, and Uwe Völker

Abstract

Proteome profiling of bacteria internalized by host cells is still a challenging task, due to low amounts of bacterial proteins in host–pathogen settings and the high amounts of contaminating host proteins. Here, we describe a workflow for the enrichment of intracellular bacteria by fluorescence activated cell sorting which in combination with highly sensitive LC-MS/MS allows monitoring of about 1200 proteins from 2 to 4 × 10^6 internalized bacterial cells as starting material.

Key words Host–pathogen interaction, *Staphylococcus aureus*, Intracellular bacteria, Epithelial cells, Internalization, Bacterial cell sorting, Mass spectrometry, Proteomics

1 Introduction

For many bacterial pathogens such as *Salmonella enterica*, *Mycobacterium tuberculosis* or *Listeria monocytogenes* the intracellular environment is a niche for protection against host immune factors or antibiotic treatments. Once inside host cells bacteria have to mount different adaptation strategies to ensure survival and proliferation, and to establish an infection [1]. Infectious diseases are still a major threat to human health, especially because of the increasing number of antibiotic resistant pathogens [2]. Therefore, it is necessary to understand the adaption of bacteria to the host cell milieu to be able to develop new drugs for the treatment of such hiding, antibiotic resistant pathogens.

Either alone or better in combination with complementing OMICS-technologies (genomics, transcriptomics, and metabolomics) proteomics can provide an unprecedented comprehensive view of the adaptive changes occurring during an infection. However, the challenging part of applying proteome analysis to host–pathogen settings is the small number of bacteria available either

Dörte Becher (ed.), *Microbial Proteomics: Methods and Protocols*, Methods in Molecular Biology, vol. 1841, https://doi.org/10.1007/978-1-4939-8695-8_15,

from cell culture and animal experiments or human specimen. Additionally, due to the larger size of the host cells, a large excess of contaminating host proteins complicates the detection and quantification of bacterial proteins [3, 4]. We developed appropriate workflows for such analyses exploring the interaction of the Gram-positive pathogen *Staphylococcus aureus* with human host cells. However, with slight modifications this workflow can also be extended to the study of other bacterial pathogens.

S. aureus is a versatile organism, which can appear as a harmless colonizer of the human body but also as an opportunistic pathogen. About 20–30% of the population is colonized by *S. aureus* without symptoms [5]. Nevertheless, *S. aureus* is the cause of severe diseases, e.g., pneumonia and endocarditis. Until recent years, *S. aureus* was exclusively regarded as an extracellular pathogen. This changed when it became more and more clear that *S. aureus* can invade nonprofessional phagocytic cells and survive intracellularly for extended periods of time [6]. Adaptation to this new milieu includes formation of subpopulations such as small colony variants (SCVs). SCVs seem to be well-adapted to the intracellular environment by displaying reduced growth rate, and this "stealth mode" is connected to long-term intracellular persistence and recurrent infections [7].

The workflow presented here (Fig. 1) provides an experimental setup for proteome analysis of intracellular *S. aureus* internalized by human epithelial cells. The combination of the enrichment of the internalized bacteria by high speed fluorescence activated cell sorting with sensitive, high accuracy MS allows identification and quantitation of a large number of proteins to comprehensively monitor adaptive responses to the host cell environment. Epithelial cells were infected with *S. aureus* HG001 carrying a plasmid encoding green fluorescent protein (GFP). Intracellular bacteria were separated from host cells after their lysis by sorting GFP-expressing bacteria directly onto a filter membrane of a 96-well plate. To achieve sorting of a few million bacteria into one well of a 96-well plate, the sheath fluid had to be removed constantly by a vacuum pump. After sorting, intact bacteria were subjected to a combination of lysostaphin treatment and tryptic digestion on the filter membrane. The complex peptide mixture was then separated via nano liquid chromatography and measured online on a Q Exactive™ Orbitrap mass spectrometer. With this workflow we were able to identify about 1200 proteins with at least one peptide per protein and about 900 proteins with at least two peptides per protein from 2 to 4×10^6 internalized *S. aureus* cells [8]. For processing of MS raw data we intentionally provide within this chapter an analysis workflow using freely available software such as Trans Proteomic Pipeline (TPP) [9, 10] in combination with Comet [11] for peptide identification and Skyline [12] for protein quantification. However, alternative data analysis workflows employing alternative software packages or commercially available software can also be envisioned [13–15].

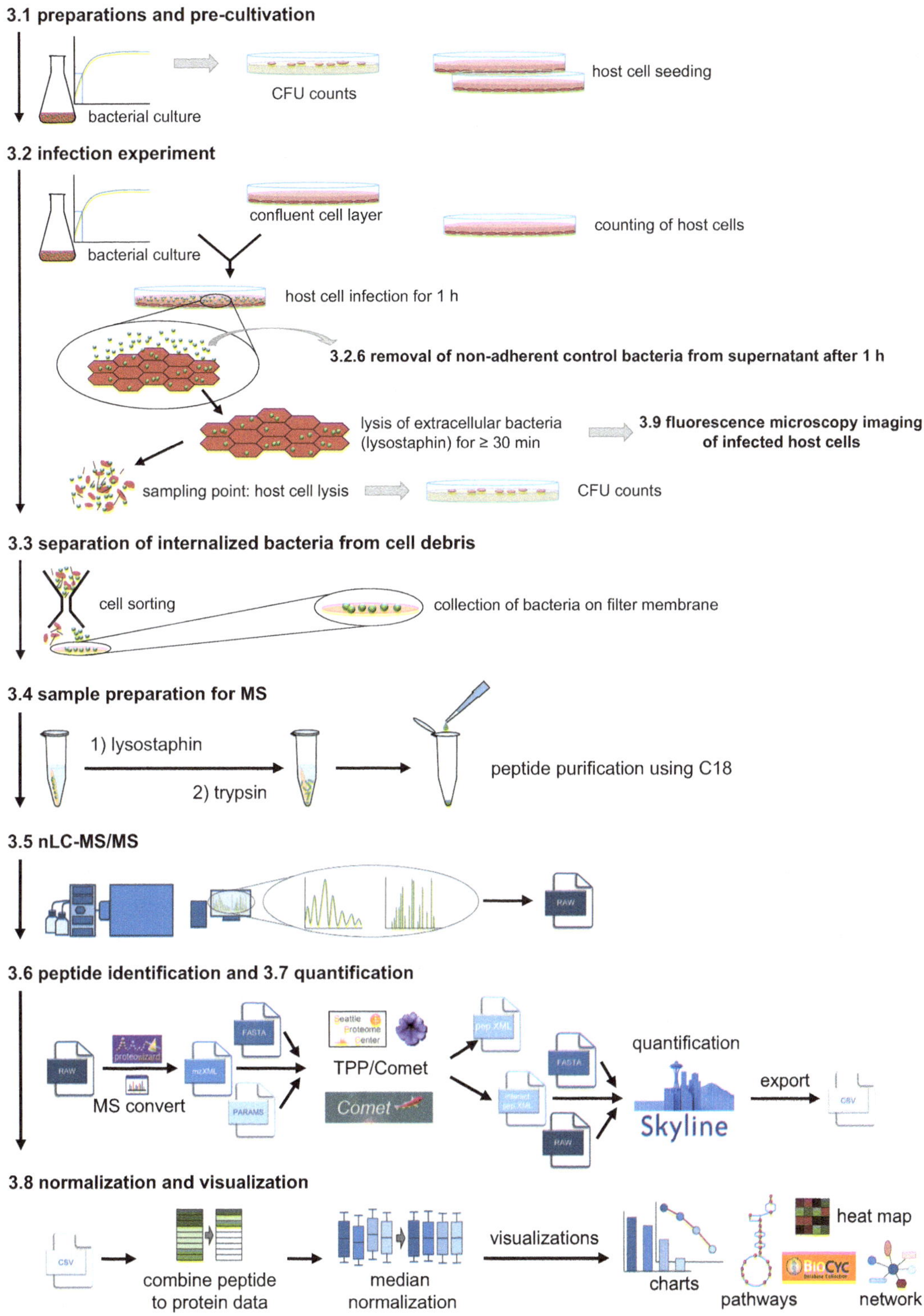

Fig. 1 Schematic presentation of the workflow for the proteome analysis of internalized bacterial pathogens. Numbered captions refer to Subheading 3

2 Materials

2.1 Infection of Eukaryotic Host Cells

Besides specific materials listed below, standard laboratory equipment is needed when performing this protocol. This includes for example equipment normally used for eukaryotic cell culture like incubators (defined temperature and humidified atmosphere with increased CO_2-content) and laminar flow cabinets for sterile working, and equipment used for cultivation of bacteria like shaking water baths. Furthermore, a cell sorter is required for the enrichment of internalized bacteria from host cell lysates.

1. 96-well filtration plates for cell sorting with a hydrophilic and low protein binding PVDF filter membrane with a pore size of 0.22 μm.
2. Amino acid stock solutions: Mix 1 (50 mM): Dissolve 4.455 g L-alanine, 5.858 g L-valine, 6.559 g L-leucine, and 6.559 g L-isoleucine in 1 L sterile HPLC-grade water. Mix 2 (50 mM): Dissolve 7.755 g L-aspartate and 8.455 g L-glutamate in 1 L sterile HPLC-grade water. Mix 3 (50 mM): Dissolve 5.255 g L-serine, 5.956 g L-threonine, and 6.058 g L-cysteine in 900 mL sterile HPLC-grade water and adjust pH to 10.3 with NaOH. Make up to 1 L with sterile HPLC-grade water. Mix 4 (50 mM): Dissolve 5.757 g L-proline, 8.260 g L-phenylalanine, and 7.758 g L-histidine in 1 L sterile HPLC-grade water. Dissolve 5.106 g L-tryptophan in 1 L sterile HPLC-grade water to obtain a 25 mM solution. Sterile filtrate all solutions using a filter with a pore size of 0.22 μm. Store the amino acid stock solutions at 4 °C.
3. Lysostaphin stock solution: 5 mg/mL lysostaphin in water. Dissolve a vial with 50 mg lyophilized lysostaphin (e.g., AMBI PRODUCTS LLC) containing >3000 U/mg in 10 mL sterile HPLC-grade water. Portion the solution in single-use aliquots of 50–200 μL and store the aliquots at −20 °C.
4. Eukaryotic Minimum Essential Medium (eMEM): Mix under sterile conditions 100 mL MEM Earl w/o L-glutamine (1×), 4 mL fetal bovine serum, 2 mL L-glutamine (200 mM), and 1 mL MEM nonessential amino acids (NEAAs, 100×). Store the eMEM at 4 °C (*see* **Note 1**).
5. eMEM/lysostaphin: eMEM with 10 μg/mL lysostaphin.
6. Prokaryotic Minimum Essential Medium (pMEM): Mix 10 mL MEM Earl w/o L-glutamine (10×), 1 mL MEM NEAAs (100×), 1 mL 4-(2-hydroxyethyl)-1-piperazineethanesulfonic acid (HEPES, 1 M), 2 mL L-glutamine (200 mM), 8 mL L-tryptophan (25 mM), 4 mL amino acid mix 1 (50 mM each), 4 mL amino acid mix 2 (50 mM each), 4 mL amino acid mix 3 (50 mM each), 4 mL amino acid mix 4 (50 mM each), and

52 mL HPLC-grade water. Adjust pH to 7.4 with NaOH, fill the volume up to 100 mL with HPLC-grade water and sterile filtrate the medium using a filter with a pore size of 0.22 μm. For preculture medium add 0.01% sterile yeast extract and respective antibiotics. Prepare the medium freshly before use (*see* **Note 2**).

7. Phosphate buffered saline (1× PBS): For this study 10× PBS was purchased (e.g., Gibco) but PBS can also be prepared (*see* **Note 3**). Dilute 10× PBS with *A. dest.* to 1× PBS and autoclave the solution. Store PBS at room temperature.
8. Sodium dodecyl sulfate solution: 0.05% (w/v) SDS in HPLC-grade water.
9. TSB agar plates: Weigh TSB powder and agar for 1 L according to manufacturer's recommendation in a 1 L glass bottle. Add 1 L *A. dest.* and autoclave the mixture. Use the liquid TSB agar to pour plates in petri dishes. Store the dishes at 4 °C and let them warm-up to room temperature before use.
10. Trypsin-EDTA: 0.05% (w/v) trypsin, 0.02% (w/v) EDTA in 1× PBS. This cell culture ready-to-use solution is purchased by our laboratory (e.g., from PAN).
11. Sodium bicarbonate solution: 7.5% sodium bicarbonate. In our laboratory, this solution is purchased for use in cell culture (e.g., from PAN).
12. Neubauer improved counting chamber.
13. Accudrop fluorescent beads (BD Biosciences).
14. Cytometer Setup and Tracking Beads (BD Biosciences).

2.2 Sample Preparation for Mass Spectrometry

1. Ammonium bicarbonate solution: 20 mM ammonium bicarbonate. Dissolve 20 mg ammonium bicarbonate in 12.5 mL HPLC-grade water. Prepare the buffer fresh for each use.
2. Lysostaphin: 0.05 μg/μL lysostaphin in water. Dilute an aliquot of the lysostaphin stock solution (*see* Subheading 2.1, **item 3**) 1:100 with sterile HPLC-grade water. Divide the solution into aliquots of 5–10 μL and store them at −20 °C.
3. Trypsin: 0.1 μg/μL trypsin in ammonium bicarbonate solution. Fill a vial containing 20 μg lyophilized trypsin (*e.g.*, Promega) with 200 μL ammonium bicarbonate solution. Divide the solution in aliquots of 10–50 μL. Store the aliquots at −20 °C (*see* **Note 4**).
4. Trifluoroacetic acid. Spectroscopy grade.
5. Acetonitrile (ACN). HPLC grade or LC-MS grade.
6. Acetic acid, 100% ultra-quality.
7. Pipette tips with C18 material and a loading capacity of 5 μg (e.g., ZipTip C18, Millipore).

8. Acetonitrile solution 80%: 80% (v/v) acetonitrile in 1% (v/v) acetic acid. Mix 800 μL 100% (v/v) acetonitrile and 200 μL 5% (v/v) acetic acid freshly before use.
9. Acetonitrile solution 50%: 50% (v/v) acetonitrile in 1% (v/v) acetic acid. Mix 500 μL 100% (v/v) acetonitrile, 200 μL 5% (v/v) acetic acid and 300 μL HPLC-grade water freshly before use.
10. Acetonitrile solution 30%: 30% (v/v) acetonitrile in 1% (v/v) acetic acid. Mix 300 μL 100% (v/v) acetonitrile, 200 μL 5% (v/v) acetic acid and 500 μL HPLC-grade water freshly before use.
11. Acetic acid solution 1%: 1% (v/v) acetic acid in water, Mix 200 μL 5% (v/v) acetic acid and 800 μL HPLC-grade water freshly before use.

2.3 Files and Software for Data Analysis

1. Database file for peptide identification in FASTA-format containing all theoretical peptides, and peptide sequences, and annotation data of the target species and annotation data (*see* **Note 5**).
2. Software for conversion of raw data files: MS convert, which is part of the ProteoWizard open-source cross-platform tools (http://proteowizard.sourceforge.net/downloads.shtml) [16, 17].
3. Database search engine for peptide identification: Trans Proteomic Pipeline (TPP) of the Seattle Proteome Center (http://www.proteomecenter.org/software.php, *see* **Note 6**) [9, 10], Comet (http://comet-ms.sourceforge.net/) [11].
4. Software for data quantification: Skyline Targeted Proteomics Environment (https://skyline.gs.washington.edu/labkey/project/home/software/Skyline/begin.view?) [12].

2.4 Fluorescence Microscopy

1. Blocking solution: 10% (v/v) fetal bovine serum in PBS.
2. Fixation solution: 2% (v/v) formaldehyde in 1× PBS.
3. Hoechst 33258 staining solution: 0.5 μg/mL and 0.2 μg/mL Hoechst 33258 in *A. dest.* Prepare a stock solution with a concentration of 10 μg/mL in sterile *A. dest.* and store aliquots of the stock solution at −20 °C. Before use, dilute Hoechst 33258 stock solution 1:20 or 1:50 in 1× PBS to obtain a concentration between 200 and 500 ng/mL.
4. Alexa Fluor® 568 Phalloidin (Thermo Fisher Scientific) staining solution: 1 U/mL Alexa Fluor® 568 Phalloidin in methanol. Dissolve a vial Alexa Fluor® 568 Phalloidin in 1.5 mL methanol to obtain a concentration of 200 U/mL. Prepare aliquots of Alexa Fluor® 568 Phalloidin solution and store them at −20 °C. Before application, dilute an aliquot of Alexa

Fluor® 568 Phalloidin in 1× PBS to obtain a final concentration of 1 U/mL.

5. Mowiol mounting medium: 33% (w/v) glycerol, 13% (w/v) Mowiol, and 0.1% (w/v) p-phenylendiamine in 0.133 mol/L Tris pH 8.5. Dissolve 6 g glycerol and 2.4 g Mowiol in 6 mL *A. dest.* by shaking at room temperature (RT) for at least 2 h. Add 12 mL of 0.2 M Tris/HCl (pH 8.5) and shake the solution at 53 °C until substances are completely dissolved. Add 0.1% *p*-phenylenediamine as antifading reagent and shake again for 3 h. Store aliquots of the solution at −20 °C (*see* **Note 7**).
6. Phosphate buffered saline (1× PBS): *see* **item** 7 in Subheading 2.1.
7. Triton X-100: 1% (v/v) Triton X-100 stock solution in sterile HPLC-grade water. Dilute the stock solution on the day of the experiment 1:10 with HPLC-grade water, and sterile filtrate the solution using a filter with a pore size of 0.22 μm (*see* **Note 8**).

3 Methods

3.1 Cultivation of Bacteria and Host Cells

1. Determine the colony forming units per mL (CFU/mL) of *S. aureus* at an optical density at 600 nm (OD_{600}) of 0.4 in pMEM in three independent experiments (*see* **Note 9**). To this end, inoculate pMEM with an exponentially growing overnight culture to an OD_{600nm} of 0.05, and monitor growth photometrically until an OD_{600nm} of 0.4 is reached. Dilute 100 μL of the bacterial culture in 1× PBS in tenfold dilution steps, and plate 100 μL of the appropriate dilutions on TSB agar plates in replicates. Incubate the plates for 24 h at 37 °C. Count the colonies, and calculate the CFU/mL considering the respective dilution.
2. Seed the eukaryotic host cells 3 days in advance of the experiment into 12-well tissue culture plates in order to obtain a confluent cell layer on the day of the infection experiment. For S9 human bronchial epithelial cells, a cell suspension in eMEM with a concentration of 9×10^4 cells per mL is prepared. Into each well, 2 mL of the cell suspension is transferred. The cells are incubated at 37 °C and 5% CO_2 in a humidified incubator.
3. Pour TSB agar in petri dishes for the determination of intracellular bacterial CFU.
4. In the evening before the experiment bacterial precultures have to be inoculated in pMEM in dilution steps to have an exponentially growing preculture for inoculation of the main culture available on the next day.

3.2 Infection Experiment

1. Inoculate the pMEM main culture with an exponentially growing overnight culture of *S. aureus* to an OD_{600nm} of 0.05. Cultivate the bacteria to an OD_{600nm} of 0.4 (*see* **Notes 9** and **10**).
2. Meanwhile, determine the number of eukaryotic cells per well in two representative wells of a 12-well tissue culture plate. To this end, remove the eMEM of two wells, wash the cell layer once with 2 mL of 1× PBS, and add 500 μL of trypsin-EDTA solution to detach cells. Incubate the plate for 7 min at 37 °C and 5% CO_2 in a humidified incubator. Stop the trypsin reaction by adding 500 μL of eMEM. Carefully separate the eukaryotic cells by pipetting them up and down. Determine the cell concentration with the help of a Neubauer improved counting chamber, and calculate the number of eukaryotic cells per well.
3. Calculate the amount of bacterial culture required to infect the eukaryotic cells with a multiplicity of infection (MOI) of 25, i.e., 25 bacteria per eukaryotic cell (*see* **Note 11**).
4. Prepare the infection master mix (IMM) when the bacteria reach the OD_{600nm} of 0.4. IMM will be used in the next step (*see* **step 5**) for the infection of eukaryotic cells. For preparation of the IMM, mix the calculated volume of bacterial culture required for an MOI of 25 with 29 μL sodium bicarbonate solution per mL of bacterial culture added to the IMM. Fill up the mixture with eMEM to the favored volume (*see* **Note 12**).
5. Remove the medium from the eukaryotic cells, and add 2 mL of IMM to each well of the 12-well tissue culture plate. Coincubate the bacteria and the eukaryotic cells for 1 h at 37 °C and 5% CO_2 in a humidified incubator. During this time, the eukaryotic cells internalize the bacteria.
6. Collect bacteria for the non-adherent control from supernatants of a 12-well cell culture dish. After 1 h of incubation, transfer the supernatant of the eukaryotic cells with the remaining extracellular bacteria into a 15 mL tube. Add SDS to a final concentration of 0.05% (w/v), and incubate the sample for at least 7 min at 37 °C and 5% CO_2 in a humidified incubator with a loosely screwed cap (*see* **Notes 13** and **14**). After incubation pass the sample to the cell sorter for sorting of the bacteria for the non-adherent control (*see* Subheading 3.3) (*see* **Note 15**).
7. Kill the remaining extracellular bacteria. Therefore, remove the remaining IMM from the eukaryotic cells, and add to each well 2 mL of prewarmed (~37 °C) eMEM/lysostaphin, to lyse the noninternalized *Staphylococcus aureus* (*see* **Note 16**). Incubate the tissue culture plates for at least 30 min at 37 °C and 5% CO_2 in a humidified incubator. If sampling of the intracellular bacteria is desired later, the medium with the lysostaphin is not

removed from the cell layer until sampling is carried out (*see* **Note 17**).

8. Sampling of intracellular bacteria. Remove the supernatant at the desired sampling points post infection (p.i.) (e.g., 1.5–8.5 h), and wash the cell layer twice with 2 mL 1× PBS. To lyse the host cells, add 500 μL 0.05% (w/v) SDS solution to each well and incubate for 7 min at 37 °C and 5% CO_2 in a humidified incubator (*see* **Note 18**). Subsequently, detach and homogenize the lysed cells by extensive pipetting (*see* **Note 19**). Collect the cell lysate of each well in a tube on ice. Rinse four wells successively with 400 μL 1× PBS, and finally add the washing aliquot to the tube.
9. Determination of intracellular bacterial titer. Take 100 μL of the cell lysate from each sampling point, and serially dilute the lysate in 1× PBS. Plate 100 μL of the appropriate dilutions on TSB agar plates in triplicates, and incubate the plates at 37 °C for 24 h (*see* **Note 20**).

3.3 Enrichment of Intracellular Bacteria

For this step a flow cytometer with sorting function is needed. The description of the cell sorting follows the protocol for the use of an Aria III cell sorter. As the settings vary with the different instruments utilized the settings described here may have to be adapted.

1. Prior to cell sorting, verify and calibrate the performance of the instrument with Cytometer Setup and Tracking Beads, and adjust the drop delay with Accudrop fluorescent beads to ≥99% in the initial mode and ≥95% in the fine tune mode.
2. Excite the bacterial GFP with the blue laser at 488 nm, and detect the emitted light between 515 and 545 nm.
3. Generate a dot plot with the side scatter area (SSC-A) on the *Y*-axis and the forward scatter area (FSC-A) on the *X*-axis. Apply a threshold for the FSC of 600 and a FSC area scaling of 0.87. Place a gate around the population in the SSC versus FSC dot plot, and display this population in a dot plot with the GFP fluorescence intensity area on the *Y*-axis and the side scatter area (SSC-A) on the *X*-axis. Place a gate around the GFP positive population, and sort these events with a 70 μm nozzle in the "4-way-purity" sort mode with a sort rate between 1500 and 2000 cells per second at 4 °C. Sort the bacteria directly into wells of a 96-well plate containing a low protein binding filter membrane with a pore size of 0.22 μm (*see* **item 1** in Subheading 2.1). For internalized *S. aureus* the desired number of sorted bacteria per filter is between 1×10^6 and 5×10^6. For a better comparability of the samples use a constant number of sorted bacteria throughout the experiment.
4. While sorting, remove the sheath fluid of the well of the 96-well plate by applying 500–550 mbar pump pressure. This

allows the sorting of a few million bacteria on the same filter in a well.

5. Rinse the filter twice with 200 μL 1× PBS after sorting. Subsequently, cut out the filter and divide it into four pieces with a scalpel. Transfer the filter pieces with tweezers into a low protein binding reaction tube (*see* **Note 21**).

3.4 Sample Preparation for Mass Spectrometry

1. Digestion of intact *S. aureus* on a filter membrane. Add to each reaction tube containing filter pieces with sorted bacteria 20 μL of ammonium bicarbonate solution and 1.5 μL lysostaphin (*see* **Note 15**). Incubate the reaction solution for 30 min at 37 °C. Subsequently, add 10 μL of trypsin, and incubate the samples for 16–18 h at 37 °C (*see* **Note 22**).
2. Transfer the digestion solution to a new reaction tube. Stop the enzymatic reaction by adding trifluoroacetic acid to a final concentration of 0.1% and incubate for 5 min at 37 °C. Remove debris by centrifugation for 10 min at room temperature and 16,000 × *g*. Transfer the supernatant into a fresh reaction tube.
3. For peptide purification and concentration use a pipette tip with C18 material and equilibrate the C18 material stepwise. To do so, pipette three times 10 μL of 100% acetonitrile up and discard it in the waste, then five times the 80, 50, and 30% acetonitrile solutions, and as a last equilibration step pipette two times 10 μL of the 1% acetic acid solution up and discard it into waste. Afterwards, load the peptides onto the tip by pipetting the peptide solution 20 times up and down. Transfer with every pipetting step the peptide solution into a fresh reaction tube. Wash the tip by pipetting five times 10 μL of 1% acetic acid solution up and discard it into waste. Elute the peptides first in 5 μL of the 50% acetonitrile solution by pipetting up and down for 10 times and then in 5 μL of the 80% acetonitrile solution. Collect both eluates in a micro vial with a conical micro insert intended for use during nLC-MS/MS analysis. Concentrate the purified peptide solution in a speed vacuum centrifuge to a volume of 2 μL to remove the acetonitrile. Fill the volume of the peptide solution up with the LC-buffer A used for HPLC to a final volume of 10 μL (*see* **Note 23**).

3.5 nLC-MS/MS

In the framework of this protocol section, only general recommendations can be made, as the nLC-MS/MS measurements will probably be performed according to the standard procedures of the actual MS facility. Table 1 summarizes the settings used for MS analysis of internalized *S. aureus* cells on a Q Exactive™ Orbitrap mass spectrometer.

Table 1
Settings for the measurement of internalized *S. aureus* on a Q Exactive™ Orbitrap mass spectrometer (Thermo Scientific)

Step	Setting
nLC-separation	
Buffer system	Binary buffer system consisting of 0.1% acetic acid, 5% ACN (LC-buffer A) and 100% ACN in 0.1% acetic acid (LC-buffer B)
Trap column	75 μm inner diameter, packed with 3 μm C18 particles (Acclaim PepMap100, Thermo Scientific)
Analytical column	25 cm analytical column packed with 2 μm C18 particles (Acclaim PepMap RSLC, Thermo Scientific)
Gradient	120 min with a total run time of 145 min. The percentage for LC buffer A decreased during the run and the percentage for LC buffer B increased steadily: 2% (v/v) for 10 min, 2% to 25% (v/v) for 120 min, 25% to 40% (v/v) for 5 min, 40% to 90% (v/v) for 2 min, and 90% (v/v) for 5 min.
Analytical column temperature	40 °C
Flow rate	300 nL/min during elution
Mass spectrometry	
Instrument	Q Exactive mass spectrometer (Thermo Scientific)
Electrospray	via TriVersa NanoMate (Advion Biosciences, Norwich, UK)
Acquisition mode	Data-dependent, profile mode with positive polarity
Full scan MS	
Full scan mass range	300–1650 m/z
Resolution on MS level	70,000
AGC target	3×10^6
Maximum ion injection time	120 ms
MS/MS acquisition	
Resolution on MS/MS level	17,500
AGC target	2×10^5
Maximum ion injection time	120 ms
MS/MS scan acquisition	8 most intense ions with charge ≥2 from the survey scan
Fragmentation mode	Higher-energy collisional dissociation (HCD)
Intensity threshold	8.3×10^4
Normalized collision energy	27.5%
Monoisotopic precursor selection	Rejecting +1 and ≥+7 charged ions
Dynamic exclusion	30 s

1. Inject 8 μL of the purified peptide solution per sample (*see* **step 3** in Subheading 3.4) for analysis.
2. Perform reversed phase nano liquid chromatography. Samples are usually loaded via a trap column, and peptides are separated on an analytical column using a binary buffer system of slightly acidified HPLC-grade water with low percentage of organic solvent (LC-buffer A) and a slightly acidified organic solvent (LC-buffer B) (*see* **Note 23**). Often, linear gradients from 2 up to 25% LC-buffer B are used with gradient lengths of 1–2 h. Operating nLC in a column oven at 40 °C is recommended.
3. Perform tandem MS analysis. MS spectra are recorded online, i.e., the MS instrument is directly coupled to the nLC and acquires MS data continuously while peptides elute from the analytical column. The use of a modern, highly accurate MS instrument like the Q Exactive™ (Thermo Scientific) is recommended. Operate the MS instrument in profile mode with data-dependent acquisition (DDA) and a positive polarity. Record survey full-scan spectra, and acquire MS/MS scans for the eight most intense peaks. Dynamically exclude already targeted ions for MS/MS scans for 30 s.

3.6 Peptide Identification

The analysis of tryptic peptides from host cell-internalized bacteria using reverse phase nLC-MS/MS results in extensive raw data files. These files comprise information about retention times (i.e., the time when a peptide elutes from the LC-column, which is directly correlated to the peptide's physicochemical characteristics like hydrophobicity and the actual percentage of organic LC-solvent), information about peptide ion size and charge (MS level), and information about peptide ion fragments' sizes and charges (MS/MS level) for peptide identification. Additionally, they contain information about intensities of the detected peptide ions and peptide ion fragments. These data have to be integrated and processed in order to yield information about peptide identity and quantity and furthermore, after combination of peptide data, about protein identity and quantity. Identification is enabled by comparing recorded size data with theoretically possible size data collected in a specific database tuned to the experimental settings employed. During the identification via a database search the peptide identifications are assigned to precursor and fragment ion data from the nLC-MS/MS raw data files. Besides commercial options like MASCOT Server (Matrix Science Ltd., London, England) [18], Sorcerer (Sage-N Research Inc., CA, USA) or SEQUEST [19], freely available tools like TPP/Comet exist. All database search tools need as input on one hand the raw data and on the other hand a database file in FASTA-format containing all theoretical peptide sequences of the species of interest and, if available, annotation data. Some search engines require data transformation into universal or generic data formats prior to database search. Quantification can either be performed with peptide or protein intensities (e.g., sum of intensities) or by calculating the area under the curve (AUC) of a spectral peak. Here, we present an example workflow which uses freely available proteomics software tools such as the TPP/Comet analysis suite for protein identification and the Skyline software package for protein quantification.

1. Provide database file in FASTA-format, and install TPP and MS convert.
2. Convert raw data into the suitable data format mzXML using the MS convert tool. Add raw data files, define output directory, and select output format mzXML. Choose default settings for the other fields, and do not apply filters.
3. Prepare database search by generating a Comet parameter file. Download MS instrument-specific parameter file at http://comet-ms.sourceforge.net/parameters/, and edit the file using a text editor, e.g., NotePad++ (http://notepad-plus-plus.org/) in order to adjust the analysis-specific settings. Include database path and name, enable concatenated decoy search if interested in FDR results, and define mass accuracy thresholds according to the MS-instrument used for analysis (reasonable thresholds are for example 10 ppm precursor tolerance and 0.02 Da fragment mass tolerance (fragment_bin_tol) for Q Exactive™ data sets). Select trypsin as enzyme used during digestion, select two enzyme termini (fully digested), and allow up to two missed cleavages or set allowed missed cleavages to 0 for a more stringent analysis. Choose for example oxidation of methionine as variable modification (*see* **Note 24**).
4. Perform database search with suitable criteria using TPP and Comet. Login to TPP using "guest" as user name and password. In the "HOME" index card menu, select "Comet" as analysis pipeline to be used. Navigate to "Analysis Pipeline (Comet) → Database Search". Add mzXML data files from MS convert (*see* **step 2**), the edited Comet parameter file (*see* **step 3**), and the database file in FASTA-format (*see* **step 1**). Subsequently, start the database search, which will produce a pep.XML output file.
5. Analyze peptides using TPP and PeptideProphet. Navigate in the TPP to "Analysis Pipeline (Comet) → Analyze Peptides". Add the previously created pep.XML output files from the database search (*see* **step 4**), and create a results file name ending with ".interact.pep.XML". Choose criteria from Filter Options, e.g., minimal peptide length and PeptideProphet probability. However, it is suggested to only use loose filter criteria at this early step of the analysis pipeline and apply strict filters later on. For a basic analysis check only "RUN

PeptideProphet" from the following options menu. Optionally select "Use decoy hits to pin down the negative distribution. ..." in the PeptideProphet options menu, and enter the decoy indicating prefix used in the FASTA-database in the second part of that line "... Decoy Protein names begin with:" (*see* **Note 25**). Run Xinteract, which will create an interact.pep. XML file.

6. Save pep.XML output files (*see* **step 4**) and the interact.pep. XML file (*see* **step 5**) for subsequent analysis steps (*see* **Note 26**).

3.7 Quantification of Peptide and Protein Intensities

Raw data containing peptide/protein intensities or areas, search results of peptide identifications, and if available, annotation information need to be combined. Furthermore, raw data need to be processed into usable quantitative values in order to interpret the proteome analysis results under a biological point of view. We propose to use the freely available software tool Skyline for this step of data analysis (*see* **Note 27**).

1. Input of FASTA-database file into Skyline. Go to "File → Import → FASTA", and browse to the storage location of the appropriate FASTA-file, select FASTA-file.
2. Define settings for shotgun data. Go to "Settings → Transition Settings" and enable "full scan" on the corresponding index card menu. MS1 filtering corresponds to a standard data-dependent-acquisition analysis. Select "count" at the field "Isotope peaks included" in combination with a number of peaks, e.g., three, choose the correct MS-instrument type and its resolution at the fields "Precursor mass analyzer" and "Resolving Power". Set "MS/MS filtering" to "Acquisition method" "none". The alternative "DIA" in this field applies for data-independent acquisition data files, while the third option "targeted" will be used for Selected Reaction Monitoring (SRM) data. Leave other fields on default values (*see* **Note 28**).
3. Define peptide settings. Go to "Settings → Peptide Settings" and select "Trypsin [KR | P]", the desired number of allowed missed cleavages on the "Digestion" index card menu, and optionally choose variable modifications used in the search engine to be included in the analysis on the "Modifications" index card menu, e.g., oxidation of methionine. Leave other fields on default values. "Prediction" and "Filter" index card menus are not needed for the analysis of shotgun MS data.
4. Build a library of peptide identifications (*see* **Note 29**). Go to "Settings → Peptide Settings" and the "Library" index card menu. Select "Build" and define a name for the new library in the newly opened window. Skyline proposes a path for saving

the new library which at least has to be confirmed (*see* **Note 30**). The default cutoff score of 0.95 can be accepted or changed if required. Do not leave the field "Lab authority" empty. Continue with "Next" to the selection of files which should be included in the library. Choose all search results files (.interact.pep.XML files) of the current experiment using the functions "Add files", "Add directories", or "Add Paths" depending on the relevant data storage structure (*see* **Note 31**).

5. Import raw data information. Go to "File → Import → Results", select "single injection replicates", browse to storage location of appropriate raw files, and select and load raw files.
6. When all data loading steps are completed, Skyline offers a chromatogram of intensity vs. retention time for each peptide and a spectrum of intensity vs. m/z for each transition. Data can also be evaluated by comparing the retention time for each single peptide between all measurements (go to "View → Retention Time → Replicate Comparison") or the retention time alignment for all peptides (go to "View → Retention Time → Alignment"), as well as by comparing the peptide areas for each single peptide between all measurements (go to "View → Peak Areas → Replicate Comparison").
7. All quantitative values are available within Skyline. Export relevant values for further analysis steps. First, define a selection of parameters for the values that will be exported. Go to "File → Export → Report" and afterwards to "Edit list... → Add...". Define a report template name in the field "Report name". We propose to include the following parameters: Replicates → Replicate name (data set names); Protein → Protein name, Protein description, Protein accession (annotation information, *see* **Note 32**); Protein → Peptide → Peptide sequence, Missed cleavages, Peptide modified sequence (peptide characteristics; modified sequence only available when modifications had already been included in the database search); Protein → Peptide → Precursor → Peptide results → Retention time (retention on the nLC-column); Protein → Peptide → Precursor → Precursor results → Total Area (quantitative value on peptide level for each data set included in the analysis). Save report template, and use it afterwards for exporting data from the current analysis to a csv-file.

3.8 Data Normalization and Visualization

1. Combine quantitative values on peptide level to protein level. A commonly accepted way for fusing peptide values into a single value per protein is the calculation of the sum of all peptide values per protein. This step can be performed in spreadsheet applications, e.g., Excel or bioinformatics platforms like R after the peptide level data have been exported to a csv-file (*see* **step 7** in Subheading 3.7).

2. Normalize data in order to make quantitative protein level values of the current analysis directly comparable. Apply a median normalization. Calculate the median of the quantitative protein level values for each sample data set. Choose a target value, which will be used for normalization. This target value is usually a calculated median of all data sets, e.g., the median of all protein areas without missing values or a fixed value, e.g., 1. Calculate for each median of a sample data set the factor which adjusts the median value to the target value. Finally, divide each quantitative protein level value by the data set-specific adjustment factor. The normalization results in equal median values for all sample data sets. Now, quantitative values per protein can be compared between the individual data sets. Relevant changes in protein abundance can be accessed by statistical testing procedures, by applying fold change and replicate variation cutoff values, or by a combination of those methods.
3. For an integrated data interpretation different kinds of data visualizations can be applied, e.g., protein levels or levels of pathway-related functional groups of proteins can be displayed in bar charts or heat maps. Proteins with significant protein level changes can be mapped onto pathways, or the freely available and highly curated KEGG/Kyoto Encyclopedia of Genes and Genomes (http://www.genome.jp/kegg/) or Bio-Cyc Database Collection (http://biocyc.org/) can be used for interpretations of the proteomics data (Fig. 2).

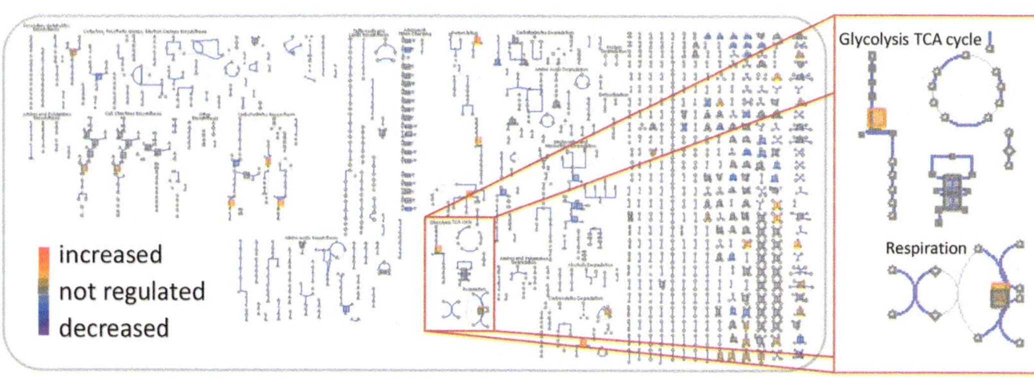

Fig. 2 BioCyc cellular pathways of *Staphylococcus aureus* HG001. The part of glycolysis, TCA cycle, and respiration (red box) is enlarged from the overview of all annotated pathways. Blue lines indicate proteins found in the current BioCyc database collection. With the toolbox "Overlay Experimental Data" also quantitative data (e.g., ratios of proteome analysis results) can be visualized. Orange rectangles indicate upregulation of the specific metabolic reaction while gray color labels nonregulated reactions. Downregulation of metabolic reactions (blue rectangles) is not included in the enlarged pathway details in this example data set

3.9 Fluorescence Microscopy of Intracellular Bacteria

For the visualization of internalized bacteria, fluorescence microscopy can be applied. As the bacteria express GFP, only the eukaryotic cells need to be stained with fluorescent dyes. To this end, F-actin is stained with Alexa Fluor® 568 Phalloidin, and DNA of the nuclei is stained with Hoechst 33258. An example of fluorescence microscopy images is shown in Fig. 3.

1. Place into each well of a 12-well tissue culture dish a sterile precision coverslip with a diameter of 18 mm. Seed the eukaryotic cells onto the coverslips as described in **step 2** of Subheading 3.1. The infection of the eukaryotic cells is carried out as described in Subheading 3.2.
2. Remove the supernatant of each well at the desired sampling points (e.g., 1.5–8.5 h p.i.), and wash the cell layer with 2 mL 1× PBS.
3. Fix the cells with 2 mL fixation solution for 20 min at room temperature, and subsequently wash the cell layer again with 2 mL 1× PBS.
4. Add 500 μL blocking solution to each well to block unspecific binding sites, and incubate the cells for 20 min at room temperature in the dark.
5. Remove the blocking solution, and add to each well 500 μL 0.1% (v/v) Triton X-100 to make the cell membrane permeable

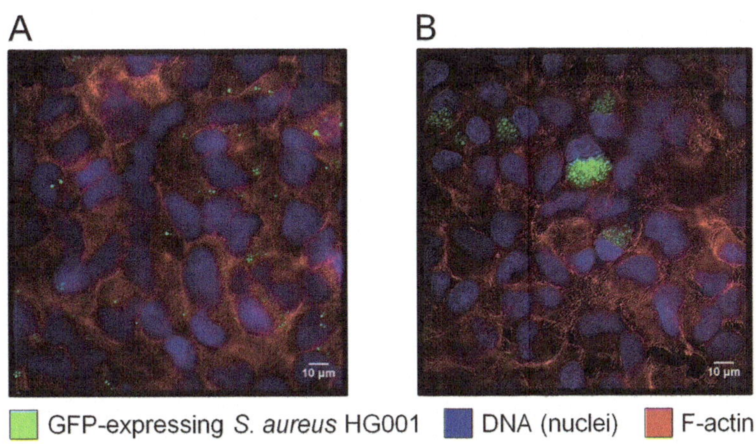

Fig. 3 Fluorescence microscopy images of *S. aureus* HG001 cells 2.5 h (**a**) and 6.5 h (**b**) post-infection in S9 human epithelial cells. *S. aureus* replicates inside the cells over time and forms cell clusters inside the epithelial cells after 6.5 h. S9 cells were grown and infected with *S. aureus* on glass coverslips. The F-actin of the S9 cells (stained with Alexa Fluor® 586 Phalloidin) is shown in red, and the DNA of the nuclei (stained with Hoechst 33258) is displayed in blue. The internalized *S. aureus* express GFP and thus, are depicted in green. Preparations were visualized on a Deltavision RT Image Restoration Workstation with an Olympus IX71 w/ Nomarski Differential Interference Contrast microscope with a 40x objective

for the following staining. Incubate the samples for 5 min at room temperature.

6. Remove the Triton X-100, and wash the cell layer twice with 2 mL 1× PBS. Subsequently, add 500 μL of the Alexa Fluor® 568 Phalloidin staining solution to each well. Incubate for 20 min at 37 °C.
7. Remove the Alexa Fluor® 568 Phalloidin staining solution, and wash the cell layer with 2 mL 1× PBS. Afterwards, add to each well 500 μL of the Hoechst 33258 staining solution and incubate for 10 min at room temperature in the dark.
8. Rinse the samples thoroughly three times with distilled water.
9. Pipette 6 μL of the Mowiol mounting medium centric on a clean microscope slide (*see* **Note 33**). Remove the coverslip with the cells from the well with tweezers and a hypodermic needle. Place the coverslip upside down in the Mowiol mounting medium, and let the slide dry for 16–18 h at room temperature in the dark.
10. To take images on a fluorescence microscope, the dyes have to be excited at their respective excitation wavelength, and the emission has to be detected at the respective emission wavelength. These settings can vary between the different microscopes used. In our fluorescence microscope Hoechst 33258 stained DNA was excited between 340 and 380 nm, and the emission was detected between 432 and 482 nm. The F-actin stained with Alexa Fluor® 568 Phalloidin was excited between 630 and 650 nm, and the emission was detected between 665 and 705 nm. The GFP of intracellular bacteria was excited between 470 and 510 nm, and the emission was detected between 509 and 547 nm.

4 Notes

1. eMEM is used for the cultivation of S9 human bronchial epithelial cells. If another cell line is used, the medium has to be adapted.
2. Please keep in mind that pMEM is a cell culture medium adapted to the requirements of *Staphylococcus aureus.* If another bacterium is used, the medium has to be adapted to its special requirements.
3. 10× PBS can also be prepared: weigh 80 g NaCl, 2 g KCl, 26.8 g $Na_2HPO_4 - 7\ H_2O$, 2.4 g KH_2PO_4, and add 800 mL A. dest. Adjust pH to 7.4 with HCl and fill up the volume to 1 L with A. dest. Sterilize PBS by autoclaving and store it at room temperature.

4. Make sure that all steps are carried out on ice, to prevent self-digestion of trypsin.
5. Annotation data help enormously for the interpretation of the data in the analysis step. For *S. aureus* a repository with detailed annotation information called AureoWiki is available at http://aureowiki.med.uni-greifswald.de/Main_Page.
6. The Seattle Proteome Center provides a help page with information about the TPP: http://tools.proteomecenter.org/wiki/index.php?title=Software:TPP
7. Alternatively to the self-prepared Mowiol mounting medium also purchased mounting medium can be used (e.g., Dako Fluorescent Mounting Medium, Agilent Technologies).
8. Triton X-100 is a viscous liquid. Either use inverse pipetting or weigh the wanted amount of Triton X-100 to obtain an exact concentration.
9. OD_{600nm} of 0.4 refers to the exponential growth phase of *S. aureus* HG001 in pMEM medium. This is the growth phase from which *S. aureus* was derived for infection of eukaryotic cells. The CFU/mL is needed to calculate the multiplicity of infection (MOI), i.e., the number of bacteria per eukaryotic cell, later in the infection experiment.
10. Do not add any antibiotics to the main culture. The antibiotics can affect the eukaryotic cells, when bacteria are added for infection.
11. This is an example how to calculate the volume of bacterial culture needed per well to reach an MOI of 25. Earlier determined parameters: CFU/mL at OD_{600nm} of 0.4 (e.g., 1×10^8) and the eukaryotic cell number per well (e.g., 6.7×10^5). To each well 2 mL of the infection master mix (IMM) are added.

    First, the required number of bacteria for an MOI of 25 is calculated: Wanted number of bacteria = MOI × eukaryotic cells per well = $25 \times 6.7 \times 10^5 = 1.68 \times 10^7$ bacteria per well. *Then, the volume of bacterial culture needed per well is calculated*: Volume of bacterial culture per well = number of bacteria per well/CFU/mL at OD_{600nm} of 0.4/volume of IMM added per well = 1.68×10^7 CFU/1×10^8 CFU/mL = 0.168 mL bacterial culture.

    This means that 2 mL of IMM must contain 168 μL of bacterial culture.
12. Example for the composition of 24 mL IMM, sufficient for one 12-well tissue culture plate, with the bacterial volume calculated in **Note 11**: Mix 2.01 mL bacterial culture (12 × 168 μL), 0.058 mL sodium bicarbonate solution (2.01 mL × 0.029 mL/mL of bacterial culture), and 21.9 mL eMEM. The sodium bicarbonate is added to stabilize

the pH of the IMM. For the preparation of the pMEM, 10× MEM is used, which does not contain sodium bicarbonate. For culturing bacteria, it is no problem that the sodium bicarbonate is not included because the HEPES fulfills the buffer function at environmental CO_2 concentrations. But when the IMM is prepared the sodium bicarbonate from the eMEM would be diluted, and the eukaryotic cells need the sodium bicarbonate for a stable pH in an atmosphere with 5% CO_2.

13. The addition of SDS to the control sample and the incubation time is to treat the control equally to the samples. The incubation time includes 7 min incubation to lyse eukaryotic cells plus the sampling time. The sampling time is a variable factor and varies with the amount of wells harvested at the different sampling points and with the experimenter.
14. We decided to use non-adherent bacteria collected after 1 h of coincubation with eukaryotic cells as control samples. Other possible controls are exponentially growing bacteria, which are used to prepare the infection master mix, or to incubate the infection master mix in a 12-well tissue culture dish without eukaryotic cells for 1 h (the time of the infection) at 37 °C and 5% CO_2 in a humidified incubator, to elucidate the influence of the serum present in the IMM and the incubation in an atmosphere with 5% CO_2.
15. Cell sorting has an influence on the bacteria. For comparison of control samples to samples of internalized bacteria, controls have to be treated equally to the internalized bacterial samples. Therefore, control samples have to be sorted as well.
16. Dilute the lysostaphin stock solution in eMEM freshly before use. Note that lysostaphin is exclusively used to lyse *Staphylococci*. If you are using another bacterium, you might take for example lysozyme to kill remaining extracellular bacteria.
17. After 30 min of incubation is the earliest time for sampling. Leave the lysostaphin medium on the cell layer and do not remove it until sampling of the intracellular bacteria is performed. Host cells may die and lyse during the progressing infection. Released bacteria would replicate in the medium, if lysostaphin would not be present. For monitoring the infection over several days, change the medium with fresh lysostaphin every 2 days.
18. Alternatively, also a 0.1% (v/v) Triton X-100 solution in HPLC-grade water can be used for the lysis of the eukaryotic cells. However, depending on the used bacteria for infection the suitability of SDS or Triton X-100 as well as the concentration of the solutions have to be tested.
19. Try to avoid the formation of bubbles while homogenizing the cell lysate.

20. As an alternative to the classical plating procedure, the intracellular bacterial titer can also be determined by flow cytometry, e.g., with Guava easyCyte (Merck Millipore) or BD Accuri C6 (BD Biosciences) cytometers. These cytometers are capable of determining absolute cell counts, volumes, and therefore, exact cell concentrations without the use of calibration beads.
21. Wear gloves while cutting out the filter and for all following steps where the samples are prepared for MS, to avoid contamination of samples with keratins from your skin.
22. Make sure that all filter pieces are covered with liquid. Otherwise the digestion might be incomplete.
23. Use the LC-buffer utilized at the local MS facility. In this example 0.1% acetic acid, 5% acetonitrile in HPLC-grade water is used as LC-buffer A and 0.1% acetic acid in 100% acetonitrile is used as LC-buffer B.
24. Modifications in fact occurring in the data set can be determined by an error-tolerant search prior to the identification search with the freely available version of Mascot (http://www.matrixscience.com/search_form_select.html). Though, there are some restrictions in the noncommercial Mascot version, e.g., searches for MS/MS data are limited to 1200 spectra and automated searching of batches of files is not permitted.
25. This decoy strategy can only be used if the FASTA-database contains decoy entries.
26. We recommend creating a file folder which contains all analysis-related data (raw-data, mzXML-files, FASTA-file, pep.XML- and interact.pep.XML-output files) which will also be used for the analysis steps in Skyline.
27. We recommend performing analysis on a 64-bit-computer because the data analysis often requires extensive computing power. All file and data import steps as well as the build library step can take considerable time.
28. Full scan settings must be enabled before loading search and raw data into the software.
29. Building a library is an essential step during shotgun data analysis with Skyline. Peptide identifications of all search result files will be available together with intensity and area data of the specific raw data sets in the subsequent analysis steps.
30. For confirming the path click on "browse" and in the next step on "save".
31. Acceptable search results file formats are *.dat, *.pep.xml, *.pepXML, *.mzid, *.xtan.xml, *.msf, *.group, *.group.xml, *.pride.xml, *.idpXML, *.sqt, *.ssl, *.perc.xml, *.results.xml, *.msms.txt, and *.final_fragment.csv. Additionally, BiblioSpec

Library files (*.blib) can be used, if a library has already been constructed before.

32. Only information already included in the database FASTA-file can be accessed at this step.
33. Try to avoid air bubbles while pipetting the mowiol mounting medium on the microscope slide.

Acknowledgments

This work was supported by the German Research Foundation Grant SFB/TR34.

References

1. Scott CC, Botelho RJ, Grinstein S (2003) Phagosome maturation: a few bugs in the system. J Membr Biol 193:137–152
2. Norrby S, Nord C, Finch R (2005) Lack of development of new antimicrobial drugs: a potential serious threat to public health. Lancet Infect Dis 5:115–119
3. Cash P (2008) Proteomics of bacterial pathogens. Expert Opin Drug Discov 3:461–473
4. Bumann D (2010) Pathogen proteomes during infection: a basis for infection research and novel control strategies. J Proteome 73:2267–2276
5. Wertheim HFL, Melles DC, Vos MC et al (2005) The role of nasal carriage in *Staphylococcus aureus* infections. Lancet Infect Dis 5:751–762
6. Garzoni C, Kelley WL (2009) *Staphylococcus aureus*: new evidence for intracellular persistence. Trends Microbiol 17:59–65
7. Sendi P, Proctor R (2009) *Staphylococcus aureus* as an intracellular pathogen: the role of small colony variants. Trends Microbiol 17:54–58
8. Pförtner H, Burian MS, Michalik S et al (2014) Activation of the alternative sigma factor SigB of *Staphylococcus aureus* following internalization by epithelial cells–an *in vivo* proteomics perspective. Int J Med Microbiol 304:177–187
9. Keller A, Eng J, Zhang N et al (2005) A uniform proteomics MS/MS analysis platform utilizing open XML file formats. Mol Syst Biol 1:2005.0017
10. Deutsch EW, Mendoza L, Shteynberg D et al (2010) A guided tour of the trans-proteomic pipeline. Proteomics 10:1150–1159
11. Eng JK, Jahan T, Hoopmann MR (2013) Comet: an open-source MS/MS sequence database search tool. Proteomics 13:22–24
12. MacLean B, Tomazela DM, Shulman N et al (2010) Skyline: an open source document editor for creating and analyzing targeted proteomics experiments. Bioinformatics 26:966–968
13. Becker D, Selbach M, Rollenhagen C et al (2006) Robust salmonella metabolism limits possibilities for new antimicrobials. Nature 440:303–307
14. Schmidt F, Scharf SS, Hildebrandt P et al (2010) Time-resolved quantitative proteome profiling of host-pathogen interactions: the response of *Staphylococcus aureus* RN1HG to internalisation by human airway epithelial cells. Proteomics 10:2801–2811
15. Surmann K, Michalik S, Hildebrandt P et al (2014) Comparative proteome analysis reveals conserved and specific adaptation patterns of *Staphylococcus aureus* after internalization by different types of human non-professional phagocytic host cells. Front Microbiol 5:392
16. Kessner D, Chambers M, Burke R et al (2008) ProteoWizard: open source software for rapid proteomics tools development. Bioinformatics 24:2534–2536
17. Chambers M, Maclean B, Burke R (2012) A cross-platform toolkit for mass spectrometry and proteomics. Nat Biotechnol 30:918–920
18. Perkins D, Pappin D, Creasy D et al (1999) Probability-based protein identification by searching sequence databases using mass spectrometry data. Electrophoresis 20:3551–3567
19. Eng JK, McCormack L, Yates JR (1994) An approach to correlate tandem mass spectral data of peptides with amino acid sequences in a protein database. J Am Soc Mass Spec 5:976–989

Part IV

Analysis of Post Translational Protein Modifications

Chapter 16

Phosphopeptide Enrichment from Bacterial Samples Utilizing Titanium Oxide Affinity Chromatography

Boumediene Soufi, Christoph Täumer, Maja Semanjski, and Boris Macek

Abstract

Mass spectrometry (MS)-based proteomics detected hundreds of phosphorylation sites on serine, threonine and tyrosine in numerous bacterial proteins, firmly establishing the presence and importance of this posttranslational modification in prokaryotes. Recent biological follow up of these results revealed that vital processes in bacterial cell, such as cell division, differentiation, spore germination and persistence, are regulated by protein phosphorylation, raising the need to study this modification on a global scale under additional physiological conditions. Due to low abundance and low stoichiometric levels of protein phosphorylation, initial protocols for phosphopeptide enrichment and analysis required relatively high amounts of starting material, extensive fractionation and MS measurement time. Here we present a protocol for phosphopeptide enrichment and detection based on TiO_2 chromatography and high resolution MS that enables in-depth detection and quantification of phosphorylation sites from significantly lower amounts of starting material and in a fraction of MS measurement time.

Key words Phosphoproteome, Phosphorylation, TiO_2 chromatography, Mass spectrometry, Phosphopeptide enrichment

1 Introduction

Phosphorylation is one of the most prominent and best studied protein modifications, widely distributed in all kingdoms of life. Proteins can be phosphorylated on reactive groups of various amino acid side chains, such as hydroxyl group (Ser, Thr, Tyr), acid group (Asp), amino group, or imidazole ring nitrogens (Lys, Arg, His). Two component systems, employing His and Asp phosphorylation, were long considered a hallmark of bacterial signalling [1]. More recently, attention has focused on phosphorylation of proteins on serine/threonine and tyrosine residues in bacteria [2, 3]. While the number of two component systems varies in proportion to genome size, Ser/Thr kinases seem to be most abundant in Actinomycetes and comparatively less abundant in other bacteria [4]. Interestingly, recent metagenomic studies

Dörte Becher (ed.), *Microbial Proteomics: Methods and Protocols*, Methods in Molecular Biology, vol. 1841,
https://doi.org/10.1007/978-1-4939-8695-8_16,

indicate that bacterial Ser/Thr protein kinases likely outnumber histidine kinases, traditionally considered as canonical bacterial kinases [5]. The importance of bacterial serine/threonine and tyrosine kinases has been established in regulation of processes such as the cell cycle [6], cell differentiation [7], stress response [8], exopolysaccharide production [9], coordination of cell division [10–12], virulence [13, 14], and persistence [15, 16]. Despite the extent and importance of His and Asp phosphorylation in bacterial signal transduction, its chemical instability at lower pH values (pH < 8) makes the analysis of this modification by mass spectrometry (MS) extremely difficult and, consequently, proteomics studies targeting Asp/His phosphorylation have so far been very rare [17, 18]. Conversely, phosphorylation on Ser/Thr/Tyr is acid-stable and therefore more accessible to global phosphoproteomics strategies. Here we provide a brief overview of the methodology used to study phosphoproteomes in a site-resolved manner.

1.1 Cell Lysis and Phosphopeptide Enrichment

Prior to phosphoproteomics analysis, bacterial cells need to be lysed in a way that enables the highest possible proteome retrieval. For comprehensive bacterial phosphoproteome analysis, proteins should be extracted and solubilized in a buffer containing enzymes that disrupt the cell wall (usually lysozyme) and a detergent. Proteins are subsequently digested in-solution, and the resulting peptide mixtures should be enriched for phosphopeptides prior to MS analysis. Efficient extraction of membrane proteins requires potent ionic detergents, such as sodium dodecyl sulfate (SDS), which inhibits commonly used proteases and is not compatible with MS. Recently, a sample preparation method was introduced that combines the advantages of the in-gel digestion (SDS compatibility) and in-solution digestion (higher peptide yields after protein digestion) [19, 20]. The method involves tissue homogenization/lysis in an SDS-containing denaturing buffer, which is followed by extensive washing, buffer exchange and protein digestion on a microfilter. Different versions of this method have recently been applied to bacterial (phospho)proteomics, and proved to be very useful in extraction of membrane proteins [21, 22].

The extracted bacterial proteome is commonly digested in solution by a protease (e.g., trypsin or endoproteinase Lys-C) and results in a complex peptide mixture. Since the phosphopeptides represent only a small fraction of the total peptides in the mixture, they need to be further enriched. This arguably presents the most important part of every global phosphoproteomics experiment. Many different strategies for up-front selective enrichment of phosphopeptides before MS analysis have been reported and extensively reviewed elsewhere [23]. Here we will briefly mention the most commonly used enrichment strategies.

1.1.1 Immobilized Metal Affinity Chromatography (IMAC)

Phosphopeptide enrichment by IMAC is based on the high-affinity coordination of the phosphate groups to certain trivalent metal ions, such as Fe^{3+}, Ga^{3+}, Al^{3+}, and Zr^{3+} [24]. Metal ions are immobilized by loading onto porous column packing material and are commercially available as ready-to-use kits.

1.1.2 TiO_2 Chromatography

Titanium dioxide (TiO_2) spheres have a high potential for capturing and selectively enriching for phosphopeptides [25]; however, they also retain acidic peptides containing glutamic and aspartic acid. To circumvent this problem, 2,5-dihydroxy benzoic acid (DHB) is usually used in the buffers for TiO_2 enrichment of phosphopeptides [26]. An excess of DHB efficiently outcompetes other carboxylic acids from interacting with TiO_2 while phosphopeptides are nonetheless retained. This significantly improves the selectivity of TiO_2 for phosphopeptides.

1.1.3 Antibody Pulldown of Tyrosine-Phosphorylated Peptides

Specific anti-p(Y) antibodies were for long successfully used in eukaryotic systems to study signal transduction based on tyrosine phosphorylation [27, 28]. Only recently, this workflow was applied to bacterial samples, revealing hundreds of tyrosine phosphorylation events in pathogenic *E. coli* strains [29].

A typical workflow for enrichment of phosphopeptides based on the methods mentioned above is depicted in Fig. 1. We note that this workflow is generic and can also be applied to eukaryotic samples. Due to higher abundance of protein phosphorylation in eukaryotes, typical enrichment efficiencies achieved in eukaryote samples are much higher than in prokaryote samples (Fig. 2), underscoring the challenges of bacterial phosphoproteomics. Strong cation exchange (SCX) chromatography was previously routinely used as a powerful phosphopeptide enrichment and fractionation step prior to IMAC or TiO_2 chromatographies [23, 30]; however, optimization of TiO_2 enrichment protocols and introduction of faster-scanning mass spectrometers have rendered this method unnecessary in current enrichment protocols. A different method, High-pH (HpH) reversed phase chromatography was recently shown to be a powerful prefractionation method in eukaryotic samples [31], but it was still not applied to prokaryotic samples.

1.2 LC-MS Measurement and Data Processing

After enrichment, phosphopeptides are mass-measured and fragmented in a mass spectrometer. Even if previously fractionated, phosphopeptide mixtures are usually very complex and have to be separated by liquid chromatography (LC) prior to MS analysis. Current LC-MS setups involve the use of nano-UHPLC columns (inner diameter 25–75 μm, length 10–50 cm and flow rates 100–200 nL/min), usually packed with reverse-phase C_{18} material (bead size 1.9 μm). These columns can be produced in form of packed fused silica emitters and directly mounted onto an LC-MS

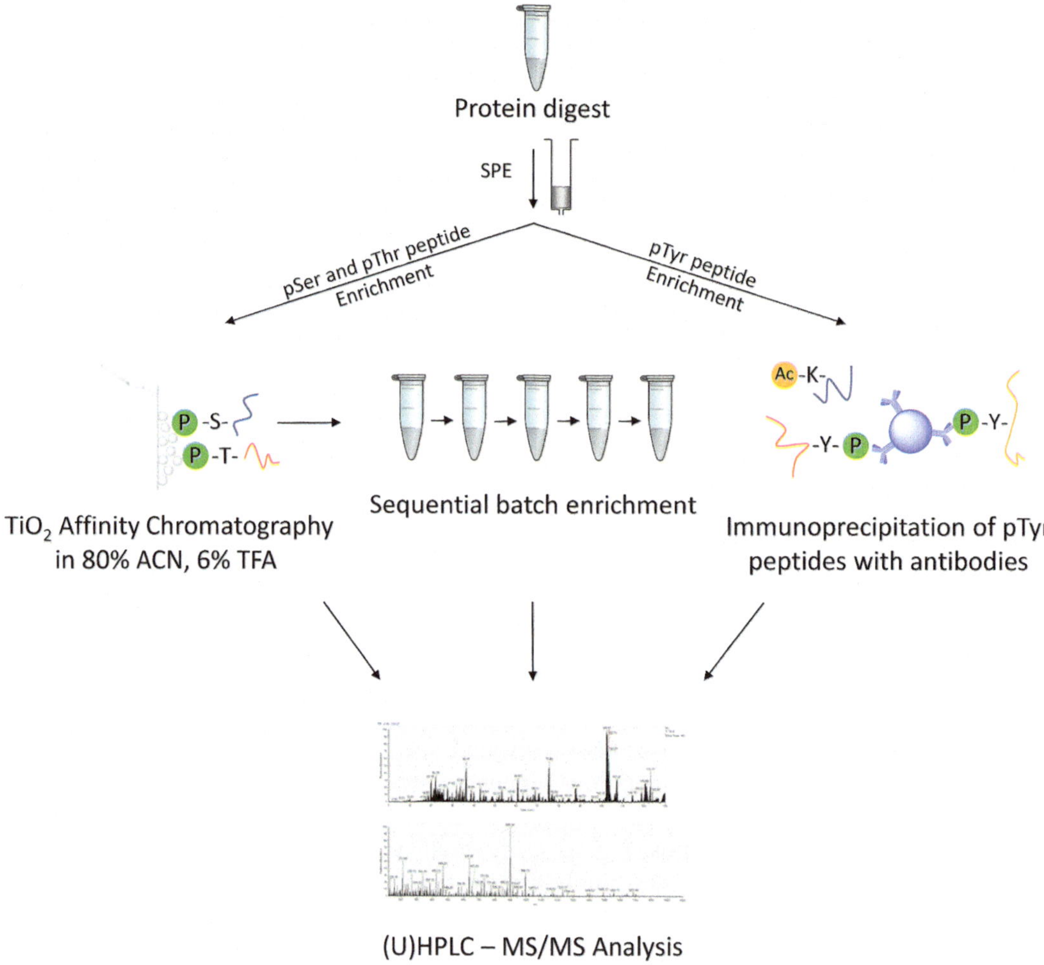

Fig. 1 Workflow for phosphopeptide enrichment. Protein digests are desalted by solid-phase extraction. Peptides phosphorylated on Ser and Thr residues are preferentially enriched using TiO_2 beads, whereas peptide phosphorylated on Tyr residues are specifically enriched using anti-(pY) antibodies (level of tyrosine-phosphorylated peptides is typically below 10% of all phosphopeptides). Alternately, pTyr peptides can first be pulled down following a buffer exchange and TiO_2 enrichment

interface. Such experimental setups enable electrospray ionization of peptides immediately upon elution from the column and therefore minimize dead volume that may result in peak broadening after separation [32]. This in turn results in a high peak capacity and chromatographic resolving power, which increases the sensitivity and dynamic range of LC-MS measurement. Alternatively, HPLC fractions may be spotted on a target plate, mixed with a matrix and ionized using MALDI ionization [33]. Different MS platforms and instruments currently in routine use in proteomics have been extensively reviewed elsewhere [34]. The most promising current development in MS is the recent introduction of fast scanning and high accuracy mass spectrometers, such as the new generation of

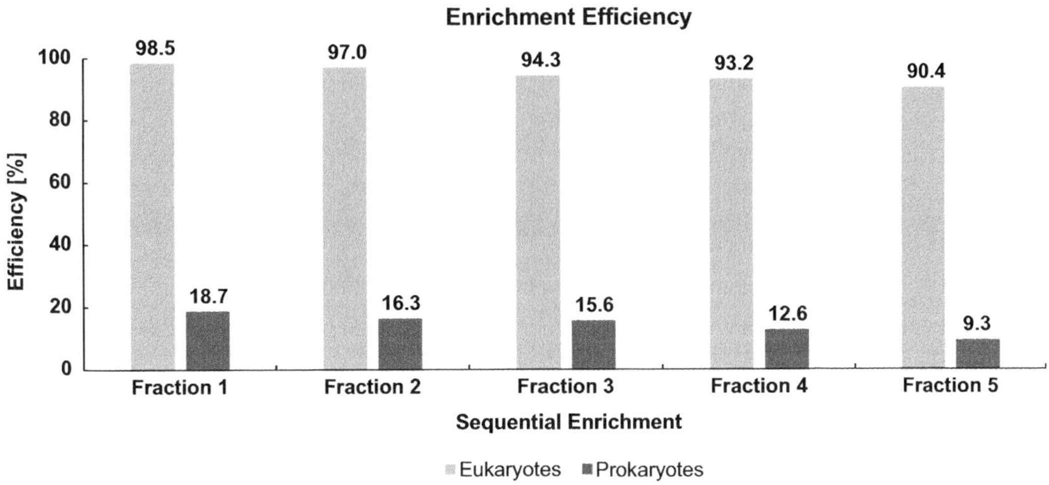

Fig. 2 Efficiency of phosphopeptide enrichment. The enrichment efficiency of phosphopeptides in eukaryotic samples can reach over 90%, whereas for prokaryotic samples the efficiency up to 20% or lower. Subsequent enrichment will increase the phosphoproteome coverage until the phosphopeptides in the sample are depleted

quadrupole-TOFs and LTQ-Orbitrap [35] mass spectrometers. The latter usually provide several fragmentation methods applicable to phosphopeptides; these include collision-induced dissociation (CID) with multistage activation (in the linear ion trap mass analyzer) [36], higher-energy CID (HCD) [37] and, if equipped with an additional chemical ionization source, electron transfer dissociation (ETD) [38]. These fragmentation methods can also be used in combination [39, 40] and each has specific advantages for analysis of protein phosphorylation and other posttranslational modifications; they are extensively discussed elsewhere [23].

MS data processing and protein database searches are equally important features of proteomic analyses. The latest generation of processing software uses high measurement mass accuracy of modern mass spectrometers to identify spectral features important for recalibration and quantification of stable isotope pairs, if these were used in the experiment [41]. Recalibrated and quantified peak lists are then submitted to a database search engine, which identifies peptides by comparing measured fragmentation spectra with theoretical mass spectra of all peptides in a protein database, such as the UniProt [42]. Most commonly used search engines in proteomics are Mascot [43], Sequest [44], and OMSSA [45]. The choice of the appropriate database is of great importance: specialized databases, consisting of forward and reversed (or randomized) protein sequences are nowadays commonly used for the estimation of false discovery rates of database searches, using a "target-decoy" approach [46]. Once identified, peptides are assembled into proteins.

A very important issue at the stage of data processing is the localization of the phosphorylation sites. Since the sequence information in the MS fragment spectra may not be sufficient to automatically assign the phosphorylation site by standard database search engines, the modern processing software suites compare spectra of identified peptides with theoretical spectra in which the phosphate group is placed at each possible sequence position. The probability scores for each position are normalized, and phosphorylation sites are then classified according to the localization score [47, 48]. The localization score provides a statistical means for assigning individual phosphorylation sites within the identified peptide sequences, and can also be applied to the analysis of other posttranslational modifications. Various resources for localization of phosphorylation sites from MS data have been reviewed elsewhere [49].

2 Materials

Prepare all solutions using ultrapure water (e.g., Milli-Q quality).

2.1 Preparation of Protein Extract and Determination of Protein Concentration

Lysis buffer (10 mL) (*see* **Note 1**): 4% (w/v) SDS (0.4 g/10 mL). 100 mM tetraethylammonium bromide (TEAB) or Tris–HCl, pH 8.6 (1 mL 1 M TEAB or Tris-HCl pH 8.6), 5 mM 2-glycerol phosphate. 5 mM sodium fluoride. 1 mM sodium orthovanadate. 10 mM EDTA (200 μL 0.5 M EDTA solution, pH 8.0). 1 tablet of cOmplete EDTA-free protease inhibitor cocktail (Roche).

Add water to 10 mL final volume.

1. Chloroform (purity >99.8% GC grade).
2. Methanol (purity >99.8% GC grade).
3. Bradford Protein Assay Dye Reagent Concentrate (e.g., Bio-Rad).
4. Filter paper (folded).
5. For protein standard measurements, bovine serum albumin (BSA) or lysozyme can be utilized.
6. Centrifuge Multifuge 1 S-R; rotor Sorvall TTH-400 (Heraeus).
7. Branson Sonifier 250; Microtip 5 (Fa. G. Heinemann).

2.2 Protein Digestion and Desalting of Peptides

1. Denaturation buffer: 6 M urea, 2 M thiourea in 10 mM Tris–HCl buffer, pH 8.0 (*see* **Note 2**).
2. Reduction buffer: 1 M dithiothreitol (DTT) in 50 mM ammonium bicarbonate.
3. Alkylation buffer: 550 mM iodoacetamide (IAA) in 50 mM ammonium bicarbonate.

4. 100 mM Tris–HCl, pH 8.0).
5. Lysyl endopeptidase LysC (e.g., Waco): stocks 0.5 μg/μL LysC in H_2O.
6. Trypsin (e.g., Promega): sequencing grade, modified, 0.5 μg/μL.
7. Methanol (purity >99.8% GC grade).
8. Solvent A: 0.1% (v/v) formic acid.
9. Loading solution: 80% (v/v) acetonitrile (ACN), 6% (v/v) trifluoroacetic acid (TFA).
10. Sep-Pak C18 Classic Cartridge, 360 mg sorbent (*see* **Note 3**).
11. Sep-Pak C18 1 cc Vac Cartridge, 100 mg sorbent (*see* **Note 4**).

2.3 Phosphotyrosine Enrichment by Immunoprecipitation

1. Cell Signaling PTM Scan® Phospho-Tyrosine Rabbit mAb (P-Tyr-1000) Kit.
2. Sigma Aldrich SigmaPrep™ Spin Columns SC1000-1KT.
3. 5× IP Buffer: 250 mM 3-(N-Morpholino)propanesulfonic acid (MOPS pH 7.2, 50 mM sodium phosphate, 250 mM sodium chloride.
4. 1× IP Buffer: 50 mM MOPS pH 7.2, 10 mM sodium phosphate, 50 mM sodium chloride.
5. 1× phosphate buffered saline (PBS Buffer): 137 mM NaCl, 2.7 mM KCl, 10 mM Na_2HPO_4, 2 mM KH_2PO_4.
6. 0.1% (v/v) TFA solution.
7. Vacuum centrifuge SpeedVac (Eppendorf).

2.4 Phospho–Enrichment and Elution Procedure

1. Zirchrom-TiO_2 5 μm, 300 Å (Zirchrom Separations Inc.) (*see* **Note 5**).
2. Solvent A: 0.1% (v/v) formic acid.
3. Solvent B: 80% (v/v) ACN, 0.1% formic acid.
4. 20% (v/v) TFA (98–100%).
5. Loading solution: 80% (v/v) ACN, 6% (v/v) trifluoroacetic acid (TFA).
6. Washing solution: 80% (v/v) ACN, 1% (v/v) TFA.
7. Elution solution 1: 1.25% (v/v) NH_4OH in water (pH 11).
8. Elution solution 2: 5% (v/v) NH_4OH, 60% (v/v) ACN, pH 11.
9. Elution solution 3: 60% (v/v) ACN, 1% (v/v) TFA.
10. Empore™ 47 mm C8 Disk (3M, Cat. No. 2214).
11. Methanol (purity >99.8% GC grade).
12. Tabletop centrifuge MiniStar (VWR).
13. Vacuum centrifuge SpeedVac (Eppendorf).

2.5 Peptide Cleanup on C18 Stage Tip

1. Methanol (purity >99.8% GC grade).
2. Solvent A*: 2% (v/v) ACN, 1% formic acid.
3. Solvent A: 0.1% (v/v) formic acid.
4. Solvent B: 80% (v/v) ACN, 0.1% (v/v) formic acid.
5. Empore™ 47 mm C18 Disk (3M, Cat. No. 2215).
6. Vacuum centrifuge SpeedVac (Eppendorf).

2.6 Liquid Chromatography Mass Spectrometry

1. HPLC solvent "A": 0.1% (v/v) formic acid.
2. HPLC solvent "B": 0.1% (v/v) formic acid, 80% (v/v) ACN in water.
3. 15 cm fused silica emitters with an inner diameter of 75 μm (new objective).
4. Reversed phase material for nano-HPLC: Reprosil-Pur C18-AQ, 3 μm resin (Dr. Maisch GmbH).
5. EASY-nLC II system (Thermo) or other HPLC system compatible with MS platform of choice.
6. High resolution MS instrument. For example, LTQ Orbitrap or Q Exactive instruments (Thermo).

2.7 Data Analysis

1. MaxQuant software suite [41, 50–52].

3 Methods

3.1 Cell Lysis

1. Spin cells until pelleted using centrifuge at approximately 5000 × *g* for 10 min at 4 °C (*see* **Note 6**).
2. Resuspend cells via slow pipetting with the appropriate amount of lysis buffer until cells are homogenized.
3. Sonicate at least three times for 30 s with a cycle of 0.4 s at a 40% amplitude to degrade DNA (*see* **Note** 7).
4. After sonication, spin intact cells down at 15,000 × *g* for 30 min and remove protein supernatant.

3.2 Chloroform/Methanol Protein Precipitation (See Note 8) and Determination of Protein Concentration

1. Mix the sample with 4 volumes of methanol and briefly vortex.
2. Add 1 volume of chloroform and briefly vortex.
3. Add 3 volumes of water and briefly vortex.
4. Spin 10 min at a minimum of 4000 × *g* in order to obtain adequate protein separation of the protein mixture.
5. Discard upper, aqueous phase (*see* **Note 9**).
6. Add 4 volumes of methanol and briefly vortex.
7. Spin 10 min at a minimum of 4000 × *g* in order to obtain adequate protein separation.

8. Remove methanol and allow protein precipitate to dry (*see* **Note 10**).
9. Dissolve the protein sample in the low volume of denaturation buffer such that a final protein concentration is up to 5 mg/mL.
10. Dilute Bradford Protein Assay-stock solution at a ratio of 1:5 using water.
11. Filter through folded filter paper (*see* **Note 11**).
12. Prepare standard calibration curve (using BSA, lysozyme etc.) and proceed with standard Bradford procedures.

3.3 Protein Digestion

1. Add reduction buffer to dissolved protein solution (denaturation buffer) to a final concentration of 1 mM DTT and incubate for 1 h at room temperature (*see* **Note 12**).
2. Add alkylation buffer to the sample to a final concentration of 5.5 mM IAA followed by a 1 h incubation period in the dark at room temperature.
3. Add 1 μg of lysyl endopeptidase LysC per 100 μg protein and incubate for 3 h at room temperature.
4. Dilute sample with 4 volumes of 100 mM Tris pH 8.0.
5. Add 1 μg trypsin (or LysC) per 100 μg protein and incubate overnight at room temperature (*see* **Note 13**).
6. After digestion is complete, acidify the sample with TFA to a pH around 2 (*see* **Note 14**).
7. Incubate the sample for 10 min at room temperature while shaking.
8. Centrifuge for 5 min to remove any participate at 15,000 × *g*.

3.4 Desalting of Peptides

1. Use a Sep-Pak Waters C18 Cartridge (*see* **Notes 14** and **15**).
2. Activate column with 1 mL of methanol. Use 5 mL of following solvents if using Sep-Pak C18 Classic Cartridge.
3. Wash/equilibrate with 1 mL of solvent A.
4. Add protein sample.
5. Wash with 1 mL of solvent A.
6. Elute sample with loading solution directly into an Eppendorf tube (*see* **Note 16**).
7. Take same eluted loading solution, and re-elute through the Sep-Pak C18 column into the same Eppendorf tube used in **step 6**.

3.5 Phospho-Tyrosine Enrichment by Immunoprecipitation (See Note 17)

1. Vacuum centrifuge (SpeedVac) samples (no heating) to a volume of 1.2 mL (*see* **Note 18**).
2. Add 2 mL of 5× IP buffer and then add water to a final volume of 10 mL (*see* **Note 19**).
3. Wash the antibody-bound beads three times with 1x PBS buffer.
4. Mix the antibody-bead slurry with 800 μL sample and transfer everything to the Falcon tube.
5. Incubate for 2 h on a spinning carousel.
6. Transfer sample to a Spin Column by pipetting 700 μL sample at a time.
7. Spin for approximately 3 s to allow liquid to flow through the column in a table-top centrifuge.
8. Collect the flow through in a falcon tube (*see* **Note 20**).
9. Repeat **steps 6** and **7** until the complete sample has been passed through Spin Column (the column is now loaded with the beads).
10. Wash the beads 3× with 800 μL of 1× IP buffer.
11. Wash the beads 1× with 800 μL of water.
12. Incubate for 5 min at room temperature with 50 μL of 0.1% (v/v) TFA solution and elute in a new Eppendorf tube (*see* **Note 21**).
13. Repeat **step 12** twice.
14. Perform MS measurements directly after elution from stage tips as described in Subheading 3.8.

3.6 Phospho-Enrichment Procedure

1. Prepare a stock solution containing TiO_2 beads in an Eppendorf tube (*see* **Note 22**).
2. Add appropriate amount of TiO_2 beads from the stock solution to the empty tube (*see* **Note 23**).
3. Incubate peptide solution with TiO_2 beads at room temperature for 10 min on a spinning carousel (*see* **Note 24**).
4. Spin TiO_2 beads down.
5. Remove the supernatant and transfer it to a new tube containing TiO_2 beads and repeat **steps 2–4** for several consecutive rounds. Meanwhile proceed with **steps 6–8** (*see* **Notes 25** and **26**).
6. Wash pelleted TiO_2 beads with 200 μL of loading solution by resuspending and spinning down. Discard the supernatant.
7. Resuspend TiO_2 beads in 100 μL of loading solution.
8. Pipet a bead solution onto a stage tip containing C8 material (2× C8 disks) (*see* **Note 26**).
9. Wash the stage tip with 100 μL of washing solution (*see* **Note 27**).
10. Remove the flow-through from the tube and perform an additional round of centrifugation.

3.7 Elution Procedure

1. Prepare a 2 mL Eppendorf tube (for each enrichment), and add 20–30 μL of 20% (v/v) TFA into each tube.
2. Elute sample with 30 μL of elution solution 1 (*see* **Note 28**).
3. Elute with 70 μL of elution solution 2.
4. Elute with 20 μL of elution solution 3.
5. Vacuum centrifuge (SpeedVac) samples (no heating) for 15–20 min to a volume of around 70 μL and add 100 μL of solvent A.
6. Adjust pH to <2 using 20% TFA and proceed with stage tipping (below).

3.8 Peptide Cleanup on C18 Stage Tip

1. Activate stage tip with 100 μL of methanol.
2. Wash/equilibrate with 200 μL of solvent A* (*see* **Note 29**).
3. Add sample.
4. Wash with 200 μL of solvent A (*see* **Note 30**).
5. Elute with 50 μL of solvent B directly into an Eppendorf tube.
6. Lower the volume in SpeedVac (no heating) for approximately 10–12 min to around 5 μL of sample.
7. Add 0.5 μL of solvent A* to the sample and adjust the volume to 5.5 μL using solvent A.

3.9 Mass Spectrometry Analysis

Peptide samples are analyzed using online nanoflow liquid chromatography tandem mass spectrometry (LC-MS/MS). In our laboratory, nanoLC-MS/MS experiments are performed on an EASY-nLC system connected to an LTQ-Orbitrap Elite (Thermo Scientific, Bremen, Germany) through a nano-electrospray ion source.

1. Load peptides directly onto a 15 cm long 75 μm-inner diameter analytical column packed with reversed-phase C18 Reprosil AQUA-Pur 3 μm particles using "Intellifow" setting (variable flow rate up to a maximum pressure of 280 bar).
2. Reduce flow rate to 200 nL/min after loading, and then separate peptides with a segmented linear gradient of solvent B from 5-33-50% (v/v) in 0.1% (v/v) formic acid for 150 min.
3. Use the following MS settings on the LTQ-Orbitrap Elite: target values 1E6 charges (Orbitrap analyzer) and 5E3 charges (Linear ion trap). Operate the instrument in the positive ion mode, with the following acquisition cycle: a full scan measured using the Orbitrap cell at resolution R 120,000 followed by MS/MS of the top 15 most intense peptide ions in the HCD collision cell and recorded at resolution R 15,000 in the Orbitrap cell.

3.10 Data Processing and Analysis

1. Use the latest version of MaxQuant software (*see* **Note 31**).

4 Notes

1. Alternative methods of bacterial cell lysis can be utilized, such as the commercially available mild detergent based YPER buffer from Thermo Scientific, which could be more suitable depending on the species of bacteria being analyzed. Care should also be taken if preferred lysis buffer contains enzymes such as DNase or lysozyme. When used in large quantities, these enzymes can lead to an over estimation of the protein concentration measurements. All solvents used throughout this section and protocol should be prepared with deionized Milli-Q water.
2. All buffers in this section can be aliquoted and stored at −20 °C for several months.
3. Sep-Pak C18 Classic Cartridges, 360 mg sorbent with a capacity of up to 5% (w/w) of the sorbent weight. They can be used for up to approximately 18 mg of protein extract.
4. Sep-Pak C18 Vac 1 cc Cartridges, 100 mg sorbent with a capacity of up to 5% (w/w) of the sorbent weight. They can be used for up to approximately 5 mg of crude protein extract.
5. As various commercially available TiO_2 beads exist, it is important to ensure that the TiO_2 beads used are the ones recommended in this protocol. Since a high percentage of TFA is employed, some beads are unstable at this low pH range, and will typically disintegrate during the enrichment process. The specific beads used here were tested and are compatible with this protocol.
6. Cell lysis can be performed immediately after centrifugation or, alternatively, cell pellet can be stored at −80 °C.
7. Sonication time and amplitude can vary depending on bacterial species. Moreover, if access to sonicator is not possible, benzonase can be used as a replacement.
8. Other forms of protein precipitation can be utilized such as via acetone precipitation or dialysis against water followed by lyophilization.
9. A visible disc of protein precipitate will be formed between hydrophobic and hydrophilic organic phases.
10. Avoid excessive drying of crude protein extract as solubility issues when dissolving in denaturation buffer may arise.

11. Store filtered Bradford reagent in a glass bottle at room temperature in the dark, check for any aggregates before using and discard if aggregates are found.
12. Due to the fact that bacteria contain a much lower frequency of phosphorylation sites with generally a low occupancy level compared to their eukaryotic counterparts, we recommend using at least 5 mg of protein starting material for this procedure. In case this protocol is applied to eukaryotic samples, we recommend using at least 1 mg of protein starting material. In case of phosphotyrosine enrichment by immunoprecipitation we recommend 15–20 mg of starting protein material.
13. Other proteases may also be used depending on the desired protein cleavage, or experiment being performed.
14. For the Sep-Pak C18 Classic Cartridge, the best way to operate the cartridge is by using a syringe (e.g., Injekt® 10 mL Luer Solo, Braun). Draw up 5 mL into the syringe and use the syringe to push the solution through the cartridge. Ensure that the cartridge does not dry out—this is critical until the sample has been loaded.
15. For the Sep-Pak C18 Vac 1 cc Cartridge, the liquid can be easily pipetted into the cartridge. To push the solution through the cartridge use a 1000 μL pipette and cut the lower end of the pipette tip so that it fits onto the cartridge. Ensure that the cartridge does not dry out—this is critical until the sample has been loaded.
16. Elution volume will depend on the initial volume of the sample prior to desalting. A typical recommended maximum elution volume is 1.8 mL in order to be compatible with Eppendorf tubes used. In case of phosphotyrosine immunoprecipitation elute with 6 mL of 80% (v/v) ACN into a 15 mL Falcon tube and continue with Subheading 3.5 Phosphotyrosine enrichment by Immunoprecipitation. Split the sample into 6× 1 mL fractions in 2 mL Eppendorf tubes for the vacuum centrifugation (SpeedVac) step.
17. The enrichment of phosphorylated tyrosine peptides is based on the commercially available antibody as part of the PTMScan Kit from Cell Signaling technology. We have slightly modified the protocol recommended by Cell Signaling due to the fact that we use different buffers for the lysis of bacterial cells, and during the optimization of this protocol, peptide lyophilization was unavailable in our laboratory. Alternatively, if one has access to a lyophilizer, the protocol that is provided in the PTMScan kit should be used and followed, otherwise the procedures described in this protocol can be utilized.
18. Sample can be split into 1 mL fractions in 2 mL Eppendorf tubes and evaporated using a SpeedVac to an approximate

volume of 200 μL/fraction. According to **Note 17** eluting with 6 mL will result in 6× 200 μL aliquots and a final total volume of 1.2 mL.

19. Collect all fractions in a 15 mL Falcon tube and once collected, the remainder of the enrichment procedure should be performed in a cold room at 4 °C.
20. Sample collected from the flow through can be used for another pull-down or alternatively can be used for the general phosphopeptide enrichment using the 6% TFA in 80% ACN approach (after desalting of peptides as listed in Subheading 3.4).
21. During the 5 min incubation step, the lower end of the column should be blocked in order to avoid loss of solution through the column.
22. In order to determine the optimal bead to peptide ratio for bacteria, we tested a broad range of possibilities. To this end, in the cases *of Bacillus subtilis* and *Escherichia coli*, the optimal ratio to use is a 1:10 bead to peptide ratio, for eukaryotes (tested using Hela cells) a 1:2 bead to peptide ratio produced the best enrichment efficiency. As the optimal ratios can vary depending on the bacteria used, we recommend performing an initial series of tests varying the bead to peptide amounts in order to ensure the best results possible. TiO_2 beads should be prepared fresh for each experiment and discarded afterward. The beads should be resuspended in a certain volume of loading solution depending on how many enrichment rounds are scheduled to be performed. For example, 20 mg of beads can be resuspended in 1 mL of loading solution. If 5 mg of peptides are present, then 25 μL of bead slurry will be added for each consecutive enrichment.
23. Bead slurry should be thoroughly mixed using a vortex, followed by immediate addition to the empty tube as TiO_2 beads settle down to bottom of the tube rapidly.
24. Ensure that the incubation time does not past 10 min. The beads will eventually become unstable the longer the incubation. Moreover, 10 min is enough time to allow for sufficient binding of the phosphopeptide to the bead material.
25. Typically a total of 5–10 consecutive rounds of enrichment are performed. For initial testing, we recommend performing at least 10 consecutive rounds of enrichment followed by MS analysis in order to determine the most practical scenario.
26. While performing the incubation steps, the C8 stage tips that are loaded with TiO_2 beads should be stored at 4 °C. It is very important to ensure that two C8 disks are used per stage tip in order to minimize the risk of TiO_2 particulates entering the tube during elution.

27. This wash step is important for the enrichment procedure as it improves the enrichment efficiency by removing nonphosphorylated peptides that have bound to the beads (for example acidic peptides). The eluted liquid from this wash step can be analyzed by MS in order to show that most peptides identified are nonphosphorylated peptides, meaning that the loss of phosphopeptides during this step is extremely low or nonexistent.
28. Elute peptides by centrifugation at lowest speed to ensure that each elution step takes approximately 10 min depending on the volume of elution buffer used.
29. Take extra caution to ensure that the stage tip does not become dry (until the sample is loaded onto the tip).
30. Stage tip can be stored at 4 °C until measurement.
31. MaxQuant Software can be downloaded at www.maxquant.org. The following publication can be used as a step-by-step guide on how to process the acquired MS files both for post-translational modification (PTM) analysis, as well as a guide for quantitative proteomic analysis using stable isotope labels [52]. It is important to note that there are other programs available that could be used as an alternative for downstream analysis.

Acknowledgments

We thank Dr. Olaf Voolstra for critical reading of the manuscript. Our work is supported by the SFB766 of the Deutsche Forschungsgemeinschaft and PRIME-XS consortium.

References

1. Mitrophanov AY, Groisman EA (2008) Signal integration in bacterial two-component regulatory systems. Genes Dev 22(19):2601–2611
2. Deutscher J, Saier MH Jr (2005) Ser/Thr/Tyr protein phosphorylation in bacteria - for long time neglected, now well established. J Mol Microbiol Biotechnol 9(3–4):125–131
3. Dworkin J (2015) Ser/Thr phosphorylation as a regulatory mechanism in bacteria. Curr Opin Microbiol 24(0):47–52
4. Wehenkel A, Bellinzoni M, Grana M, Duran R, Villarino A, Fernandez P, Andre-Leroux G, England P, Takiff H, Cervenansky C, Cole ST, Alzari PM (2008) Mycobacterial Ser/Thr protein kinases and phosphatases: physiological roles and therapeutic potential. Biochim Biophys Acta 1784(1):193–202
5. Kannan N, Taylor SS, Zhai Y, Venter JC, Manning G (2007) Structural and functional diversity of the microbial kinome. PLoS Biol 5(3): e17
6. Petranovic D, Michelsen O, Zahradka K, Silva C, Petranovic M, Jensen PR, Mijakovic I (2007) *Bacillus subtilis* strain deficient for the protein-tyrosine kinase PtkA exhibits impaired DNA replication. Mol Microbiol 63 (6):1797–1805
7. Shah IM, Laaberki MH, Popham DL, Dworkin J (2008) A eukaryotic-like Ser/Thr kinase signals bacteria to exit dormancy in response to peptidoglycan fragments. Cell 135 (3):486–496
8. Klein G, Dartigalongue C, Raina S (2003) Phosphorylation-mediated regulation of heat

shock response in *Escherichia coli*. Mol Microbiol 48(1):269–285

9. Lacour S, Bechet E, Cozzone AJ, Mijakovic I, Grangeasse C (2008) Tyrosine phosphorylation of the UDP-glucose dehydrogenase of *Escherichia coli* is at the crossroads of colanic acid synthesis and polymyxin resistance. PLoS One 3(8):e3053
10. Molle V, Kremer L (2010) Division and cell envelope regulation by Ser/Thr phosphorylation: mycobacterium shows the way. Mol Microbiol 75(5):1064–1077
11. Fleurie A, Lesterlin C, Manuse S, Zhao C, Cluzel C, Lavergne JP, Franz-Wachtel M, Macek B, Combet C, Kuru E, VanNieuwenhze MS, Brun YV, Sherratt D, Grangeasse C (2014) MapZ marks the division sites and positions FtsZ rings in *Streptococcus pneumoniae*. Nature 516(7530):259–262
12. Fleurie A, Manuse S, Zhao C, Campo N, Cluzel C, Lavergne JP, Freton C, Combet C, Guiral S, Soufi B, Macek B, Kuru E, VanNieuwenhze MS, Brun YV, Di Guilmi AM, Claverys JP, Galinier A, Grangeasse C (2014) Interplay of the serine/threonine-kinase StkP and the paralogs DivIVA and GpsB in pneumococcal cell elongation and division. PLoS Genet 10 (4):e1004275
13. Morona JK, Miller DC, Morona R, Paton JC (2004) The effect that mutations in the conserved capsular polysaccharide biosynthesis genes cpsA, cpsB, and cpsD have on virulence of *Streptococcus pneumoniae*. J Infect Dis 189 (10):1905–1913
14. Basell K, Otto A, Junker S, Zuhlke D, Rappen GM, Schmidt S, Hentschker C, Macek B, Ohlsen K, Hecker M, Becher D (2014) The phosphoproteome and its physiological dynamics in *Staphylococcus aureus*. Int J Med Microbiol 304(2):121–132
15. Schumacher MA, Piro KM, Xu W, Hansen S, Lewis K, Brennan RG (2009) Molecular mechanisms of HipA-mediated multidrug tolerance and its neutralization by HipB. Science 323(5912):396–401
16. Germain E, Castro-Roa D, Zenkin N, Gerdes K (2013) Molecular mechanism of bacterial persistence by HipA. Mol Cell 52(2):248–254
17. Lapek JD, Tombline G, Friedman AE (2010) Mass spectrometry detection of histidine phosphorylation on NM23-H1. J Proteome Res 10 (2):751–755
18. Zu XL, Besant PG, Imhof A, Attwood PV (2007) Mass spectrometric analysis of protein histidine phosphorylation. Amino Acids 32 (3):347–357
19. Wisniewski JR, Zougman A, Nagaraj N, Mann M (2009) Universal sample preparation method for proteome analysis. Nat Methods 6 (5):359–362
20. Manza LL, Stamer SL, Ham AJ, Codreanu SG, Liebler DC (2005) Sample preparation and digestion for proteomic analyses using spin filters. Proteomics 5(7):1742–1745
21. Soufi B, Macek B (2015) Global analysis of bacterial membrane proteins and their modifications. Int J Med Microbiol 305(2):203–208
22. Soufi B, Macek B (2014) Stable isotope labeling by amino acids applied to bacterial cell culture. Methods Mol Biol 1188:9–22
23. Macek B, Mann M, Olsen JV (2009) Global and site-specific quantitative phosphoproteomics: principles and applications. Annu Rev Pharmacol Toxicol 49:199–221
24. Posewitz MC, Tempst P (1999) Immobilized gallium(III) affinity chromatography of phosphopeptides. Anal Chem 71(14):2883–2892
25. Pinkse MWH, Uitto PM, Hilhorst MJ, Ooms B, Heck AJR (2004) Selective isolation at the femtomole level of phosphopeptides from proteolytic digests using 2D-nanoLC-ESI-MS/MS and titanium oxide precolumns. Anal Chem 76(14):3935–3943
26. Larsen MR, Thingholm TE, Jensen ON, Roepstorff P, Jorgensen TJD (2005) Highly selective enrichment of phosphorylated peptides from peptide mixtures using titanium dioxide microcolumns. Mol Cell Proteomics 4 (7):873–886
27. Blagoev B, Ong SE, Kratchmarova I, Mann M (2004) Temporal analysis of phosphotyrosine-dependent signaling networks by quantitative proteomics. Nat Biotechnol 22(9):1139–1145
28. Schmelzle K, Kane S, Gridley S, Lienhard GE, White FM (2006) Temporal dynamics of tyrosine phosphorylation in insulin signaling. Diabetes 55(8):2171–2179
29. Hansen AM, Chaerkady R, Sharma J, Diaz-Mejia JJ, Tyagi N, Renuse S, Jacob HK, Pinto SM, Sahasrabuddhe NA, Kim MS, Delanghe B, Srinivasan N, Emili A, Kaper JB, Pandey A (2013) The *Escherichia coli* phosphotyrosine proteome relates to core pathways and virulence. PLoS Pathog 9(6):e1003403
30. Beausoleil SA, Jedrychowski M, Schwartz D, Elias JE, Villen J, Li JX, Cohn MA, Cantley LC, Gygi SP (2004) Large-scale characterization of HeLa cell nuclear phosphoproteins. Proc Natl Acad Sci U S A 101(33):12130–12135
31. Kelstrup CD, Jersie-Christensen RR, Batth TS, Arrey TN, Kuehn A, Kellmann M, Olsen JV

(2014) Rapid and deep proteomes by faster sequencing on a benchtop quadrupole ultra-high-field Orbitrap mass spectrometer. J Proteome Res 13(12):6187–6195
32. Ishihama Y, Rappsilber J, Andersen JS, Mann M (2002) Microcolumns with self-assembled particle frits for proteomics. J Chromatogr A 979(1–2):233–239
33. Hillenkamp F, Karas M, Beavis RC, Chait BT (1991) Matrix-assisted laser desorption/ionization mass spectrometry of biopolymers. Anal Chem 63(24):1193A–1203A
34. Ahmed FE (2008) Utility of mass spectrometry for proteome analysis: part I. Conceptual and experimental approaches. Expert Rev Proteomics 5(6):841–864
35. Olsen JV, de Godoy LM, Li G, Macek B, Mortensen P, Pesch R, Makarov A, Lange O, Horning S, Mann M (2005) Parts per million mass accuracy on an Orbitrap mass spectrometer via lock mass injection into a C-trap. Mol Cell Proteomics 4(12):2010–2021
36. Schroeder MJ, Shabanowitz J, Schwartz JC, Hunt DF, Coon JJ (2004) A neutral loss activation method for improved phosphopeptide sequence analysis by quadrupole ion trap mass spectrometry. Anal Chem 76(13):3590–3598
37. Olsen JV, Macek B, Lange O, Makarov A, Horning S, Mann M (2007) Higher-energy C-trap dissociation for peptide modification analysis. Nat Methods 4(9):709–712
38. Chi A, Huttenhower C, Geer LY, Coon JJ, Syka JE, Bai DL, Shabanowitz J, Burke DJ, Troyanskaya OG, Hunt DF (2007) Analysis of phosphorylation sites on proteins from *Saccharomyces cerevisiae* by electron transfer dissociation (ETD) mass spectrometry. Proc Natl Acad Sci U S A 104(7):2193–2198
39. Good DM, Wirtala M, McAlister GC, Coon JJ (2007) Performance characteristics of electron transfer dissociation mass spectrometry. Mol Cell Proteomics 6(11):1942–1951
40. Frese CK, Altelaar AF, Hennrich ML, Nolting D, Zeller M, Griep-Raming J, Heck AJ, Mohammed S (2011) Improved peptide identification by targeted fragmentation using CID, HCD and ETD on an LTQ-Orbitrap Velos. J Proteome Res 10(5):2377–2388
41. Cox J, Mann M (2008) MaxQuant enables high peptide identification rates, individualized p.p.b.-range mass accuracies and proteome-wide protein quantification. Nat Biotechnol 26(12):1367–1372
42. The Universal Protein Resource (UniProt) (2009) Nucleic Acids Res 37(Database issue): D169–D174
43. Perkins DN, Pappin DJ, Creasy DM, Cottrell JS (1999) Probability-based protein identification by searching sequence databases using mass spectrometry data. Electrophoresis 20 (18):3551–3567
44. Yates JR 3rd, Eng JK, McCormack AL, Schieltz D (1995) Method to correlate tandem mass spectra of modified peptides to amino acid sequences in the protein database. Anal Chem 67(8):1426–1436
45. Geer LY, Markey SP, Kowalak JA, Wagner L, Xu M, Maynard DM, Yang X, Shi W, Bryant SH (2004) Open mass spectrometry search algorithm. J Proteome Res 3(5):958–964
46. Elias JE, Gygi SP (2007) Target-decoy search strategy for increased confidence in large-scale protein identifications by mass spectrometry. Nat Methods 4(3):207–214
47. Olsen JV, Blagoev B, Gnad F, Macek B, Kumar C, Mortensen P, Mann M (2006) Global, in vivo, and site-specific phosphorylation dynamics in signaling networks. Cell 127 (3):635–648
48. Beausoleil SA, Villen J, Gerber SA, Rush J, Gygi SP (2006) A probability-based approach for high-throughput protein phosphorylation analysis and site localization. Nat Biotechnol 24(10):1285–1292
49. Ravikumar V, Macek B, Mijakovic I (2016) Resources for assignment of phosphorylation sites on peptides and proteins. Methods Mol Biol 1355:293–306
50. Cox J, Matic I, Hilger M, Nagaraj N, Selbach M, Olsen JV, Mann M (2009) A practical guide to the MaxQuant computational platform for SILAC-based quantitative proteomics. Nat Protoc 4(5):698–705
51. Cox J, Neuhauser N, Michalski A, Scheltema RA, Olsen JV, Mann M (2011) Andromeda: a peptide search engine integrated into the MaxQuant environment. J Proteome Res 10 (4):1794–1805
52. Tyanova S, Mann M, Cox J (2014) MaxQuant for in-depth analysis of large SILAC datasets. In: Warscheid B (ed) Stable isotope labeling by amino acids in cell culture (SILAC), Methods in molecular biology, vol 1188. Springer, New York, pp 351–364

Chapter 17

Phosphoproteomics in Microbiology: Protocols for Studying *Streptomyces coelicolor* Differentiation

Angel Manteca, Beatriz Rioseras, Adelina Rogowska-Wrzesinska, and Ole N. Jensen

Abstract

The extension and biological role of Ser/Thr/Tyr phosphorylation in prokaryotes have been only scarcely studied. In this chapter, we describe the state of the art of microbial phosphoproteomics, focusing on protocols used for studying the phosphoproteome of *Streptomyces coelicolor*, one of the bacteria encoding the largest number of eukaryote-like Ser/Thr/Tyr kinases.

Key words *Streptomyces*, Phosphoproteomics, Differentiation, Sporulation, LC-MS/MS

1 Introduction

Reversible protein phosphorylation at serine, threonine, and tyrosine residues is a well-known dynamic posttranslational modification with stunning regulatory and signaling potential in eukaryotes [1]. In contrast, the extension and biological function of Ser/Thr/Tyr protein phosphorylation in bacteria are in most cases only poorly defined. Histidine/aspartate phosphorylations are common in bacterial two-component systems [2, 3], but its extension and biological role remains basically unknown. This is a direct consequence of the high instability of histidine phosphorylation in the acidic conditions used in phosphoproteomic protocols [4]. This chapter focus on bacterial Ser/Thr/Tyr phosphorylation.

During the last years, large-scale Ser/Thr/Tyr phosphoproteome studies were reported for *E. coli* [5, 6], *Streptococcus pneumonia* [7], *Klebsiella pneumoniae* [8], *Lactococcus lactis* [9], *Pseudomonas* [10], *Bacillus subtilis* [11], *Halobacterium salinarum* [12], *Clostridium acetobutylicum* [13], *Streptomyces coelicolor* [14, 15], *Mycobacterium tuberculosis* [16], *Acinetobacter baumanii* [17], *Listeria monocytogenes* [18], *Rhodopseudomomas palustris* [19], *Thermus thermophilus* [20], *Helicobacter pylori* [21], or *Staphylococcus*

Dörte Becher (ed.), *Microbial Proteomics: Methods and Protocols*, Methods in Molecular Biology, vol. 1841, https://doi.org/10.1007/978-1-4939-8695-8_17,

aureus [22] (Table 1). The extent of protein phosphorylation in bacteria is dramatically low in comparison to eukaryotes, making bacterial phosphoproteomics a challenge. Most bacterial phosphoproteomic studies only analyzed the vegetative stages, due to the fact that differentiation is uncharacterized for most bacteria. The only two studies analyzing different developmental stages in bacteria, demonstrated that phosphorylation is dramatically higher at the differentiated stages of *S. coelicolor* [15] and *E. coli* [6]. Quantitative phosphoproteomics is challenging, and to our knowledge, there are only two published gel-free quantitative phosphoproteomic studies in bacteria: the label-free quantitative phosphoproteomics performed in *S. coelicolor* [15], and the stable isotope labeling by amino acids in cell culture (SILAC) performed in *E. coli* [6]. All published mass spectrometry based bacterial phosphoproteomics studies applied TiO_2 for phosphopeptide enrichment, because the low phosphorylation levels of prokaryotes made other methods, such as immobilized metal affinity chromatography (IMAC), inadequate. Most of the bacterial phosphoproteomic studies used rather large amounts of protein starting material (milligrams) obtained during the vegetative growth state to detect relatively low numbers of phosphopeptides (Table 1).

In this chapter we describe the mass spectrometry based experimental workflow and analytical methods used for studying the *Streptomyces coelicolor* phosphoproteome [15], which consists of prefractionation of peptide samples by means of calcium phosphate precipitation (CPP) of phosphopeptides [23] prior to titanium dioxide (TiO_2) affinity-enrichment of phosphopeptides. CPP enhances dramatically the number of Ser/Thr/Tyr phosphopeptides identified.

Streptomycetes produce most of the biologically active compounds used in biomedicine [24–27]. They have complex life cycles (reviewed in [28]), and represent the evolutionary origin of several protein domains that are known to regulate eukaryotic signaling pathways, including eukaryotic-like protein kinases [29]. *Streptomyces coelicolor*, the model *Streptomyces* strain [30], encodes 47 predicted eukaryote-like protein kinases, twice the number of kinases predicted from genomes of other well characterized bacteria, including *E. coli* and *Bacillus subtilis* (Table 1), and it has been demonstrated to be a good model for the study of bacterial Ser/Thr/Tyr phosphorylation [15].

2 Materials

1. GYM medium [31]: 5 g/L glucose, 4 g/L yeast extract, 5 g/L malt extract, 0.5 g/L $MgSO_4.7H_2O$, 20 g/L agar. Before plating, supplement with 0.5 g/L K_2HPO_4 (add 2 mL per liter of a stock solution of 0.25 g/mL).

Table 1
Comparison of *S. coelicolor* phosphoproteome with other published prokaryotic and eukaryotic phosphoproteomes

	Bacterium	Protein (mg)[a]	Phosphoproteins	Phosphopeptides	Phosphorylation sites	Reference
Gram +	*S. coelicolor*	0.3	127	260	289[b]	Manteca et al. [15]
	S. coelicolor	50	40	44	46	Parker et al. [14]
	B. subtilis	10	78	103	78	Macek et al. [11]
	C. acetobutylicum	2	61	82	107	Bai et al. [13]
	L. lactis	20	63	102	79	Soufi et al. [9]
	M. tuberculosis	n.r.	301	380	500	Pisic et al. [16]
	S. pneumoniae	1	84	102	163	Sun et al. [7]
	L. monocytogenes	10	112	155	143	Misra et al. [18]
	S. aureus	50	108	n.r.	76	Bäsell et al. [22]
Gram −	*E. coli*	20	79	105	81	Macek et al. [5]
	K. pneumoniae	30	81	117	93	Lin et al. [8]
	P. aeruginosa	1.2	39	57	61	Ravichandran et al. [10]
	P. putida	1.2	59	56	55	Ravichandran et al. [10]
	H. pylori		67	82	126	Ge et al. [21]
	R. palustris (Ch)	2	54	100	63	Hu et al. [19]
	R. palustris (Ph)	2	42	74	59	Hu et al. [19]
	T. thermophilus	100	48	52	46	Takahata et al. [20]
	E. coli	9.8	133	n.r.	108	Soares et al. [6]
	A. baumanii Abh12O-A2	9	70	n.r.	80	Soares et al. [17]
	A. baumanii ATCC 17879	9	41	n.r.	48	Soares et al. [17]
Archaea	*H. salinarum*	20	26	42	31	Aivaliotis et al. [12]
Eukarya	*H. sapiens*	6	7832	>50,000	38,229	Sharma et al. [32]

The amount of protein used for the phosphoproteomic experiments, the number of genome ORFs, and number of phosphoproteins/phosphopeptides/phosphosites are shown.
n.r. not reported
[a]Total amount used in the all LTQ-Orbitrap or 2D gel runs
[b]This is one example out of numerous human phosphoproteome studies

2.1 *Streptomyces* Strain and Developmental Conditions

2. Cellophane disks: Use cellophane from bookshop. Cut the cellophane into discs of the diameter of a petri plate, before autoclaving.
3. Incubator (30 °C).

2.2 Protein Extraction and Sample Preparation

1. Lysis buffer: 2% SDS, 50 mM Tris–HCl pH 7, 150 mM NaCl, 10 mM $MgCl_2$, 1 mM EDTA, 7 mM ß-mercaptoethanol, EDTA-free Protease Inhibitor Cocktail Tablets (e.g., Roche) and 1% phosphatase inhibitor (e.g., mixtures 1 and 2 from Sigma).
2. Centrifuge.
3. Vacuum centrifuge.
4. Acetone, pure.
5. Ethanol (EtOH), absolute.
6. Washing solution 1: 50% (v/v) H_2O, 25% (v/v) EtOH, 25% (v/v) acetone.
7. Distilled water.
8. 12,000 Da dialysis tubing (e.g., Sigma).
9. Bradford reagent or an alternative method for protein concentration determination.
10. Dithiothreitol (DTT): 0.5 M DTT stock solution.
11. Ammonia in water (NH_3-H_2O): 2 M NH_3 in water.
12. Iodoacetamide (IAA): 150 mM IAA stock.
13. Trypsin.

2.3 Calcium Phosphate Precipitation (CPP)

1. Solution 1: 0.5 M Na_2HPO_4.
2. Solution 2: 2 M NH_3-H_2O.
3. Solution 3: 2 M $CaCl_2$.
4. Washing solution 2: 80 mM $CaCl_2$.
5. Formic acid: 5% (v/v) formic acid.
6. Acetonitrile (ACN): 50% (v/v) ACN.
7. POROS® R3 (Applied Biosystems) reversed-phase media.

2.4 Enrichment of Phosphopeptides Using TiO_2 and Desalting

1. C8 disks (3 M™ Empore™ High Performance Extraction Disk).
2. ACN.
3. TiO_2: Titansphere, 5 μm (e.g., GL Sciences).
4. Loading solution: 1 M glycolic acid, 5% (v/v) trifluoroacetic acid (TFA) (sequencing grade), 80% (v/v) ACN (HPLC grade).
5. Washing solution 3: 80% 8(v/v) ACN (HPLC grade), 1% (v/v) TFA (sequencing grade).

6. Ammonia in water (NH_3-H_2O): 2 M NH_3 in water.
7. Elution solution 1: 40% (v/v) can–1 M NH_3-H_2O.
8. Formic acid: 3.5% (v/v) formic acid.
9. POROS® R3 (Applied Biosystems) reversed-phase media.
10. Equilibration solution: 0.5% (v/v) formic acid.
11. Elution solution 2: 50% (v/v) ACN, 0.5% (v/v) formic acid.
12. Elution solution 3: 70% (v/v) ACN, 0.5% (v/v) formic acid.

2.5 LC-MS/MS Analysis

1. LC-ESI-MS/MS system: For example, a nano EasyLC system (Thermo Fisher Scientific, Odense, Denmark) interfaced by an ESI source to a LTQ-Orbitrap XL or an Q-Exactive tandem mass spectrometer (Thermo Fisher Scientific, Bremen, Germany).
2. Formic acid: 0.5% (v/v) formic acid.
3. Fused silica column: in-house packed fused silica column, 16 cm length, 100 μm inner diameter, 375 μm outer diameter, filled with ReproSil, C18 AQ 3 μm; (e.g., Dr. Maisch, Ammerbuch, Germany).
4. HPLC solvent A: 0.1% (v/v) formic acid.
5. HPLC solvent B: 90% (v/v) ACN, 0.1% (v/v) formic acid.
6. ProteomeDiscoverer Software version 1.2 (Thermo Scientific).
7. Mascot server version 2.3.02 (Matrix Science).

3 Methods

Experimental workflow is detailed in the following paragraphs and outlined in Fig. 1.

3.1 *Streptomyces* Strain and Developmental Conditions

Use the model *Streptomyces coelicolor* M145 strain [30], or another *Streptomyces* whose genome is sequenced. *Streptomyces coelicolor* developmental cycle and culture conditions are well known [28]. As introduced above, one of the keys to increase the number of detected phosphorylation sites in *Streptomyces* is to analyze different developmental stages: vegetative, MI; and reproductive, MII/sporulation [15]. In the case of *S. coelicolor*, solid sporulating cultures can be prepared as described by Manteca et al. [15]: inoculate plates of GYM medium, covered with cellophane disks, with 100 μL of a spore suspension (1×10^7 viable spores/mL); incubate them at 30 °C; analyze three key developmental stages (12 h, MI; 24 h, MII; and 72 h, sporulating MII).

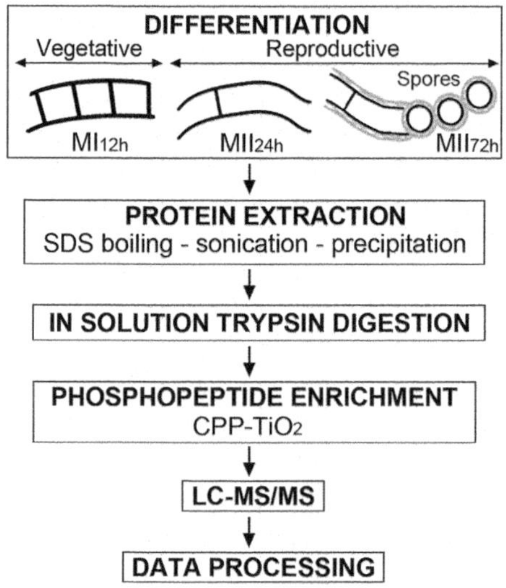

Fig. 1 Overview of the *S. coelicolor* phosphoproteomics workflow. Mycelium from three developmental stages is lysed by boiling in SDS and sonication; proteins are precipitated with acetone–ethanol; trypsin digestion is followed by phosphoenrichment using CPP-TiO_2. Phosphopeptides are separated on nano-HPLC, mass-measured (MS) and fragmented (MS/MS) in the high performance LTQ-Orbitrap mass spectrometer

3.2 Protein Extraction and Sample Preparation

1. Collect the mycelial lawns of *S. coelicolor* M145 (or alternative species) growing on cellophane disks at different time points (12, 24, and 72 h) using a plain spatula. Process as many biological replicates as possible, ideally, those necessary to ensure rigorous statistical analysis. Multiple replicates add to the workload, specially MS instrumentation time, and may not always be feasible. Process at least two biological replicates for each developmental stage.
2. Lyse the cells boiling the samples in lysis buffer (10 min). Use 65 mg of fresh weight mycelium per mL of lysis buffer (*see* **Note 1**).
3. Centrifuge at 20,000 × *g* for 10 min.
4. Collect the supernatant.
5. Reduce sample viscosity by sonication (4 cycles of 10 s, on ice).
6. Precipitate the protein with acetone–ethanol (1 volume sample–4 volumes ethanol–4 volumes acetone). Store over night at −20 °C (*see* **Note 2**).
7. Centrifuge at 20,000 × *g* for 10 min at 4 °C.
8. Discard the supernatant.
9. Wash the pellet with precooled (−20 °C) washing solution 1.
10. Repeat **steps** 7 and **8**.
11. Resuspend the pellet in precooled (4 °C) distillated water.

12. Dialyze the sample against large volumes of water (1 h at 4 °C, change water every 15 min) (*see* **Note 3**).
13. Determine protein concentration (e.g., Bradford method [28] or any other alternative).
14. Lyophilize samples in aliquots (e.g., 100 μg of protein per aliquot).
15. Store the lyophilized samples at −80 °C (*see* **Note 4**).

3.3 In-Solution Trypsin Digestion

1. Resuspend the protein in distilled water (add 25 μL of water to 50 μg of protein) (*see* **Note 5**).
2. Add DTT to a final concentration of 5 mM (0.25 μL of the 0.5 M DTT stock to 25 μL). Adjust the pH to 8 with 2 M NH_3-H_2O (*see* **Note** 6).
3. Incubate at 37 °C for 45 min.
4. Take out the sample from 37 °C, leave it at room temperature until it cool down.
5. Add IAA to a final concentration of 15 mM (2.5 μL of the 150 mM IAA stock), check the pH (*see* **Note 6**).
6. Incubate at room temperature for 45 min, in the dark.
7. Add the same amount of DTT as in **step 2**.
8. Add trypsin (0.5 μg of trypsin to the 25 μg of protein).
9. Incubate at 37 °C overnight.

3.4 Phosphopeptide Enrichment: CPP Prefractionation/TiO_2

1. Adjust the volume of the peptide solution (50 μg) to 100 μL with distilled water.
2. Adjust the pH of the sample solution to 9–10 using 2 M NH_3-H_2O (*see* **Note 6**).
3. Add 8 μL of precipitation solutions 1 and 8 μL of precipitation solutions 2 to the 100 μL sample solution and mix.
4. Add 8 μL of precipitation solution 3 and vortex.
5. Centrifuge at 20,000 × *g* for 10 min at room temperature.
6. Discard the supernatant (this supernatant contains the non-phosphorylated peptides and can be stored to compare it with the phosphopeptide-enriched sample).
7. Add 120 μL of washing solution 2 to the pellet, vortex until it is suspended.
8. Centrifuge at 20,000 × *g* for 5 min.
9. Repeat **steps 6–8**.
10. Discard the supernatant.
11. Dissolve the pellet with 40 μL of 5% (v/v) formic acid.

12. Desalt the resulting solution by reverse phase chromatography as detailed below (Subheading 3.5); store the dried phosphopeptides at −20 °C (*see* **Notes 4** and 7).
13. Prepare TiO_2 microcolumns putting a plug of the C8 disk into the narrow end of the tip and load the TiO_2 material on top (*see* **Note 8**).
14. Wash the column with 30 μL of 100% ACN.
15. Resuspend the dried CPP pre-enriched sample (**step 12**) in 25 μL of loading solution.
16. Load the sample on the TiO_2 microcolumn.
17. Collect the flow through and load it a second time on the same column.
18. Wash with 25 μL of loading solution.
19. Wash with 25 μL of washing solution 3 (two times).
20. Elute the peptides with 10 μL of 2 M NH_3-H_2O (two times).
21. Elute with 2 μL elution solution 1 to elute the phosphopeptides that bind to the C8-disk.
22. Vacuum-dry the sample, resuspend it in 30 μL 3.5% (v/v) formic acid (FA), and desalt it using the protocol detailed below (Subheading 3.5).

3.5 Sample Desalting

1. Prepare an oligoR3 RP column into a p200 pipette tip with C18 plug. Equilibrate the column with 50 μL equilibration solution. (*see* **Note 9**).
2. Load the acidified peptide mixture onto the column.
3. Wash the column with 50 μL equilibration solution.
4. Elute peptides in a two-step elution of peptides with 30 μL elution solution 2 and 20 μL elution solution 3.
5. Combine the eluates and dry down in a vacuum centrifuge. Store at −20 °C until further analysis (*see* **Note** 7).

3.6 LC-MS/MS Analysis

1. For LC-MSMS analysis resuspend samples in 0.5% (v/v) formic acid.
2. Analyze samples using nanoliter flow chromatography (nanoLC) interfaced to a high mass resolution electrospray tandem mass spectrometer (LC-ESI-MS/MS). For example a nanoliter flow EasyLC system (Thermo Fisher Scientific, Odense, Denmark) coupled to a Q-Exactive ESI tandem mass spectrometer (Thermo Fisher Scientific, Bremen, Germany) can be used. (*see* **Note 10**).
3. Separate peptides using an in-house packed fused silica.
4. Inject 5 μL of peptide/phosphopeptide sample onto the column.

5. Separate peptides using a linear 50 min gradient from 0% to 34% of solvent B with a constant flow of 250 nL/min.
6. Record mass spectra in in the positive ion mode with data dependent acquisition. Acquire survey scans in the m/z range 400–2000 in the Orbitrap at a mass resolution of 60,000 FWHM. Perform collision induced dissociation MS/MS of 5–10 most intense ions in the linear ion trap with an activation time of 15 s.
7. Convert raw data into peak lists using for example Proteome Discoverer software. Generate mgf files, perform database searches using for example Mascot server (version 2.3.02; Matrix Science), search against the NCBInr database. Use the following parameters to filter the results: precursor mass tolerance 10 ppm, fragment ion mass tolerance of 0.8 Da, fixed modifications: carbamidomethylation of Cys residues; variable modifications: phosphorylation of Ser/Thr/Tyr, oxidation of Met; enzyme: trypsin, two missed cleavages allowed (*see* **Note 11**).

3.7 Criteria to Validate Candidate Phosphopeptides

Candidate phosphopeptides are validated by manual inspection of their respective MS/MS spectra using the next criteria (*see* Manteca et al. [15] for examples). Peptides should have:

1. At least four consecutive y or b ions.
2. An intense signal should be assigned to ions produced by fragmentation at peptide bond N-terminal to proline if proline is present in the sequence.
3. Phosphoserine sites were assigned by the appearance of a 69 Da (dehydroSer) and 167 Da (phosphoserine, 87 + 80 Da) distance between fragment ions.
4. Phosphothreonine sites were assigned by the appearance of 83 Da (dehydroThr) and 181 Da (phosphothreonine, 80 + 101 Da) distance for phosphothreonine.
5. Phosphotyrosine containing peptides were validated by the observation of a mass increase of 80 Da to unmodified peptide and the presence of the immonium ion at m/z 216 (singly charged) or a mass difference of 243 Da between fragment ions in spectra.

3.8 Label-Free Quantitative Analysis

The relative abundance of the phosphopeptides at different developmental stages is estimated based in the averaged counts per second following the next criteria (*see* Manteca et al. [15] for examples):

1. Phosphopeptides were only considered for quantification if they were detected and sequenced in two biological replicates.
2. Average abundance was calculated in the 24 LC-MS/MS runs.

3. For the LC-MS/MS runs in which the precursor ion was not sequenced, it was searched manually using a tolerance of 0.14 min for retention time and 5 ppm as mass tolerance.
4. Relative abundances were considered as significant when the average counts per second ±SD were not overlapping between developmental stages.

4 Notes

1. Collect the sample in lysis buffer in ice. Boil the samples in the SDS-lysis buffer as soon as possible to inactive phosphatases/proteases and to stabilize the phosphoproteome.
2. Ethanol–acetone precipitation is critical to remove SDS.
3. Dialysis cleans rests of SDS and salts, but it is not critical.
4. In order to minimize methodological variation, prepare single use aliquots, process them as soon as possible, and use samples with the same storage times.
5. Do not use a buffer during trypsin digestion, because it interferes with CPP. The stock of IAA (150 mM) should be prepared freshly for each digestion. Prepare it in water, and adjust the pH to 8 with 2 M NH_3-H_2O (*see* also **Note 1**).
6. Adjust the pH to 9–10 is critical for CPP. pH should be checked adding minimal volumes (around 0.2 μL) to pH strips.
7. Take into account that CPP is a phosphopeptide pre-enrichment, and most peptides in the CPP pre-enriched sample are not phosphorylated.
8. In order to improve reproducibility of the label free quantification between replicates, TiO_2 microcolumns should be as homogeneous as possible.
9. OligoR3 RP microcolumn should be big enough to retain all the peptides. This can be checked analyzing the absence of peptides in the column flow through by MALDI-TOF.
10. It is recommended to perform 2–5 replicate analysis.
11. Take into account that most of the peptides identified in the final CPP-TiO_2 enriched sample are not phosphorylated due to the low level of Ser/Thr/Tyr phosphorylation in bacteria.

Acknowledgments

We wish to thank the European Research Council (ERC Starting Grant; Strp-differentiation 280304), the Spanish "Ministerio de Economía y Competitividad" (MINECO; BIO2015-65709-R), and the VILLUM Foundation (VILLUM Center for Bioanalytical Sciences at University of Southern Denmark) for financial support.

References

1. Pawson T, Scott JD (2005) Protein phosphorylation in signaling—50 years and counting. Trends Biochem Sci 30(6):286–290
2. Hoch JA (2000) Two-component and phosphorelay signal transduction. Curr Opin Microbiol 3(2):165–170
3. Galperin MY, Nikolskaya AN, Koonin EV (2001) Novel domains of the prokaryotic two-component signal transduction systems. FEMS Microbiol Lett 203(1):11–21
4. Kleinnijenhuis AJ, Kjeldsen F, Kallipolitis B, Haselmann KF, Jensen ON (2007) Analysis of histidine phosphorylation using tandem MS and ion-electron reactions. Anal Chem 79 (19):7450–7456
5. Macek B, Gnad F, Soufi B, Kumar C, Olsen JV, Mijakovic I, Mann M (2008) Phosphoproteome analysis of *E. coli* reveals evolutionary conservation of bacterial Ser/Thr/Tyr phosphorylation. Mol Cell Proteomics 7 (2):299–307
6. Soares NC, Spat P, Krug K, Macek B (2013) Global dynamics of the *Escherichia coli* proteome and phosphoproteome during growth in minimal medium. J Proteome Res 12 (6):2611–2621
7. Sun X, Ge F, Xiao CL, Yin XF, Ge R, Zhang LH, He QY (2010) Phosphoproteomic analysis reveals the multiple roles of phosphorylation in pathogenic bacterium *Streptococcus pneumoniae*. J Proteome Res 9(1):275–282
8. Lin MH, Hsu TL, Lin SY, Pan YJ, Jan JT, Wang JT, Khoo KH, Wu SH (2009) Phosphoproteomics of *Klebsiella pneumoniae* NTUH-K2044 reveals a tight link between tyrosine phosphorylation and virulence. Mol Cell Proteomics 8(12):2613–2623
9. Soufi B, Gnad F, Jensen PR, Petranovic D, Mann M, Mijakovic I, Macek B (2008) The Ser/Thr/Tyr phosphoproteome of *Lactococcus lactis* IL1403 reveals multiply phosphorylated proteins. Proteomics 8(17):3486–3493
10. Ravichandran A, Sugiyama N, Tomita M, Swarup S, Ishihama Y (2009) Ser/Thr/Tyr phosphoproteome analysis of pathogenic and non-pathogenic Pseudomonas species. Proteomics 9(10):2764–2775
11. Macek B, Mijakovic I, Olsen JV, Gnad F, Kumar C, Jensen PR, Mann M (2007) The serine/threonine/tyrosine phosphoproteome of the model bacterium *Bacillus subtilis*. Mol Cell Proteomics 6(4):697–707
12. Aivaliotis M, Macek B, Gnad F, Reichelt P, Mann M, Oesterhelt D (2009) Ser/Thr/Tyr protein phosphorylation in the archaeon *Halobacterium salinarum*—a representative of the third domain of life. PLoS One 4(3):e4777
13. Bai X, Ji Z (2012) Phosphoproteomic investigation of a solvent producing bacterium *Clostridium acetobutylicum*. Appl Microbiol Biotechnol 95(1):201–211
14. Parker JL, Jones AM, Serazetdinova L, Saalbach G, Bibb MJ, Naldrett MJ (2010) Analysis of the phosphoproteome of the multicellular bacterium *Streptomyces coelicolor* A3 (2) by protein/peptide fractionation, phosphopeptide enrichment and high-accuracy mass spectrometry. Proteomics 10(13):2486–2497
15. Manteca A, Ye J, Sanchez J, Jensen ON (2011) Phosphoproteome analysis of Streptomyces development reveals extensive protein phosphorylation accompanying bacterial differentiation. J Proteome Res 10(12):5481–5492
16. Prisic S, Dankwa S, Schwartz D, Chou MF, Locasale JW, Kang CM, Bemis G, Church GM, Steen H, Husson RN (2010) Extensive phosphorylation with overlapping specificity by *Mycobacterium tuberculosis* serine/threonine protein kinases. Proc Natl Acad Sci U S A 107 (16):7521–7526
17. Soares NC, Spat P, Mendez JA, Nakedi K, Aranda J, Bou G (2014) Ser/Thr/Tyr phosphoproteome characterization of *Acinetobacter baumannii*: comparison between a reference strain and a highly invasive multidrug-resistant clinical isolate. J Proteome 102:113–124
18. Misra SK, Milohanic E, Ake F, Mijakovic I, Deutscher J, Monnet V, Henry C (2011) Analysis of the serine/threonine/tyrosine phosphoproteome of the pathogenic bacterium *Listeria monocytogenes* reveals phosphorylated proteins related to virulence. Proteomics 11 (21):4155–4165
19. Hu CW, Lin MH, Huang HC, Ku WC, Yi TH, Tsai CF, Chen YJ, Sugiyama N, Ishihama Y, Juan HF, Wu SH (2012) Phosphoproteomic analysis of *Rhodopseudomonas palustris* reveals the role of pyruvate phosphate dikinase phosphorylation in lipid production. J Proteome Res 11(11):5362–5375
20. Takahata Y, Inoue M, Kim K, Iio Y, Miyamoto M, Masui R, Ishihama Y, Kuramitsu S (2012) Close proximity of phosphorylation sites to ligand in the phosphoproteome of the extreme thermophile *Thermus thermophilus* HB8. Proteomics 12(9):1414–1430
21. Ge R, Sun X, Xiao C, Yin X, Shan W, Chen Z, He QY (2011) Phosphoproteome analysis of the pathogenic bacterium *Helicobacter pylori* reveals over-representation of tyrosine

phosphorylation and multiply phosphorylated proteins. Proteomics 11(8):1449–1461

22. Bäsell K, Otto A, Junker S, Zühlke D, Rappen GM, Schmidt S, Hentschker C, Macek B, Ohlsen K, Hecker M, Becher D (2014) The phosphoproteome and its physiological dynamics in Staphylococcus aureus. Int J Med Microbiol 304(2):121–132
23. Zhang X, Ye J, Jensen ON, Roepstorff P (2007) Highly efficient phosphopeptide enrichment by calcium phosphate precipitation combined with subsequent IMAC enrichment. Mol Cell Proteomics 6(11):2032–2042
24. Omura S (1992) The expanded horizon for microbial metabolites—a review. Gene 115 (1–2):141–149
25. Tamaoki T, Nakano H (1990) Potent and specific inhibitors of protein kinase C of microbial origin. Biotechnology (N Y) 8(8):732–735
26. Umezawa K (1997) Induction of cellular differentiation and apoptosis by signal transduction inhibitors. Adv Enzym Regul 37:393–401
27. Katz L, Baltz RH (2016) Natural product discovery: past, present, and future. J Ind Microbiol Biotechnol 43:155–176
28. Yague P, Lopez-Garcia MT, Rioseras B, Sanchez J, Manteca A (2013) Pre-sporulation stages of Streptomyces differentiation: state-of-the-art and future perspectives. FEMS Microbiol Lett 342(2):79–88
29. Perez J, Castaneda-Garcia A, Jenke-Kodama H, Muller R, Munoz-Dorado J (2008) Eukaryotic-like protein kinases in the prokaryotes and the myxobacterial kinome. Proc Natl Acad Sci U S A 105(41):15950–15955
30. Bentley SD, Chater KF, Cerdeno-Tarraga AM, Challis GL, Thomson NR, James KD, Harris DE, Quail MA, Kieser H, Harper D, Bateman A, Brown S, Chandra G, Chen CW, Collins M, Cronin A, Fraser A, Goble A, Hidalgo J, Hornsby T, Howarth S, Huang CH, Kieser T, Larke L, Murphy L, Oliver K, O'Neil S, Rabbinowitsch E, Rajandream MA, Rutherford K, Rutter S, Seeger K, Saunders D, Sharp S, Squares R, Squares S, Taylor K, Warren T, Wietzorrek A, Woodward J, Barrell BG, Parkhill J, Hopwood DA (2002) Complete genome sequence of the model actinomycete Streptomyces coelicolor A3(2). Nature 417(6885):141–147
31. Novella IS, Barbes C, Sanchez J (1992) Sporulation of *Streptomyces antibioticus* ETHZ 7451 in submerged culture. Can J Microbiol 38 (8):769–773
32. Sharma K, D'Souza RC, Tyanova S, Schaab C, Wisniewski JR, Cox J, Mann M (2014) Ultra-deep human phosphoproteome reveals a distinct regulatory nature of Tyr and Ser/Thr-based signaling. Cell Rep 8(5):1583–1594

Chapter 18

Thiol-Redox Proteomics to Study Reversible Protein Thiol Oxidations in Bacteria

Martina Rossius, Falko Hochgräfe, and Haike Antelmann

Abstract

Thiol-redox proteomics methods are rapidly developing tools in redox biology. These are applied to identify and quantify proteins with reversible thiol oxidations that are formed under normal growth and oxidative stress conditions inside cells. The proteins with reversible thiol oxidations are usually prepared by alkylation of reduced thiols, subsequent reduction of disulfide bonds followed by a second differential alkylation of newly released thiols. Here, we describe two methods for detection of protein *S*-thiolations in Gram-positive bacteria using the direct shotgun approach and the fluorescent-label thiol-redox proteomics method that have been successfully applied in our previous work.

Key words Fluorescent-based redox proteomics, Thiol trapping, Mass spectrometry, Reversible thiol oxidations, Protein *S*-thiolations

1 Introduction

Bacteria are exposed to reactive oxygen species (ROS) during aerobic respiration or under infection conditions by the oxidative burst encountered by activated neutrophils [1, 2]. ROS can damage all cellular macromolecules, but the most susceptible target is the thiol group of cysteine in proteins. The thiol group is subject to reversible oxidation by ROS to form intramolecular or intermolecular protein disulfides or mixed protein disulfides with low molecular weight (LMW) thiols (termed as *S*-thiolations). Highly reactive ROS, such as hydroxyl radicals are formed in the Fenton reaction and can lead to irreversible overoxidation of protein thiols to sulfinic or sulfonic acids leading to a loss of protein function [3] (Fig. 1).

Protein *S*-thiolations protect the thiol groups against overoxidation and function as thiol-redox switches to control protein activities. In eukaryotes and Gram-negative bacteria, the LMW

The original version of this chapter was revised. The correction to this chapter is available at https://doi.org/10.1007/978-1-4939-8695-8_23

Dörte Becher (ed.), *Microbial Proteomics: Methods and Protocols*, Methods in Molecular Biology, vol. 1841, https://doi.org/10.1007/978-1-4939-8695-8_18, © The Author(s) 2018

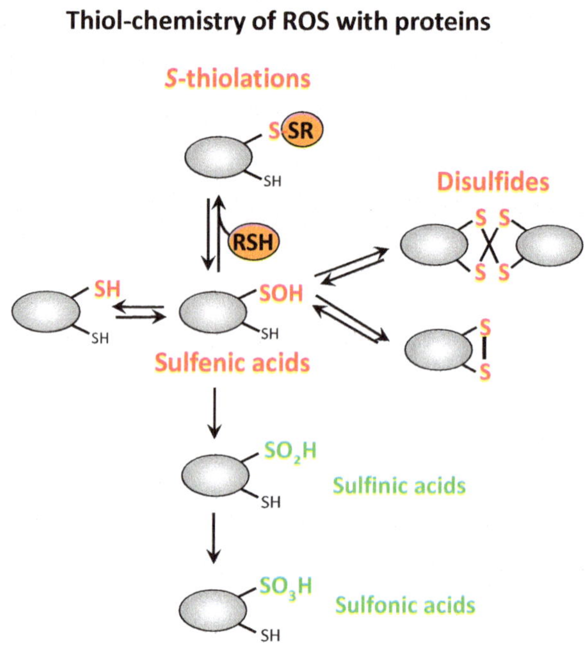

Fig. 1 Thiol chemistry of reactive oxygen species (ROS). Reversible thiol oxidation by ROS leads first to a Cys sulfenic acid intermediate (R-SOH) that is unstable and reacts further to form intramolecular and intermolecular disulfides or mixed disulfides with LMW thiols, such as glutathione, bacillithiol, cysteine, or CoASH, termed as *S*-thiolations. The Cys sulfenic acid can be also overoxidized to Cys sulfinic and sulfonic acids which are irreversible thiol oxidations

thiol glutathione (GSH) is used for protein *S*-glutathionylation. Protein *S*-glutathionylation controls numerous cellular functions and is involved in many physiological and pathophysiological processes [4]. However, most Gram-positive bacteria lack the ability to synthesize GSH and alternative redox buffers are used for protein *S*-thiolations. Actinomycetes utilize the alternative LMW thiol mycothiol (MSH, Acetyl-Cys-GlcN-myoinositol) and Firmicutes produce bacillithiol (BSH, Cys-GlcN-malate) [5]. Thus far, the knowledge about the targets and the physiological role of protein *S*-thiolations in bacteria is limited. Recent advances in redox proteomics and mass spectrometry have facilitated the detection of various forms of protein *S*-thiolations in bacteria, including *S*-cysteinylation, *S*-bacillithiolations, and *S*-mycothiolations [6–10]. Here, we provide two protocols that were applied in our lab to identify these different forms of protein *S*-thiolations in Gram-positive bacteria, such as *Staphylococcus carnosus, Bacillus subtilis*, and *Corynebacterium glutamicum*. We describe first the fluorescent-label based thiol-redox proteomics method that was developed to quantify and visualize reversibly oxidized protein thiols using the two-dimensional gel electrophoresis method (2D PAGE) [7, 9]. In this approach, all reversibly oxidized proteins are reduced and labeled with the fluorescent dye BODIPY FL C_1-IA

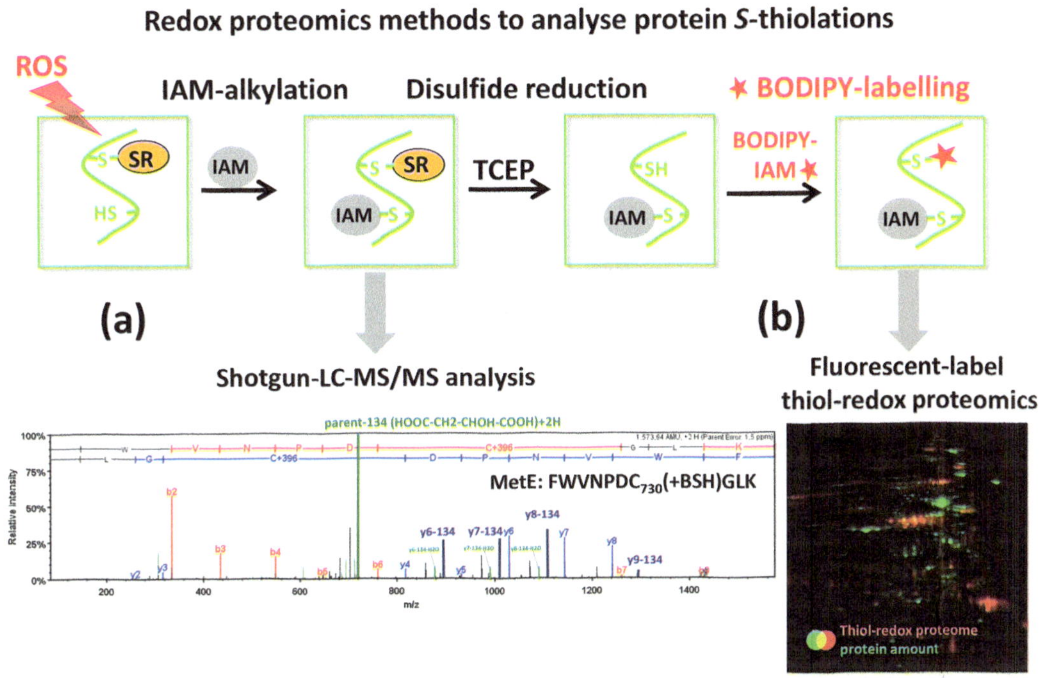

Fig. 2 Workflow for preparation of proteins for shotgun LC-MS/MS analyses of protein *S*-thiolations (**a**) and fluorescent-label thiol-redox proteomics (**b**). (**a**) Bacterial cells are harvested and reduced thiols are alkylated with IAM and subjected to non-reducing SDS-PAGE and in-gel tryptic digestion. LC-MS/MS analysis of the tryptic peptides is performed using a LTQ-Orbitrap-Velos mass spectrometer. The MS/MS spectrum shows as example the *S*-bacillithiolated MetE-Cys730 peptide. Bacillithiol (BSH) is composed of cysteine, glucosamine and malate. The abundant neutral loss precursor ion is characteristic for the loss of malate (−134 Da) and serves as indicator for the *S*-bacillithiolated peptide since malate of BSH is lost during fragmentation. (**b**) For the thiol redox proteome analysis, reduced protein thiols are alkylated with IAM and reversibly oxidized proteins (-S-SR) are reduced with TCEP and labeled with the fluorescent dye BODIPY FL C_1-IA. Proteins are separated using 2D PAGE, scanned for BODIPY fluorescence and stained with Colloidal Coomassie blue

(*N*-(4,4-difluoro-5,7-dimethyl-4-bora-3a,4a-diaza-*s*-indacene-3-l) methyl)iodoacetamide). The BODIPY-labeled protein extracts are separated by 2D-PAGE and scanned for BODIPY-fluorescence. The gels are subsequently stained for total protein amounts by Coomassie blue or a compatible fluorescent protein dye. Using the Decodon Delta 2D software an overlay image is generated of the fluorescent-labeled thiol-redox proteome and the Coomassie stained protein amount image. Finally, the level of oxidation is quantified in comparison to the protein amount and the oxidized proteins can be identified using MALDI-TOF mass spectrometry (Figs. 2b and 3).

The direct identification of the different forms of protein *S*-thiolations is based on a bottom-up proteomics approach using shotgun LC-MS/MS analysis of alkylated proteins samples under non-reducing conditions [6, 7, 10]. The reduced thiols in the protein extracts are blocked using iodoacetamide (IAM) or N-ethylmaleimide (NEM) during cell disruption and the extracts

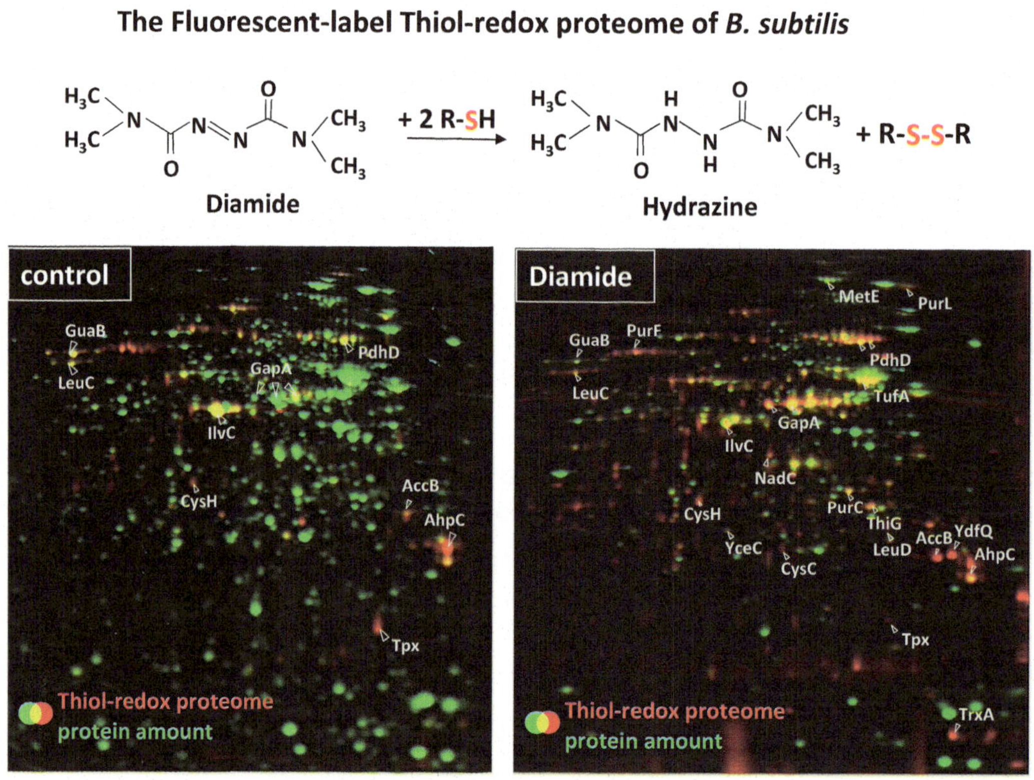

Fig. 3 Fluorescence-label thiol redox proteomics to visualize reversibly oxidized proteins in *B. subtilis* before (control) and after diamide stress. Diamide is a thiol-reactive azo compound that causes disulfide stress in cells as shown by the chemistry above. For the thiol redox proteome analysis, cells were harvested at control and diamide stress conditions and prepared as described in Fig. 2b. The overlay image shows the BODIPY fluorescence image (redox proteome, red image) compared with the Coomassie-stained protein amount image (green image). Quantitative image analysis is performed using the DECODON Delta 2D software (http:/www.decodon.com) to calculate fluorescence (redox) ratios versus protein amounts. Proteins which appear in red in the overlay images are highly oxidized proteins with low protein levels, yellow spots are abundant oxidized proteins and green proteins are reduced proteins. The strongly increased reversible protein thiol oxidation after diamide stress is visualized which includes also protein *S*-thiolations

are directly applied for tryptic in-gel digestion and mass spectrometry to identify for example *S*-bacillithiolated or *S*-mycothiolated peptides (Fig. 2a). The 2D gel-based method and gel-free shotgun approach use similar protocols for thiol-trapping by IAM or NEM during sample preparation leading to a significant overlap between the identified reversibly oxidized proteins with those identified with *S*-thiolations [7–9]. However, even in combination both methods cover only a small fraction of protein *S*-thiolations and have several limitations. For example, false-positive redox-regulated proteins may be identified using the 2D gel-based approach and *S*-thiolated proteins cannot be enriched using both methods. Hence, these 2D gel-based method and shotgun-approach should be combined with more sensitive mass spectrometry-based methods of redox proteomics, such as OxICAT or the NEM-Biotin-switch assay for comprehensive quantification of protein *S*-thiolations [11–14]. These mass spectrometry-based thiol-trapping assays could be even further developed to use purified bacilliredoxins or mycoredoxins for selective reduction of *S*-bacillithiolations and *S*-mycothiolations, respectively. This would facilitate the enrichment and more comprehensive identification and quantification of the various protein *S*-thiolation forms in bacteria. Similarly, the glutaredoxin-coupled NEM-Biotin-switch assay has been successfully applied for global identification of *S*-glutathionylations in different eukaryotic cells [11, 13]. Further advances in mass spectrometry and the design of novel chemical probes will provide leads for a more comprehensive description of the dynamics of the complete *S*-thiolome in various bacterial cells.

2 Materials

Centrifugation of protein samples in Eppendorf tubes is performed using a microfuge. For concentration of protein samples, a vacuum concentrator (SpeedVac) is used. Cells are disrupted using a Precellys-homogenizer (Peqlab). The rehydration chamber is required for rehydration of the IPG-strips. The Multiphor electrophoresis instrument including a power supply (e.g., EPS3500-XL from Pharmacia) is used for the IEF run. The 2D electrophoresis unit and a power supply are used for the subsequent second dimension (SDS-PAGE). In addition, casting trays must be provided for casting the 1D and 2D gels. Scanning of the 2D redox gels is performed using the Typhoon Scanner.

2.1 Components for Fluorescent-Label Thiol-Redox Proteomics

Prepare solutions for thiol-trapping, reduction and BODIPY-IAM-fluorescent-labeling of thiols using double distilled water.

2.1.1 Components for Preparation of Fluorescent-Labeled Protein Samples

1. TE-IAM buffer (for washing bacterial cell pellets): 10 mM Tris–HCl, pH 8.0, 1 mM EDTA, 100 mM iodoacetamide (IAM).
2. UCE buffer (for protein denaturation): 8 M urea, 1% (w/v) CHAPS (3-((3-cholamidopropyl) dimethylammonio)-1-propanesulfonate), 1 mM EDTA, 200 mM Tris–HCl, pH 8.0. Store in Falcon tubes at −20 °C.
3. UCE-IAM buffer (for thiol-blocking): 8 M urea, 1% (w/v) CHAPS, 1 mM EDTA, 200 mM Tris–HCl, pH 8.0, 100 mM IAM. Store in Falcon tubes at −20 °C.
4. 80% acetone (v/v) in Aqua dest (for removal of excess IAM after thiol-blocking).

2.1.2 Components for 2D-PAGE and Staining Used in Fluorescent-Label Thiol-Redox Proteomics

5. Rehydration solution (RH) for IPG strips: 8 M urea, 2 M thiourea, 0.5% (v/v) CHAPS, 20 mM DTT, 0.5% (v/v) Pharmalytes pH 3–10. Add fresh DTT, CHAPS and Pharmalytes to 20 mL urea/thiourea solution before usage for isoelectric focusing (IEF).
6. SERVA IPG Blue Strips pH 4–7 NL/18 cm (SERVA Electrophoresis GmbH).
7. Electrode wicks.
8. Mineral oil for IEF.
9. TCEP stock solution: 10 mM Tris(2-carboxyethyl)phosphine hydrochloride (TCEP HCl) should be prepared in Aqua dest.
10. 15 mM BODIPY-IAM (Invitrogen D-6003) stock solution: Dissolve one flask of BODIPY FL C1-IA in 800 μL 100% acetonitrile and store as 80 μL aliquots at −20 °C in dark tubes (*see* **Note 1**).
11. Micro Bio-Spin six Chromatography Columns (e.g., Bio-Rad 732-6221) (for removal of excess BODIPY-IAM from the protein samples).
12. Separating gel buffer: 1.5 M Tris–HCl, pH 8.8. Dissolve 181.71 g Tris base in 800 mL distilled water, adjust with HCl to pH 8.8 and fill up to 1 L.
13. Stacking gel buffer: 0.5 M Tris–HCl, pH 6.8. Dissolve 60.57 g Tris base in 800 mL distilled water, adjust with HCl to pH 6.8 and fill up to 1 L.
14. Sodium dodecyl sulfate (SDS): 10% (w/v) SDS in distilled water.
15. Ammonium persulfate (APS): 10% (w/v) Ammonium persulfate in distilled water. Store at 4 °C.
16. N, N, N, N-tetramethyl-ethylenediamine (TEMED).
17. Acrylamide solution: 40% (w/v) acrylamide solution (e.g., SERVA).
18. Bisacrylamide solution: 2% (w/v) bisacrylamide solution (e.g., SERVA).
19. Separating gel solution (12.5%): acrylamide solution, bisacrylamide solution, separating gel buffer, SDS, APS, TEMED. Mix 266 mL distilled water with 304 mL 40% (w/v) acrylamide solution, 170 mL 2% (w/v) bisacrylamide solution, 250 mL separation gel buffer, and 10 mL 10% (w/v) SDS using a magnetic stirrer. Add 2.5 mL 10% (w/v) APS and 500 μL TEMED to initiate the polymerization process of the separating gel.
20. Stacking gel solution: acrylamide solution, bisacrylamide solution, stacking gel buffer, 10% (w/v) SDS. Mix 60 mL distilled

water, 9 mL 40% (w/v) acrylamide solution, 4.5 mL 2% (w/v) bisacrylamide solution, 25 mL stacking gel buffer, and 1 mL 10% (w/v) SDS using a magnetic stirrer.

21. SDS–Tris–Glycine running buffer (tenfold): Dissolve 30 g Tris base, 144 g glycine, and 10 g SDS in distilled water and fill up to 1 L. The tenfold running buffer is stored at RT and used as a dilution of 1:10 in distilled water.
22. Equilibration solution for IPG strips: To prepare 1 L of the equilibration solution, dissolve the following components in Aqua dest and fill up to 1 L: 100 mL 0.5 M Tris–HCl pH 6.8 (final conc. 50 mM), 360 g urea (final conc. 6 M), 300 mL 99% (v/v) glycerol (final conc. 30% (v/v)), 40 g SDS (final conc. 4% (w/v)). Divide the equilibration solution into two parts (A, B). Add 1.75 g DTT to 500 mL solution (A) corresponding to a final DTT concentration of 3.5 mg/mL and store as 50 mL aliquots in Falcon tubes at −20 °C. Add 22.5 g IAM and traces of bromophenol blue to 500 mL solution (B) corresponding to a final IAM concentration of 45 mg/mL and store as 50 mL aliquots in Falcon tubes at −20 °C.
23. Gel fixation solution: 40% (v/v) ethanol, 10% (v/v) acetic acid.
24. Colloidal Coomassie staining solution: Prepare 1 L Colloidal Coomassie staining solution by dissolving 100 g ammonium sulfate, 12 mL orthophosphoric acid 85%, and 20 mL of 5% (w/v) Coomassie Brilliant Blue stock in distilled water and fill up to 1 L. Add methanol directly before the staining of the gels.
25. Methanol, purity >99.95%.

2.2 Components for Mass Spectrometry-Based Identification of Protein S-Thiolations

2.2.1 Components for Preparation of Protein Samples

1. TE-IAM buffer (for washing bacterial cell pellets) is described in Subheading 2.1.1, **item 1**.
2. UCE-buffer (for denaturation of proteins) is described in Subheading 2.1.1, **item 2**.
3. Bradford reagent (e.g., Roti-Nanoquant, Fa Roth) for protein determination.

2.2.2 Components for SDS-PAGE

4. Separating and stacking gel buffer, SDS 10% (w/v), APS 10% (w/v) and TEMED are described in Subheading 2.1.2, **items 12–16**.
5. Nonreducing SDS-PAGE sample buffer: Stacking gel buffer, SDS, glycerol, Bromophenol Blue, Aqua dest. To 3 mL Aqua dest add 1.2 mL stacking gel buffer, 4.0 mL 10% (w/v) SDS, 1.0 mL glycerol, and a small amount (tip of the spatula) Bromophenol Blue and fill up to 10 mL with distilled water. This buffer is used as sample buffer to dissolve proteins with *S*-thiolations before non-reducing SDS-PAGE.

6. Acrylamide–bisacrylamide solution: 30:0.8% (w/v) acrylamide–bisacrylamide solution (e.g., Applichem).
7. 1D Separating gel solution: Separating gel buffer, acrylamide–-bisacrylamide solution, SDS, APS, TEMED, Aqua dest. For preparation of the separation gel solution mix 4.7 mL distilled water, 5 mL separating gel buffer, 10 mL acrylamide–bisacrylamide solution, and 0.2 mL 10% (w/v) SDS. Add 100 μL APS and 10 μL TEMED to initiate the polymerization process.
8. 1D Stacking gel solution: Stacking gel buffer, acrylamide–bisacrylamide solution, SDS, APS, TEMED, Aqua dest. For preparation of stacking gel solution mix 6.1 mL distilled water, 2.5 mL Stacking gel buffer, 1.3 mL 30:0.8% (w/v) acrylamide–bis, and 100 μL 10% (w/v) SDS. Add 50 μL 10% (w/v) APS and 5 μL TEMED to initiate the polymerization process.
9. SDS–Tris–Glycine running buffer: is described in Subheading 2.1.2, **item 21**.
10. Colloidal Coomassie staining solution: is described in Subheading 2.1.2, **item 24**.
11. Gel fixation solution: 40% (v/v) ethanol, 10% (v/v) acetic acid.
12. Digester solutions for in-gel tryptic digestion: Digester-A: 50 mM NH_4HCO_3, pH 8; 50% (v/v) acetonitrile. Digester-B: 75% (v/v) acetonitrile. Spotter solution: 0.5% (v/v) TFA; 50% (v/v) acetonitrile.
13. Trypsin solution: 20 μg trypsin (e.g., Promega) is dissolved in 1 mL distilled water.

3 Methods

Fluorescent-label thiol-redox proteomics to visualize reversibly oxidized proteins in bacteria in response to oxidative stress.

3.1 Fluorescent-Label Thiol-Redox Proteomics: Growth, Harvesting of Cells, and Blocking of Reduced Thiols by IAM

Fluorescent-label thiol-redox proteomics is carried out to visualize reversibly oxidized proteins in bacteria in response to oxidative stress.

1. Inoculate an overnight culture of bacterial cells into two flasks each containing 100 mL defined minimal medium to an optical density at 500 nm (OD_{500}) of 0.05.
2. Grow bacterial cells until the mid-exponential phase (e.g., OD_{500} of 0.4). Flask 1 is used as control and flask 2 is exposed to sublethal NaOCl concentration which allows the cells to recover from growth arrest. The sublethal NaOCl concentration has to be determined for the specific bacterium before the stress experiment.

3. Harvest bacterial cultures in flask 1 before the stress (control) and that in flask 2 after 30 min of NaOCl stress exposure in 2 × 50 mL Falcon-like tubes on ice by rapid centrifugation (ca 9500 × *g*, 5 min, 4 °C).
4. Wash cell pellets immediately in 1–2 mL TE-IAM buffer to remove the oxidant, transfer the cells to 2 mL Eppendorf-like tubes, repeat centrifugation using a microfuge (19,400 × *g*, 5 min, 4 °C).
5. Resuspend washed cell pellet on ice immediately in 400 μL UCE buffer with 100 mM IAM. Do not vortex to avoid air bubbles and resuspend the cell pellet slowly using a pipette tip (*see* **Note 2**).
6. Break cells in UCE-IAM buffer immediately using a homogenizer in the presence of glass beads and remove glass beads by short centrifugation in a microfuge (19,400 × *g*, 5 min, 4 °C). Thiol alkylation is performed for 15 min in the dark (*see* **Note 2**).
7. Add four parts pure acetone to one part of protein extract in 2 mL Eppendorf-like tubes (e.g., add 0.4 mL protein extract to 1.6 mL acetone), precipitate proteins for 1 h on ice and centrifuge protein pellet in a microfuge (19,400 × *g*, 20 min, room temperature (RT)).
8. Wash protein pellet with 80% (v/v) acetone by mixing up the pellet mechanically with a pipette tip to ensure the proteins are suspended and any trace of IAM is removed. Remove supernatant after centrifugation using a microfuge (19,400 × *g*, 20 min, 20 °C) and repeat washing and centrifugation with 80% (v/v) acetone four times. Dry the washed protein pellet in a vacuum centrifuge and store proteins frozen at −20 °C until reduction and BODIPY labeling **(*see* Note 3).**

3.2 Fluorescent Labeling of Reversibly Oxidized Proteins by BODIPY-IAM and Isoelectric Focusing of the Labeled Protein Extracts

1. Dissolve dried protein pellet from a 50 mL cell culture in 150 μL UCE buffer (without IAM) by shaking for 30–60 min. Remove cell waste by centrifugation in a microfuge (20 min, 17,900 × *g*, RT), transfer the supernatant into a new Eppendorf tube and determine protein concentration using Bradford reagent according to manufacturer instructions.
2. Transfer 250 μg protein extract into a new Eppendorf tube and fill up to 50 μL with UCE buffer. Reduce protein disulfides with 1 mM TCEP for 30 min at RT in the dark (add 5 μL 10 mM TCEP to 50 μL protein extract). Perform fluorescence labeling of the protein extract by addition of 6 μL BODIPY-IAM stock solution and incubate for 15 min at RT in the dark (*see* **Note 4**).
3. Equilibrate spin columns with 500 μL RH solution, centrifuge twice (1 min, 1020 × *g*, RT) and discard the flow-through.

Add BODIPY-IAM-labeled protein extract to the spin columns, centrifuge (4 min, 1020 × *g*, RT) and collect the protein containing flow-through into a new Eppendorf-like tube. The excess BODIPY-IAM will be retained in the spin columns. Add 280 μL RH solution to the BODIPY-IAM-labeled proteins, centrifuge (15 min, 17,900 × *g*, RT) and use the supernatant for rehydration of IPG strips in the dark overnight (*see* **Note 5**).

4. For rehydration of IPG strips use SERVA IPG Blue Strips pH 4–7 NL/18 cm. Add 360 μL of RH solution containing the BODIPY-IAM-labeled proteins to the rehydration chamber. Place IPG strips into the rehydration solution within the chamber and remove air bubbles below the IPG strips. Cover the rehydration chamber with a parafilm to avoid drying out of IPG strips. Protect rehydration chamber from light. Reswelling of the IPG strips is performed overnight at RT (*see* **Note 6**).
5. Place the IPG strips in the isoelectric focusing chamber. On top at both ends of the IPG strips place electrode strips that are soaked with distilled water. Place the electrodes on top of these electrode strips and connect these with the cathode and anode. Overlay the IPG strips with mineral oil, start running the IEF using the power supply EPS3500-XL according to the IEF following parameters (*see* **Note 7**):

1.	Step: 150 V	1 mA	5 W	150 Vh	1 h
2.	Step: 300 V	1 mA	5 W	300 Vh	1 h
3.	Step: 600 V	1 mA	5 W	600 Vh	1 h
4.	Step: 1500 V	1 mA	5 W	1500 Vh	1 h
5.	Step: 3000 V	1 mA	5 W	57,000 Vh	19 h

6. Remove IPG strips from the chamber and soak the mineral oil from the surface of the strips by a filter paper. Store IPG strips at −20 °C until separation by SDS-PAGE.

3.3 Second Dimension of 2D SDS-PAGE of Fluorescent-Labeled Proteins

1. For 12.5% acrylamide–bisacrylamide 2D gels prepare the separating gel solution as described in Subheading 2.1.2, **item 19**. Immediately after initiating the polymerization process cast the separating gel solution into the funnel of the multicasting chamber. Remove bubbles by pressing the flexible tube which connects the funnel and the multicasting chamber. Open the valve to cast slowly the gel solution through the flexible tube into the multicasting chamber. Make sure to avoid air bubbles in the chamber. Cover the separating gels with butanol to ensure a homogeneous polymerization of the gel. Acrylamide polymerization is finished within 2 h.

2. Prepare 100 mL stacking gel solution as described in Subheading 2.1.2, **item 20**. Keep 30 mL of the stacking gel solution for embedding the IPG strips at 4 °C. Rinse the surface of the gels with distilled water to remove the butanol. Add 380 μL 10% (w/v) APS and 62.5 μL TEMED to the stacking gel solution to initiate the polymerization process and mix gently. Load stacking gel solution with APS and TEMED immediately on top of the separating gel using a pipette. Cover the stacking gel with butanol to ensure a homogeneous polymerization of the gel. Polymerization is finished within 1 h.
3. After stacking gel polymerization, remove the 2D gels from the multicasting chamber and insert it into the running gel chamber. Fill the running gel chamber with 1× SDS–Tris–glycine running buffer. Equilibrate each IPG strip with 4.5 mL equilibration solution (A) for 15 min. Remove solution (A) and equilibrate each IPG strip with 4.5 mL solution (B) for 15 min. Remove solution (B), and load the IPG strips onto the 2D gels. Mix 30 mL stacking gel solution (from **step 2** in this section) with 460 μL 10% (w/v) APS and 75 μL TEMED using a magnetic stirrer and cover IPG with this stacking gel solution. Finally, fill the upper gel chamber with 1× SDS–Tris–glycine running buffer. Perform the SDS-PAGE run according to the following parameters: 300 V and 300 mA; 10 min 20 W per gel and overnight 2 W per gel. Stop the run when the bromphenol blue front has reached the bottom of the gel. After the run is finished, remove the gels from the glass plates. Scan the gels to observe fluorescence images immediately using a Typhoon scanner according to the following parameters: Fluorescence setup: 520 nm, BP40, 488 nm; Filter: Blue FAM; Mode: platen.
4. Fix 2D-gels in gel fixation solution for at least 1 h. Mix 200 mL Colloidal Coomassie stock solution with 50 mL methanol and use this solution for staining of the gels overnight. Next day, rinse the gels with distilled water twice and scan the Coomassie image. 2D gel images are quantified using the Decodon Delta 2D software by calculating ratios of fluorescence/protein amounts (*see* **Note 8**) (Figs. 2b and 3).

3.4 Bacterial Cultivation, Harvesting of Cells, and Preparation of Protein Extract for Identification of Protein S-Thiolation Using ESI-LC-MS/MS Analysis

1. Inoculate three flasks each containing 100 mL minimal medium with overnight cultures to an optical density of OD_{500} of 0.07 and grow cells until the mid-exponential growth phase (OD_{500} of 0.5–1.0). Flask 1 is used as untreated control. Flask 2 is exposed to diamide, and flask 3 is treated with NaOCl at sublethal concentrations each for 30 min. These sublethal NaOCl and diamide concentrations should reduce the growth rate and have to be determined before in detailed growth analyses.

2. Harvest the three bacterial cultures from untreated cells (control) and cells exposed for 30 min to diamide and NaOCl stress in 50 mL Falcon tubes on ice by rapid centrifugation (9500 × *g*, 10 min, 4 °C).
3. Wash bacterial cell pellets in TE-IAM buffer, transfer to Eppendorf tubes, centrifuge again (17,900 × *g*, 5 min, 4 °C) and resuspend the cell pellets in 1 mL UCE-IAM buffer on ice.
4. Disrupt cells using the homogenizer on ice and centrifuge cell extracts twice for 30 min at 4 °C to remove the cell debris. Alkylate the proteins containing free thiol-groups by incubation of the protein extract for 15 min at RT in the dark.
5. Determine the protein concentration by using Bradford reagent according to manufacturer instructions. Concentrate aliquots of 200 μg protein extract in the vacuum centrifuge to 15 μL and dissolve proteins in 15 μL non-reducing SDS-sample buffer (without DTT). Continue with protein separation using SDS gels.

3.5 Protein Fractionation by Separation on 15% SDS-PAGE for Identification of Protein S-Thiolation Using ESI-LC-MS/MS Analysis

1. For preparation of two 15% non-reducing SDS-PAGE separating gels, prepare 1D separating gel solution freshly (Subheading 2.2.2, **item** 7) and immediately cast the separating gels between two glass plates, cover the gel surface with 1 mL butanol and allow polymerization for 30–45 min.
2. For preparation of 1D SDS-PAGE stacking gel, first remove butanol and rinse the gel surface with Aqua dest. Prepare 1 D stacking gel solution (Subheading 2.2.2, **item 8**) and cast the stacking gel solution on top of the separation gel. Place the comb into the stacking gels and allow polymerization for 40 min.
3. Place the gels into the running chamber and fill up with running buffer. For protein fractionation, load 2 wells each with 200 μg protein extract in non-reducing SDS-sample buffer for control, diamide, and NaOCl extract, respectively. Separate proteins in 15% non-reducing SDS-PAGE according to manufacturer's description. Fix the SDS gels in gel fixation solution for 1 h. Mix 200 mL of Colloidal Coomassie stock with 50 mL methanol and stain the gels overnight. Next day, rinse the gels with distilled water twice. At this point gel can be stored at 4 °C covered in for example transparent film to avoid drying till in-gel digestion. Each gel lane is cut into ten slices and used for in-gel tryptic digestion and subsequent LC-MS/MS-analysis for identification of protein *S*-thiolations (*see* **Note 9**).

3.6 In-Gel Tryptic Digestion and Peptide Preparation for Mass Spectrometry

1. Destain the gel pieces using Digester-A solution under vigorous agitation for 20 min at 37 °C. Discard the supernatant and repeat destaining several times until the gel pieces are completely destained.
2. Shrink the destained gel pieces in Digester-B solution under vigorous agitation for 30 min at RT. Discard the supernatant and dry the gel pieces in a vacuum centrifuge for 10–20 min.
3. For the tryptic in-gel digestion add 5–20 μL trypsin solution to the dried gel pieces until the gel is completely reswollen. Perform tryptic digestion overnight at 37 °C.
4. Extract tryptic peptides with 100 μL spotter solution under vigorous agitation for 60 min at RT. Concentrate the peptides in the vacuum centrifuge and store at −20 °C until LC-MS/MS analysis.

4 Notes

1. The BODIPY FL C1-IA stock solution can be stored as frozen aliquots for 6 month and should be discarded when precipitates appear in the stock solution.
2. Cell sampling, washing, and blocking of reduced thiols by IAM for the fluorescent-label Thiol-Redox proteomics method needs to be done at the same day until the acetone-precipitation step is completed and the IAM-alkylated protein extracts is dried (**step 8**). If the cell pellet is not quickly dissolved and broken in UCE-IAM buffer, artificial oxidation can occur since reduced thiols are not blocked by IAM. The acetone precipitated protein pellet can be stored for several days at −20 °C.
3. Washing of the IAM-alkylated protein extract in acetone must be done carefully by using a pipette tip to mechanically disturb the protein pellet. Traces of remaining IAM must be avoided since these prevent the BODIPY-IAM labeling.
4. For BODIPY-IAM labeling, we recommend to spin down the BODIPY-IAM stock solution for removal of precipitated fluorescent dye which affects the thiol labeling.
5. Equilibrate the Biorad spin columns directly before the removal of excess BODIPY-IAM dye from the samples.
6. Reswelling of the IPG strips, IEF and the SDS-PAGE should be performed in the dark. The IPG strips should be subjected to SDS-PAGE directly after the IEF run is finished to avoid loss of the fluorescence label.

7. The IEF running parameters are used according to the instructions as recommended for IPG BlueStrips from SERVA (http://www.scie-plas.com/documents/SCIE-PLAS/distributors/445.pdf.)
8. The scanning of the fluorescence images is performed using a Typhoon scanner and requires to wet the surface of the scanner with distilled water. After the gel run, the 2D gel is removed from the glass plates and placed immediately to the scanner surface in the water by avoiding air bubbles under the gel. For gel scanning, the PMT voltage needs to be adjusted according to the fluorescence intensity of the spots. We further recommend for the fluorescence quantification to run all gels of one experiment (e.g., gels of controls and stress samples) at the same day in one 2D gel run.
9. For preparation of protein extracts for direct identifications of *S*-thiolations using Orbitrap LC-MS/MS analysis, the reduced thiols need to be IAM-alkylated directly after cell harvesting and alkylated proteins are subsequently used for fractionation by non-reducing SDS-PAGE. Protein sample processing at the same day avoids precipitation of IAM-alkylated proteins which often occurs after freezing the IAM-alkylated protein extracts. To maintain the thiol-modifications, it is required that cells need to be broken, alkylated, and processed by SDS-PAGE at the same day. However, it is possible to store cells frozen at −70 °C before preparation of protein extracts for SDS-PAGE and mass spectrometry.

Acknowledgments

This work was supported by grants from the Deutsche Forschungsgemeinschaft (DFG) AN746/4-1 and AN746/4-2 within the DFG priority program SPP1710, by the DFG Research Training Group GRK1947, project [C1] and by an ERC Consolidator grant (GA 615585) MYCOTHIOLOME to H.A.

References

1. Winterbourn CC, Kettle AJ (2013) Redox reactions and microbial killing in the neutrophil phagosome. Antioxid Redox Signal 18 (6):642–660
2. Forman HJ, Torres M (2001) Redox signaling in macrophages. Mol Asp Med 22 (4–5):189–216
3. Antelmann H, Helmann JD (2011) Thiol-based redox switches and gene regulation. Antioxid Redox Signal 14(6):1043–1063
4. Dalle-Donne I, Rossi R, Colombo G, Giustarini D, Milzani A (2009) Protein S-glutathionylation: a regulatory device from bacteria to humans. Trends Biochem Sci 34 (2):85–96
5. Fahey RC (2013) Glutathione analogs in prokaryotes. Biochim Biophys Acta 1830 (5):3182–3198
6. Chi BK, Busche T, Van Laer K, Bäsell K, Becher D, Clermont L, Seibold GM, Persicke M, Kalinowski J, Messens J,

Antelmann H (2014) Protein *S*-mycothiolation functions as redox-switch and thiol protection mechanism in *Corynebacterium glutamicum* under hypochlorite stress. Antioxid Redox Signal 20(4):589–605

7. Chi BK, Gronau K, Mäder U, Hessling B, Becher D, Antelmann H (2011) *S*-bacillithiolation protects against hypochlorite stress in *Bacillus subtilis* as revealed by transcriptomics and redox proteomics. Mol Cell Proteomics 10 (11). https://doi.org/10.1074/mcp.M111.009506
8. Chi BK, Roberts AA, Huyen TT, Bäsell K, Becher D, Albrecht D, Hamilton CJ, Antelmann H (2013) *S*-bacillithiolation protects conserved and essential proteins against hypochlorite stress in firmicutes bacteria. Antioxid Redox Signal 18(11):1273–1295
9. Hochgräfe F, Mostertz J, Albrecht D, Hecker M (2005) Fluorescence thiol modification assay: oxidatively modified proteins in *Bacillus subtilis*. Mol Microbiol 58(2):409–425
10. Hochgräfe F, Mostertz J, Pöther DC, Becher D, Helmann JD, Hecker M (2007) *S*-cysteinylation is a general mechanism for thiol protection of *Bacillus subtilis* proteins after oxidative stress. J Biol Chem 282 (36):25981–25985
11. Kehr S, Jortzik E, Delahunty C, Yates JR 3rd, Rahlfs S, Becker K (2011) Protein *S*-glutathionylation in malaria parasites. Antioxid Redox Signal 15(11):2855–2865
12. Leichert LI, Gehrke F, Gudiseva HV, Blackwell T, Ilbert M, Walker AK, Strahler JR, Andrews PC, Jakob U (2008) Quantifying changes in the thiol redox proteome upon oxidative stress in vivo. Proc Natl Acad Sci U S A 105(24):8197–8202
13. Lind C, Gerdes R, Hamnell Y, Schuppe-Koistinen I, von Lowenhielm HB, Holmgren A, Cotgreave IA (2002) Identification of *S*-glutathionylated cellular proteins during oxidative stress and constitutive metabolism by affinity purification and proteomic analysis. Arch Biochem Biophys 406 (2):229–240
14. Thamsen M, Jakob U (2011) The redoxome: proteomic analysis of cellular redox networks. Curr Opin Chem Biol 15(1):113–119

Part V

Metaproteomics

Chapter 19

Sequential Isolation of DNA, RNA, Protein, and Metabolite Fractions from Murine Organs and Intestinal Contents for Integrated Omics of Host–Microbiota Interactions

Pranjul Shah, Emilie E. L. Muller, Laura A. Lebrun, Linda Wampach, and Paul Wilmes

Abstract

The gastrointestinal microbiome plays a central role in health and disease. Imbalances in the microbiome, also referred to as dysbiosis, have recently been associated with a number of human idiopathic diseases ranging from metabolic to neurodegenerative. However, to causally link specific microorganisms or dysbiotic communities with tissue-specific and/or systemic disease-associated phenotypes, systematic in vivo studies are fundamental. Gnotobiotic mouse models have proven to be particularly useful for the elucidation of microbiota-associated characteristics as they provide a means to conduct targeted perturbations followed by analyses of induced localized and systemic effects. Here, we describe a methodology in the framework of systems biology which allows the comprehensive isolation of high quality biomolecular fractions (DNA, RNA, proteins and metabolites) from limited and/or heterogeneous sample material derived from murine brain, liver, and colon tissues, as well as from intestinal contents (fecal pellets and fecal masses). The obtained biomolecular fractions are compatible with current high-throughput genomic, transcriptomic, proteomic, and metabolomic analyses. The resulting data fulfills the premise of systematic measurements and allows the detailed study of tissue-specific and/or systemic effects of host–microbiota interactions in relation to health and disease.

Key words Biomolecular isolation, Host–microbe interactions, Integrated omics, Limited sample material, Microbiota, Mouse model, Sample heterogeneity

1 Introduction

Microbiota of the gastrointestinal tract (GIT) play pivotal roles in governing health and disease in humans and other mammals [1]. Imbalances in the gastrointestinal microbiome (dysbiosis) for example due to the overgrowth of facultative pathogenic strains have emerged as disease-associated traits, yet whether these alterations represent cause or consequence remains largely unknown. Examples of diseases in which the GIT microbiome has been implicated include diabetes [2], cancer [3, 4], obesity [5], inflammatory

Dörte Becher (ed.), *Microbial Proteomics: Methods and Protocols*, Methods in Molecular Biology, vol. 1841, https://doi.org/10.1007/978-1-4939-8695-8_19,

diseases [6] and neurological disorders [7, 8]. Based on large-scale meta-omic analyses, hypotheses linking gastrointestinal dysbiosis and potential pathological changes in different organs such as the brain, liver or colon have emerged [4, 8–12]. In this respect, experimental approaches involving the conventionalization of germ-free mice with defined microbiota formulations provide a compelling experimental approach to study localized as well as systemic effects of the gastrointestinal microbiota in relation to host phenotypes [13]. Currently, targeted perturbations in germ-free animal models represent the sole means to validate hypotheses relating to the role of microbial dysbiosis in the pathogenesis of disease [13].

Through the use of gnotobiotic mouse models, it was demonstrated that effects of the gastrointestinal microbiota are extensive and therefore should be assessed systemically at a multiorgan level [14]. Examples include investigations of the gut–liver [15, 16] and gut–brain [7, 8, 17, 18] axes to establish the fundamental molecular interactions between the gastrointestinal microbiota and these organs. However, in order to comprehensively model and elucidate the complex host–microbial molecular exchanges and their fundamental role in relation to the host health and disease status [19], there is an imminent need to expand current host–microbial interaction models largely obtained from applying reductionist methods to systemic multiorgan and multi-omic models [20]. Such models will reflect the complexities of the interactions between mixed microbial communities in the gastrointestinal tract and different tissues/organs. Recently, computational approaches have been proposed to model the molecular interactions within microbial communities which provide a basis to model the overall dynamics of the complex microbiota and their metabolic potential [21, 22]. Such approaches can be complemented with other computational approaches which allow for example dynamic modeling of the host–microbe interactions [20]. Recently, this in silico approach has been applied to investigate the effects of host–microbial interactions between a mouse and a single bacterium under different feeding regimes [23]. However, although such computational models allow formulation and testing of hypotheses in rapid succession, they require validation using high-resolution and high-fidelity multi-omic analyses in vivo especially since community- and organism-wide interaction networks are expected to have emergent properties and be more than merely sums of their respective parts [24].

To study the role of the gastrointestinal microbiome in the pathogenesis of idiopathic medical conditions, large-scale multi-omic studies are yielding unprecedented datasets, which are allowing novel insights into microbe–microbe and host–microbe molecular interactions. Such models encompassing (meta)genomic, (meta)transcriptomic, (meta)proteomic, and (meta)metabolomic datasets [24] will allow inclusion of the dynamic community-wide interactions of the gastrointestinal microbiota and the host, allowing multiscale modeling of multiorganism and multiorgan interactions thereby providing fundamental mechanistic insights into the processes involved in disease pathogenesis.

For the seamless generation and subsequent integration of multi-omic datasets, it is essential to: (1) isolate ample quantities of high-quality biomolecular fractions encompassing DNA, RNA, proteins, and metabolites from often limited amounts of sample material, (2) isolate all biomolecular fractions from single unique samples to account for inherent sample heterogeneity due to multiple cell types being present in tissues as well as in the mixed microbial communities from intestinal contents [24–26], and (3) utilize a single standardized extraction protocol on different sample types to eliminate protocol-specific biases in the resulting multi-omic data and allow comprehensive comparative analyses. To address these requirements, we describe a methodology, which allows the comprehensive isolation of high quality biomolecular fractions (DNA, RNA, proteins and metabolites) from limited and/or heterogeneous sample material derived from murine brain, liver, and colonic tissues, as well as from intestinal contents (fecal pellets and fecal masses).

2 Materials

2.1 Chemicals

1. Pure ethanol (−20 °C).
2. Lysis buffer consists of RLT buffer from the AllPrep DNA/RNA/Protein Mini kit (Qiagen) supplemented with 10 μL/mL of 2-β-mercaptoethanol (room temperature).
3. RNA*later*-ICE® at −80 °C (Life Technologies).
4. Isopropanol.
5. Ice-cold chloroform (−20 °C).
6. Ice-cold methanol–ultrapure water (1:1, v/v), (−20 °C) (*see* **Note 1**).

2.2 Consumables

1. AllPrep DNA/RNA/Protein Mini kit (Qiagen) (*see* **Note 2**).
2. QIAshredder columns (Qiagen).
3. Sterile nucleic-acid-free 2, 4, and 5 mm stainless steel milling beads (*see* **Note 3**).
4. Sterile pipette tips.
5. Sterile 2 mL, 1.5 mL, and 0.25 mL microcentrifuge tubes.
6. Serological pipettes.
7. Parafilm.

2.3 Equipment

1. Oscillating Mill MM 400 (Retsch).
2. Shaking block heater.
3. Microcentrifuge.
4. High-speed centrifuge.
5. Micropipettes.
6. Sterile spatula.
7. Liquid nitrogen and carrier.
8. Ice and ice carrier.
9. Refrigerated centrifugal vacuum concentrator.
10. Spectrophotometer.
11. RNA quality control equipment.

2.4 Animals

All animal experimentation had been approved by the University of Luxembourg and the appropriate Luxembourg governmental agencies (Ministry of Health, Ministry of Agriculture). All mice were housed under a 12 h light–dark cycle with *ad libitum* access to water and food until euthanasia in a conventional animal housing facility [27]. At euthanasia, mice were deeply anesthetized with a ketamine–medetomidine mix (100 and 1 mg/kg, respectively), then transcardially perfused with PBS to remove the blood [27]. The organs were quickly dissected and then snap-frozen on crushed dry ice, and stored at −80 °C until use [27].

3 Methods

The workflow provided below allows for the purification of DNA, RNA, and protein fractions from murine brain (left hemisphere) tissue, liver tissue, colonic tissue, and intestinal contents (fecal pellets and fecal masses). Following comprehensive cell lysis, small molecules are extracted first using specific solvents followed by biomacromolecular isolations using chromatographic spin columns (Fig. 1). Given the labile nature of the RNA which can be isolated from colonic tissues and intestinal contents, pretreatment of these samples with an RNA stabilization solution is required and, consequently, small molecule fractions should be extracted from separate sample material.

All procedures have to be carried out in cold conditions except for the biomacromolecular extractions as detailed below. The purified fractions have to be preserved on ice until the assessment of their respective qualities. Extracted biomolecules can be flash frozen and stored at −80 °C until further analysis. The preparation of the individual reagents can be carried out at room temperature according to the manufacturer's instructions unless specified otherwise. Reagents should be stored at the specified temperatures. To

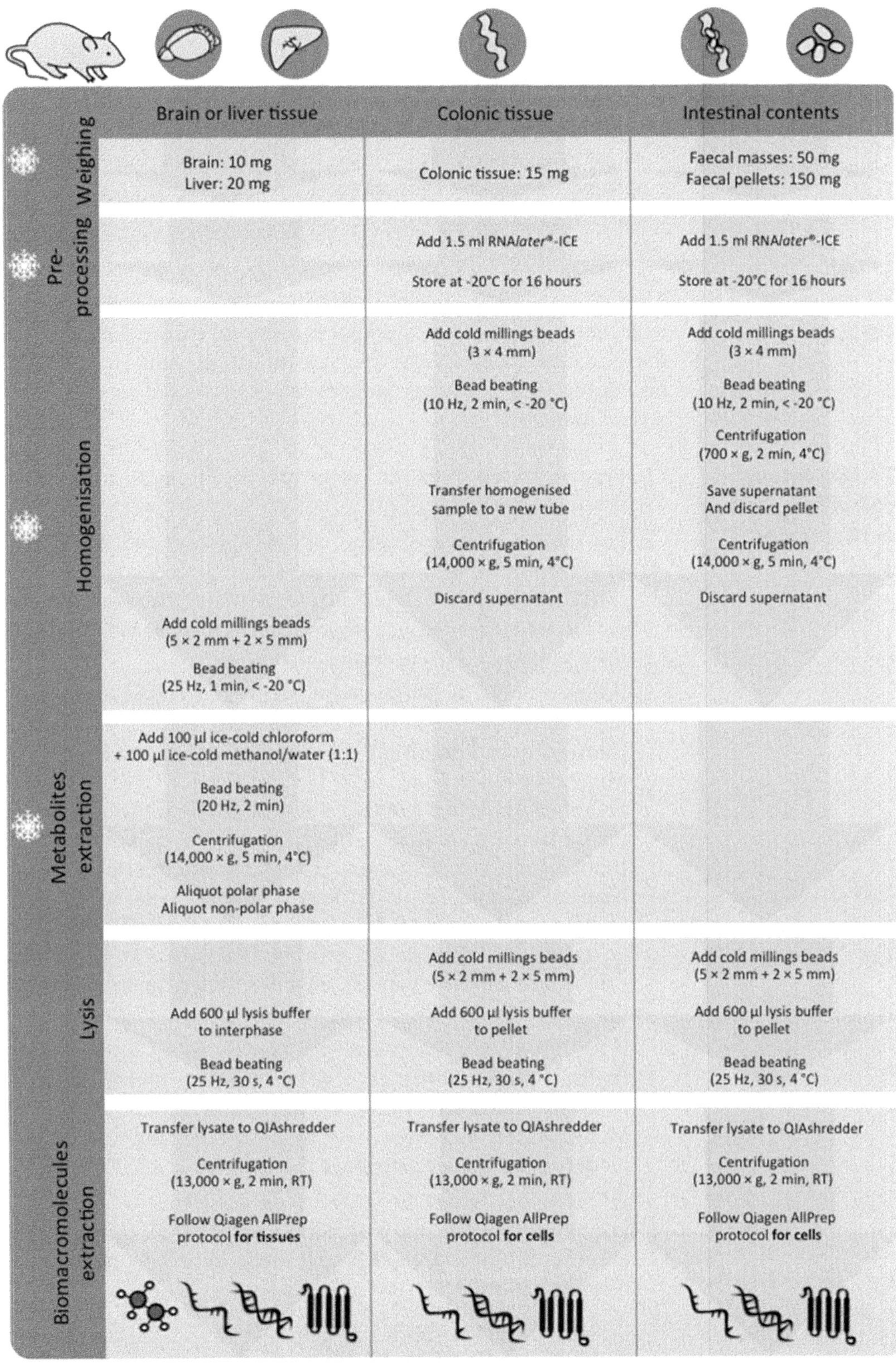

Fig. 1 Diagram summarizing the methodological workflow used to isolate the biomacromolecular fractions from the respective samples. The steps where it is essential that the samples are handled in cold conditions are depicted using an ice crystal. *Hz* Hertz, *RT* room temperature

ensure high-quality biomolecular extracts, bench space should be cleaned to remove any RNases, gloves should be worn during all sample-handling steps and disposable sterile, nucleic acid- and nuclease-free pipette tips, tubes, and other materials should be employed during the extractions. Some reagents used during the extraction protocol are toxic and must be handled and disposed of appropriately.

3.1 Sample Collection

Snap-freeze samples immediately in liquid nitrogen at the point of collection. This is the most important step to ensure the quality of the extracted biomolecular fractions and to prevent any degradation of the sample. Frozen samples can be preserved at −80 °C for several months.

3.2 Biomolecular Isolation from Brain and Liver Tissues

1. Prepare 10 mg of frozen brain tissue or 20 mg of frozen liver tissue (*see* **Note 4**).
2. For sample homogenization, add 5 × 2 mm and 2 × 5 mm diameter sterile nucleic-acid-free stainless steel milling beads (previously placed in liquid nitrogen) to the frozen samples and perform a cold bead beating in the Oscillating Mill MM400, for 1 min at 25 Hz (*see* **Notes 5** and **6**).
3. For metabolite extraction, sequentially add 100 μL of ice-cold methanol–ultrapure water (1:1,v/v) and 100 μL of ice-cold chloroform and perform a cold bead beating in the Oscillating Mill MM400 for 2 min at 20 Hz (*see* **Notes 5** and 7). Centrifuge the sample at 14,000 × *g* at 4 °C for 5 min to fractionate the polar (top) and nonpolar (bottom) metabolite phases. Immediately proceed to **step 4** with the interphase pellet (*see* **Note 8**). Aliquot 25 μL of the polar and nonpolar metabolites fractions into separate 0.5 mL glass vials (*see* **Note 9**). Dry polar fractions in a refrigerated centrifugal vacuum concentrator at −4 °C and air-dry nonpolar fractions under fume hood. Cap the glass vials and store them at −80 °C until further analysis (*see* **Note 10**).
4. For cell lysis, add 600 μL of lysis buffer to the interphase pellet. Perform a cold bead beating for 30 s at 25 Hz in the Oscillating Mill MM400 (*see* **Notes 5–7**). Transfer the lysate to a QIAshredder column and centrifuge the column at 13,000 × *g* for 2 min.
5. For biomacromolecular extraction from the flow-through, follow the AllPrep DNA/RNA/Protein Mini kit protocol for tissue (*see* **Note 2**).
6. Proceed with the respective quality controls of the isolated biomolecular fractions (*see* **Note 11**).

3.3 Biomolecular Isolation from Colonic Tissue

1. Prepare 15 mg of frozen colonic tissue (*see* **Note 4**).
2. For RNA stabilization, preprocess the sample by adding 1.5 mL of RNA*later*-ICE® to the tissue and store at −20 °C for at least 16 h prior to extraction (*see* **Note 12**).
3. For sample homogenization, add 3 × 4 mm diameter sterile nucleic acid-free stainless steel milling beads (liquid nitrogen cooled) to the sample and perform a cold bead beating in the Oscillating Mill MM400 for 2 min at 10 Hz (*see* **Notes 5** and 7). Transfer the RNA*later*-ICE containing the homogenized tissue to a new tube. Centrifuge the homogenized tissue for 5 min at 14,000 × *g* and 4 °C to pellet the sample. Discard the supernatant and proceed immediately to **step 4**.
4. For cell lysis, add 5 × 2 mm and 2 × 5 mm diameter sterile nucleic acid-free stainless steel milling beads (liquid nitrogen cooled) as well as 600 μL of lysis buffer to the pellet and perform a cold bead beating for 30 s at 25 Hz in the Oscillating Mill MM400 (*see* **Notes 5**–7). Transfer the lysate to a QIAshredder column and centrifuge the column at 13,000 × *g* for 2 min.
5. Perform sequential extractions of RNA, DNA and proteins from the flow-through, following the AllPrep DNA/RNA/Protein Mini kit protocol for tissue (*see* **Note 2**).
6. Proceed with the respective quality controls of the isolated biomolecular fractions (*see* **Note 11**).

3.4 Biomolecular Isolation from Intestinal Contents (Fecal Masses and Fecal Pellets)

1. Prepare 50 mg of frozen fecal masses or 150 mg of frozen fecal pellets (*see* **Note 4**).
2. For RNA stabilization, preprocess the sample by adding 1.5 mL of RNA*later*-ICE® to the sample and store at −20 °C for at least 16 h prior to extraction (*see* **Note 12**).
3. For sample homogenization, add 3 × 4 mm diameter sterile nucleic-acid-free stainless steel milling beads (liquid nitrogen cooled) and perform a cold bead beating in the Oscillating Mill MM400 for 2 min at 10 Hz (*see* **Notes 5**–7 and **13**). Centrifuge the homogenized samples at 700 × *g* for 2 min at 4 °C and transfer the supernatant to a new 2 mL tube. Centrifuge the supernatant for 5 min at 14,000 × *g* and 4 °C to pellet the sample. Discard the supernatant and proceed immediately to **step 4**.
4. For cell lysis add 5 × 2 mm and 2 × 5 mm diameter sterile nucleic-acid-free stainless steel milling beads (cooled in liquid nitrogen) as well as 600 μL of lysis buffer to the pellet and perform a cold bead beating for 30 s at 25 Hz in the Oscillating Mill MM400 (*see* **Notes 6** and 7). Transfer the lysate to a

QIAshredder column and centrifuge the column at 13,000 × *g* for 2 min.

5. Perform sequential extractions of RNA, DNA and proteins from the flow-through, following the AllPrep DNA/RNA/Protein Mini kit protocol for tissue (*see* **Note 2**).
6. Proceed with the respective quality controls of the isolated biomolecular fractions (*see* **Notes 11** and **14**).

4 Notes

1. In order to compare the total ion chromatogram (TIC) obtained following metabolomic analyses of the metabolites originating from multiple, histologically different sample types, it is essential to have a means to normalize the data. For this, it is typically recommended to include internal standards. For example, ^{13}C-ribitol (12 μL/mL of aqueous stock solution 0.02 mg/mL) can be added to the polar metabolite fractions to allow subsequent normalization. Furthermore, an internal standard also provides means to assess the variability between technical replicates.
2. Special attention should be paid to the preparation of all the solutions required for the extractions. In particular, remember to add supplemental components to the extraction buffers as requested by the manufacturer of the kit. In order to obtain high-quality biomolecular extracts, we recommend following all the optional steps suggested in the Qiagen AllPrep kit including additional drying of the columns and collecting biomolecules following second elutions from the columns in separate collection tubes. Similar kits allowing the sequential extraction of DNA, RNA and proteins are provided by others companies: their utilization may require further optimization.
3. After use, the beads are cleaned to remove any nucleic acid residues by successive washing steps with water, followed by 70% ethanol treatment. Finally, the beads are washed in concentrated bleach and rinsed with nucleic acid-free water before additional sterilization in an autoclave. To speed up the extraction protocol procedure, the beads can be prealiquoted into sterile tubes. To cool the beads, before transferring them to the sample tube, the tubes containing the beads can be immersed directly in liquid nitrogen.
4. It is very important to ensure that the samples are maintained in a frozen state during the entire weighing procedure. To guarantee the frozen state of the sample, the tubes and the different tools to cut the sample should be regularly placed in liquid nitrogen to cool them down. For the entire weighing step, we

recommend using plastic weighing cups, which can float at the surface of the liquid nitrogen, to maintain the sample material in the frozen state.

5. To avoid any degradation of the biomolecules due to thawing of the sample during the bead beating steps, the racks holding the samples in the Oscillating Mill MM400 should be immersed in liquid nitrogen for a short duration (10–15 s) prior to the bead beating step. The cold bead beating procedure and settings are highlighted for Oscillating Mill MM400. However, the use of a different cryomill could be envisaged but may require further optimizations of settings for ensuring proper sample homogenization, metabolite extraction and/or cell lysis.
6. To ensure the appropriate isolation of small molecules by the extraction solvents, the sample should be completely powdered at the end of this cold bead beating step. In case the sample is not completely pulverized, we recommend repeating the bead beating again after refreezing the sample in liquid nitrogen. Different bead size combinations can also be optimized for recalcitrant samples.
7. To prevent solvent spillage and loss of sample material due to lack of tightness of the tube caps, the rim of the sample tubes/caps should be sealed with Parafilm. The Parafilm should be removed before the centrifugation step in order to facilitate the efficient loading and removal of the samples on the centrifuge.
8. If required, the polar and nonpolar metabolite fractions can be stored at −20 °C for a few hours before aliquoting them into glass vials for drying. We recommend waiting until the RNA quality has been determined before aliquoting the metabolites, as a good quality RNA fraction is an indicator of a successful biomolecular extraction. The aliquoted and dried vials can be stored at −80 °C until further analysis.
9. When analyzing a large number of metabolite samples using mass spectroscopy, it is recommended to prepare separate pool samples by combining equal proportions of the polar and the nonpolar metabolite fractions of the individual samples, respectively. Pool samples serve the dual purpose of providing an additional means to normalize the metabolite profiles, and to check the quality of the measurement by scanning for instrument drift using the retention indices of the identified metabolite peaks in the pool samples.
10. Metabolites can be analyzed using many approaches and the volume of the metabolite fractions to be aliquoted for the analysis might need to be optimized based on the instrument as well as the type of metabolomic analysis. Here, the isolated polar and nonpolar metabolite fractions were analyzed using

GC-MS. The polar fractions were analyzed using split 10 procedure, while the nonpolar samples were analyzed using splitless mode. The metabolite peaks were identified using the Metabolite detector software [28].

11. The success of the biomolecular extraction can be assessed from the quantity and quality of the isolated biomolecules [25]. To confirm the quality of DNA extracts, gel electrophoresis using 1% agarose gel is recommended (Fig. 2). High molecular weight bands should be visible with minimal smear

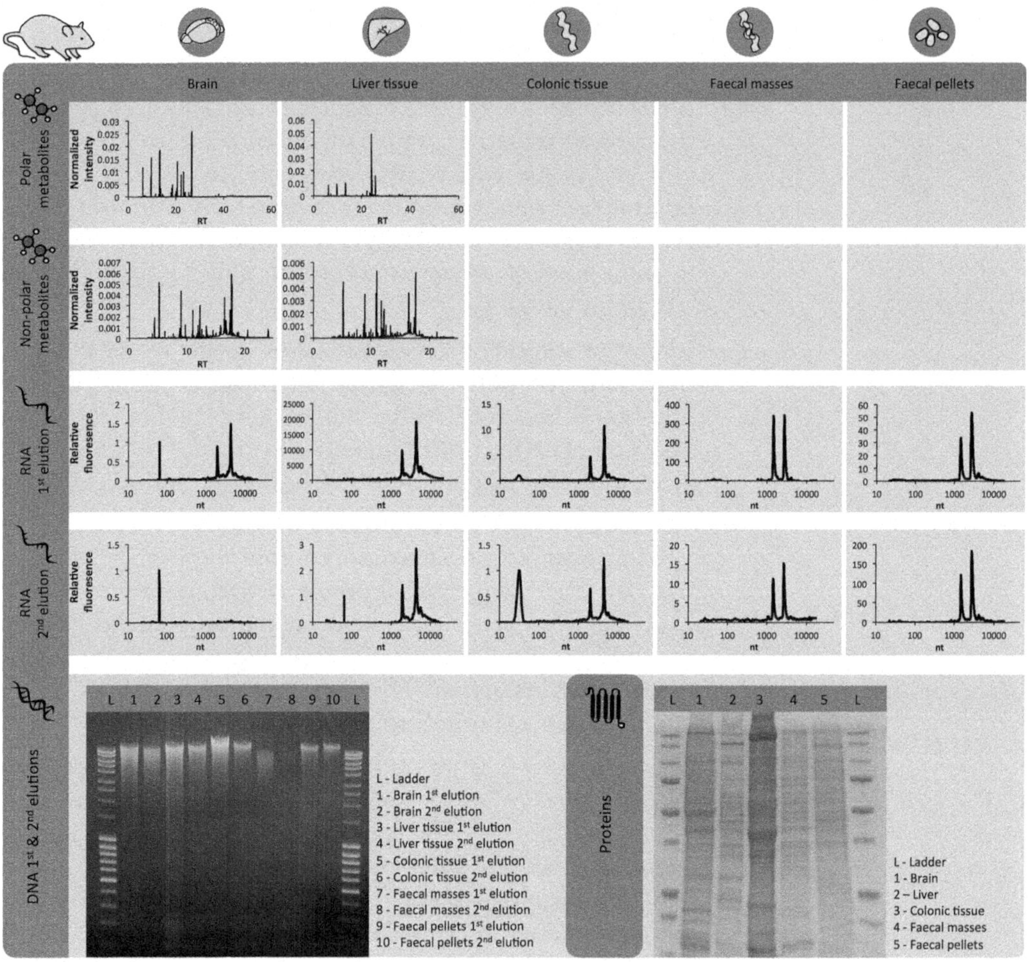

Fig. 2 Representative results highlighting the quality of the isolated biomolecular fractions from the different samples. Depicted are representative total ion chromatograms of the polar and nonpolar metabolite fractions analyzed by GC-MS, electropherograms of the RNA fractions obtained by capillary electrophoresis (Bioanalyzer), DNA fractions separated using agarose gel electrophoresis and protein fractions separated by 1D SDS-PAGE. GC-MS total ion chromatogram intensity values have been sum normalized. To correlate the RNA migration time to nucleotide size, marker and ladder peaks were included as reference [25] to fit the curve to a fourth order polynomial. The DNA and protein gels were prepared by loading similar concentrations of the different fractions. *RT* retention time, *nt* nucleotides

and RNA contamination. In case of RNA contamination in the DNA fraction (typically visible in the gel by two lower molecular weight bands corresponding to the rRNA fragments) RNA can be removed by digestion with an RNase.

RNA quality and quantity can be ascertained using for example a Agilent Bioanalyzer. A minimum RIN (Ribosomal Integrity Number) of 7 is usually required for subsequent transcriptomic analyses for example employing RNA sequencing (Fig. 2). The typical RIN values obtained on the RNA extracts of the samples analyzed in the present study were in the range of 8.8–10 ($n = 3$). For the assessment of the quality and quantity of protein extracts, 1D SDS-PAGE is utilized.RNA, DNA and protein fractions can be quantified using NanoDrop (Thermo) or Qubit® 2.0 Fluorometer (Life Technologies) measurements (Table 1). The clarity and quality of the total ion chromatograms (TICs) provide an indication of the quality and the purity of the representative metabolite fractions (Fig. 2).

12. In order to obtain high quality RNA from frozen samples of colonic tissue and intestinal contents, it is essential to stabilize the sample in RNA*later*-ICE® solution. The RNA*later*-ICE® solution has to be stored at −80 °C for at least 3 h prior to application to the sample. The intestinal contents should ideally be fragmented into smaller pieces allowing thorough penetration of the RNA*later*-ICE® to the entire sample. Store the samples in horizontal position at −20 °C for at least 16 h prior to extraction.
13. The intestinal contents are often difficult to pulverize for ensuring successful isolation of high quality extracts. It is therefore essential to ensure that these samples are completely pulverized and, if necessary, the cold bead beating step has to be repeated to pulverize the tissue entirely following the refreezing of the samples and racks.
14. The fecal masses from the intestinal contents may contain epithelial cells leading to isolation of mammalian as well as microbial biomolecular fractions. This typically results in the appearance of two series of peaks in the RNA electropherograms representing 16S and 18S rRNA as well as 23S and 28S rRNA, respectively. This circumstance needs to be taken into account in the downstream bioinformatic analyses, e.g., the filtering out of human sequences.

Table 1
Qualitative and quantitative evaluation of isolated nucleic acid fractions

Sample type	Biomolecules	Elution	Average concentration (ng/μL/)	Concentration standard deviation	Absorbance ratio range	
					A260/A230	A260/A280
Brain	RNA	1	132.73	82.55	1.09–2.04	2.09–2.11
Brain	RNA	2	2.87	0.15	0.02–0.5	2.14–2.7
Liver	RNA	1	811.53	382.88	1.09–1.81	2.05–2.09
Liver	RNA	2	111.43	47.41	0.27–1.58	2.04–2.1
Colonic tissue	RNA	1	377.87	303.59	0.59–1.46	2.08–2.1
Colonic tissue	RNA	2	43.10	26.59	0.43–0.84	2.01–2.06
Fecal masses	RNA	1	154.33	225.43	0.13–1.41	1.93–2.1
Fecal masses	RNA	2	12.70	6.60	0.13–0.69	1.51–1.97
Fecal pellets	RNA	1	531.80	86.30	1.38–2.25	2.12–2.16
Fecal pellets	RNA	2	27.87	6.20	0.23–0.52	1.98–2.07
Brain	DNA	1	85.47	5.92	0.11–0.21	1.95– 2.01
Brain	DNA	2	29.53	13.66	0.08–0.09	1.94–2.09
Liver	DNA	1	175.80	25.53	0.69–1.22	1.89–1.93
Liver	DNA	2	39.43	7.84	0.57–1.05	1.92–1.98
Colonic tissue	DNA	1	337.03	187.63	0.34–1.31	1.85–1.93
Colonic tissue	DNA	2	90.60	52.33	0.55–1.51	1.9–1.96
Fecal masses	DNA	1	107.93	144.25	0.38–0.85	1.9–2.02
Fecal masses	DNA	2	13.27	17.09	0.16–0.58	1.92–2.75
Fecal pellets	DNA	1	77.50	23.25	0.09–0.78	1.96–2.08
Fecal pellets	DNA	2	55.07	14.41	0.23–1.43	0.67–2.07
Brain	Protein	–	2.47	0.95	–	–
Liver	Protein	–	0.67	0.13	–	–
Colonic tissue	Protein	–	3.10	0.35	–	–
Fecal masses	Protein	–	1.92	0.70	–	–
Fecal pellets	Protein	–	3.03	0.15	–	–

Acknowledgments

This work was supported by an ATTRACT fellowship (ATTRACT/A09/03), CORE grant (CORE11/BM/1186762) and a European Union Joint Programming in Neurodegenerative Diseases grant (INTER/JPND/12/01) to P.W., and Aide à la Formation Recherche (AFR) grants to L.W. (PHD-2013-1/BM) and E.E.L.M. (PRD-2011-1/SR), all funded by the Luxembourg

National Research Fund (FNR). We are grateful to Anna Heintz-Buschart for SDS analysis and her valuable comments on the manuscript; Christian Jaeger and Karsten Hiller for GC-MS analysis; the LCSB mouse facility, especially, Djalil Coowar and Ludivine Treuillot for the provision of fecal pellets, and Manuel Buttini for the provision of mouse tissue samples.

References

1. Flint HJ, Scott KP, Louis P et al (2012) The role of the gut microbiota in nutrition and health. Nat Rev Gastroenterol Hepatol 9:577–589
2. Piñero-Piloña A, Raskin P (2001) Idiopathic Type 1 diabetes. J Diabetes Complicat 15:328–335
3. Zhu Q, Gao R, Wu W et al (2013) The role of gut microbiota in the pathogenesis of colorectal cancer. Tumour Biol 34:1285–1300
4. Irrazábal T, Belcheva A, Girardin SE et al (2014) The multifaceted role of the intestinal microbiota in colon cancer. Mol Cell 54:309–320
5. Villanueva-Millán MJ, Pérez-Matute P, Oteo JA (2015) Gut microbiota: a key player in health and disease. A review focused on obesity. J Physiol Biochem:1–17
6. Bellaguarda E, Chang EB (2015) IBD and the gut microbiota—from bench to personalized medicine. Curr Gastroenterol Rep 17:439
7. Hsiao EY, McBride SW, Hsien S et al (2013) Microbiota modulate behavioral and physiological abnormalities associated with neurodevelopmental disorders. Cell 155:1451–1463
8. Scheperjans F, Aho V, Pereira PAB et al (2014) Gut microbiota are related to Parkinson's disease and clinical phenotype. Mov Disord 30:350–358
9. Cryan JF, O'Mahony SM (2011) The microbiome-gut-brain axis: from bowel to behavior. Neurogastroenterol Motil 23:187–192
10. Minemura M, Shimizu Y (2015) Gut microbiota and liver diseases. World J Gastroenterol 21:1691–1702
11. Malaguarnera G, Giordano M, Nunnari G et al (2014) Gut microbiota in alcoholic liver disease: pathogenetic role and therapeutic perspectives. World J Gastroenterol 20:16639–16648
12. Sobhani I, Tap J, Roudot-Thoraval F et al (2011) Microbial dysbiosis in colorectal cancer (CRC) patients. PLoS One 6:e16393
13. Fritz JV, Desai MS, Shah P et al (2013) From meta-omics to causality: experimental models for human microbiome research. Microbiome 1:14
14. Claus SP, Tsang TM, Wang Y et al (2008) Systemic multicompartmental effects of the gut microbiome o mouse metabolic phenotypes. Mol Syst Biol 4:219–232
15. Del Chierico F, Gnani D, Vernocchi P et al (2014) Meta-omic platforms to assist in the understanding of NAFLD gut microbiota alterations: tools and applications. Int J Mol Sci 15:684–711
16. Szabo G (2015) Gut-liver axis in alcoholic liver disease. Gastroenterology 148:30–36
17. Duca F, Gérard P, Covasa M et al (2014) How gut and brain control metabolism. Front Horm Res 42:73–82
18. Mayer EA, Tillisch K, Gupta A (2015) Gut/brain axis and the microbiota. J Clin Invest 125:926–938
19. Joyce AR, Palsson BO (2006) The model organism as a system: integrating "omics" data sets. Nat Rev Mol Cell Biol 7:198–210
20. Thiele I, Heinken A, Fleming RMT (2013) A systems biology approach to studying the role of microbes in human health. Curr Opin Biotechnol 24:4–12
21. Harcombe WR, Riehl WJ, Dukovski I et al (2014) Metabolic resource allocation in individual microbes determines ecosystem interactions and spatial dynamics. Cell Rep 7:1104–1115
22. Faust K, Raes J (2012) Microbial interactions: from networks to models. Nat Rev Microbiol 10:538–550
23. Heinken A, Sahoo S, Fleming RMT et al (2013) Systems-level characterization of a host-microbe metabolic symbiosis in the mammalian gut. Gut Microbes 4:28–40
24. Muller EE, Glaab E, May P et al (2013) Condensing the omics fog of microbial communities. Trends Microbiol 21:325–333
25. Roume H, Muller EE, Cordes T et al (2012) A biomolecular isolation framework for eco-systems biology. ISME J 7:110–121
26. Roume H, Heintz-Buschart A, Muller EE et al (2013) Sequential isolation of metabolites, RNA, DNA, and proteins from the same unique sample. Methods Enzymol 531:219–236
27. Jaeger C, Glaab E, Michelucci A et al (2015) The mouse brain metabolome: region-specific signatures and response to excitotoxic neuronal injury. Am J Pathol 185:1699–1712
28. Hiller K, Hangebrauk J, Jäger C et al (2009) MetaboliteDetector: comprehensive analysis tool for targeted and nontargeted GC/MS based metabolome analysis. Anal Chem 81:3429–3439

Chapter 20

Utilization of a Detergent-Based Method for Direct Microbial Cellular Lysis/Proteome Extraction from Soil Samples for Metaproteomics Studies

Karuna Chourey and Robert L. Hettich

Abstract

Soil metaproteomics is a rapidly developing and rather complex field aimed at understanding the functionalities of soil microbial communities. One of the main challenges of such an approach is the availability of a robust and efficient protocol to extract proteins from soil microbes inhabiting this complex matrix. The wide range of soil types and the innumerable variations in soil properties confound this experimental goal. Here we present a detergent based, heat-assisted cellular lysis method coupled with trichloroacetic acid (TCA) precipitation of soil microbial proteins that has been developed in our lab and found to be reasonably robust and unbiased in extracting microbial proteins from a broad range of soils for downstream mass spectrometric characterizations of microbial metabolic activities in natural ecosystems.

Key words Soil proteomics, Metaproteomics, Protein extraction, Biological mass spectrometry, Soil protein extraction, Cell lysis

1 Introduction

The field of metaproteomics has emerged as an important component of a systems biology approach aimed at an in-depth study of the structure and function of microbial communities [1–3]. Redundancies in protein sequences and functionality are common in complex natural communities; however, recent advances in metaproteomics can connect the expression of certain proteins or protein pathways to the appropriate microbial members in the community [2, 3]. The information provided by metaproteomics will help harness the power of naturally occurring microbial members in soil for improvements in areas of human health, the environment, and biotechnology [1]. The success of metaproteomics hinges on the (1) ability to efficiently extract proteins from a complex community sample such as soil (2) the type of mass spectrometer used for measurements, as this impacts the quality and depth of the information, and (3) the availability of a well

Dörte Becher (ed.), *Microbial Proteomics: Methods and Protocols*, Methods in Molecular Biology, vol. 1841,
https://doi.org/10.1007/978-1-4939-8695-8_20,

assembled metagenome database [3]. The composition of the microbiome being studied also largely influences efficient protein extraction from a complex matrix. Microbial communities residing in human tissues and biofilms are usually embedded in environments made up of extracellular polymeric substances, but are usually relatively easy to extract and often are present at significant biomass quantities for extraction optimization. However, the application of metaproteomics for studying soil microbial communities is hindered by many factors, mainly involving the inherent complexity of the soil matrices [4], along with the fact that extracted proteins tend to get irreversibly adsorbed to soil particles or natural humic substances and are difficult to recover for metaproteomics based studies [5–7]. This complexity is further compounded by the presence of extracellular proteins from other living sources, the high diversity of microbial members (many of which are uncultivable and often underrepresented in the genomic database), and the presence of dormant or dead cells, spores, and fungi [5–8]. Efficient soil protein extraction therefore hinges on efficient cellular lysis and proteome extraction of proteins, with minimum recovery of interferences such as humic acids [9, 10].

Many methods have been proposed for soil protein extraction, but can basically be divided into two general approaches—either direct or indirect ([8, 10], and references within). In the direct method, cellular lysis is accomplished using the native unprocessed soil sample, and usually affords a more comprehensive sampling of the soil microbial community, although additional steps are often required to remove coeluting interferences [5, 7, 8, 10–13]. For the indirect method, cells are preferentially and physically separated from the soil matrix and then subjected to cellular lysis [8, 10, 14, 15]. While the indirect method avoids the simultaneous extraction of humics and other undesirable interferences, this protocol does run the risk of only sampling a subset of the microbial community (mostly the planktonic cells) from soil, while missing cells that are adherent to soil particles [8, 10]. Here we present a method for direct in-situ extraction of microbial proteins from soil using a detergent based buffer coupled with heat-assisted cellular lysis and selective precipitation of proteins using trichloroacetic acid (TCA) (*see* Fig. 1). Addition of the detergent, sodium dodecyl sulfate (SDS), disrupts the phospholipids in cellular membrane, thus lysing the cells. Also, the EDTA in the buffer chelates Ca^{+} and Mg^{+} ions, deactivating metal-requiring enzymes while disrupting cellular adhesion to soil or other matrix. Heating the soil slurry disrupts already weakened cellular membranes, thereby enhancing cellular lysis. This protocol is designed to disperse the detergent via lysis buffers throughout the soil sample to enhance effective lysis of

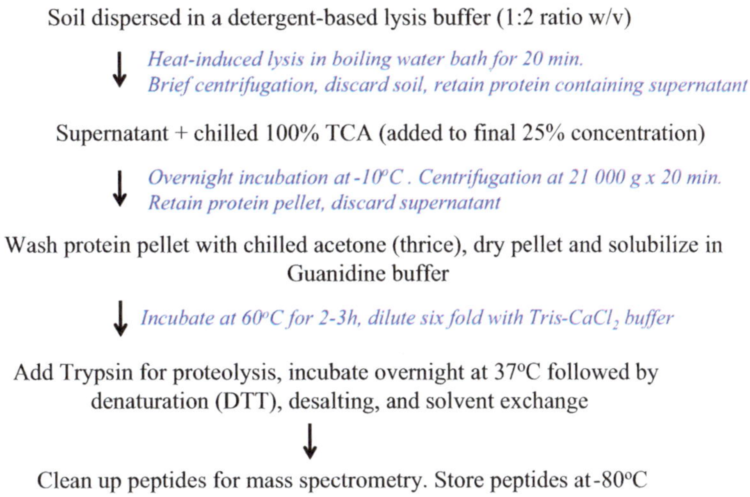

Fig. 1 Work flow showing the main steps in protein extraction from soil and sample preparation for mass spectrometry

adherent cells and inaccessible microbial cells trapped in the soil matrix. Once the cellular membranes are ruptured and the proteins are solubilized, the detergent helps them stay in solution rather than adhere to soil particles and other hydrophobic surfaces. Moreover, the addition of lysis buffer before the heating step inactivates the cellular machinery upon contact, thereby halting metabolic activities so that the heat treatment does not induce any heat-shock response in the microbial sample. Addition of TCA to the protein-containing soil supernatant selectively precipitates proteins, which are then collected via centrifugation and subjected to series of chilled acetone washes to remove detergent and soluble contaminants prior to air drying. The dried protein pellet can be solubilized in different buffers for various downstream analyses, such as SDS-PAGE or mass spectrometry analysis of proteins.

In this chapter we focus on a sample preparation protocol for mass spectrometry-based soil metaproteomics. The method was developed and tested on extraction of microbial proteins from sandy soil (Hanford, WA) and sandy, clay loam soil (Hopland Field station in California) [10]. The characteristics of the soil from California have been described elsewhere [16]. The protocol has been successfully employed in metaproteomics study of Arctic permafrost soil [17], deep sea methane seep sediment [18], subsurface sediments from Rifle, CO, USA [19], and other soil samples [20]. The method can also be employed for cellular lysis and protein extraction from microbial isolates [21, 22] and groundwater samples [23], enabling comparisons of lab-based proteomics studies with metaproteomics of environmental samples.

2 Materials

All buffers and solutions are prepared using water or acetonitrile (HPLC grade) with no further purification. Heat resistant gloves are used during sample heating process. Normal gloves are worn for all other steps to minimize contamination of the sample by keratin or other human proteins. Most of the buffers and solvents can be made and stored in advance, unless mentioned otherwise.

2.1 Cellular Lysis and Protein Extraction

1. Soil protein extraction buffer: 5% (w/v) sodium dodecyl sulfate (SDS) dissolved in 50 mM Tris–HCl buffer, pH 8.5, 150 mM sodium chloride (NaCl), 0.1 mM EDTA; 1 mM magnesium chloride ($MgCl_2$), fresh 50 mM dithiothretiol (DTT). Store at 4 °C and thaw to room temperature before use (*see* **Note 1**).
2. Trichloroacetic acid (TCA):100%, store at 4 °C.
3. Chilled acetone. The required amount of acetone is transferred to glass bottles and chilled at −20 °C for at least an hour before use.
4. Tris-$CaCl_2$ buffer: 50 mM Tris, 10 mM $CaCl_2$, pH 7.6. The buffer can be stored at room temperature.
5. Guanidine buffer: 6 M guanidine HCl, 10 mM dithiothreitol (DTT) in Tris-$CaCl_2$ buffer pH 7.6. Make fresh before use.
6. Modified sequencing grade trypsin (e.g., Promega or Sigma).
7. Formic acid.
8. Poly(vinylpolypyrrolidone) (PVPP) (e.g., Sigma).
9. Water bath with gently boiling water. (We typically use a beaker with boiling water on a hot plate).
10. 100 mL glass bottles (heat proof—i.e., pyrex) for sample heating
11. 1.5 or 2 mL Eppendorf tubes.
12. High speed centrifuge (up to 21,000 × *g*).
13. Freezer (−10 °C or −20 °C and − 80 °C).
14. Falcon tubes (15 mL or 50 mL).

2.2 Total Protein Estimation

1. RC/DC protein assay kit (Bio-Rad).
2. UV–visible spectrophotometer.

2.3 Desalting and Cleanup of Peptide Sample

1. C18 Sep-Pak solid phase extraction cartridges—Lite or Plus (Waters, Milford, MA).
2. Acidified water: HPLC grade water amended with 0.1% (v/v) formic acid. Store at room temperature.
3. Acidified acetonitrile: HPLC grade acetonitrile supplemented with 0.1% (v/v) formic acid. Store at room temperature.

4. Durapore PVDF filters, 0.45 μm (Millipore).
5. Vacuum centrifuge.

3 Methods

3.1 Protein Extraction

1. Weigh out required amount of soil (e.g., 5 g soil) in a glass bottle (*see* **Note 2**).
2. Add twice the amount (w/v) of soil protein extraction buffer (e.g., 10 mL for 5 g of soil) and disperse the soil particles; break up clumps by intermittent vigorous vortexing (5 min) (*see* **Note 3**).
3. Place the soil slurry in a boiling water bath for 20 min to lyse the cells and solubilize the proteins. After heating, remove the glass bottle from water bath and let cool for 5 min (*see* **Notes 4** and **5**).
4. Transfer the soil slurry to a fresh Falcon tube, vortex for 2 min, and then centrifuge at 2100 × *g* for 10 min to precipitate the soil particles (*see* **Note 6**).
5. Disperse the supernatant (cell lysate) into fresh Eppendorf tubes and add chilled TCA to final concentration of 25% (v/v) in protein solution. Cap the tubes and gently mix it by inverting the tube. Keep the tubes in freezer (−10 °C or −20 °C) for 10–12 h (overnight) to precipitate the proteins preferentially relative to the other soluble components of the sample.
6. Centrifuge the samples at 21,000 × *g* for 20 min (*see* **Note** 7).
7. Discard the supernatant and wash the pellet with 1 mL of chilled acetone. Centrifuge again at 21,000 × *g* for 10 min. Discard the acetone, retain pellet, and repeat the acetone wash step twice (*see* **Note 8**).
8. After the final acetone wash, discard the acetone and air-dry the protein pellet (*see* **Note 9**).
9. Dissolve the protein pellet in 1 mL of guanidine buffer and incubate at 60 °C for 1 h with intermittent vortexing to denature and reduce the proteins for subsequent effective proteolytic digestion with trypsin (*see* **Note 10**).
10. For samples where soil debris or humics (yellow solution color) are obvious, add 100 mg PVPP to pooled protein sample and gently mix the solution by keeping it on a nutator for 5 min. Centrifuge the solubilized protein solution at 20,800 × *g* for 10 min. The PVPP and any soil debris will precipitate, and the supernatant will appear clearer, since most of the humics will adhere to PVPP and be removed from the soluble fraction. The presence of observable debris or dark solution color suggests

insufficient centrifugation after TCA precipitation step or incomplete acetone washes.

11. Aliquot 25 μL of sample for total protein estimation. Use RC/DC protein assay kit and follow the manufacturer's protocol. Use Bovine Serum Albumin (provided in the kit) as protein standard to quantify the recovered amount of protein (*see* **Note 11**).

3.2 Protein Digestion

1. Dilute the soluble sample six-fold using Tris–$CaCl_2$ buffer and amend it with trypsin in the ratio of 1:25 w/w (trypsin–total protein), followed by overnight incubation at 37 °C with gentle mixing.
2. Following proteolysis, add DTT (10 mM final concentration) to keep disulfide bonds from forming in the peptide mixture, and incubate at room temperature for half an hour. Centrifuge the samples at 2100 × *g* for 5 min to pellet any debris.

3.3 Peptide Desalting and Cleanup

1. A common and convenient route is to desalt the supernatant with a Sep-Pak C18 cartridge (*see* **Note 12**). Depending on the sample size and composition, other routes for desalting (on-column cleanup, filtration, etc.) may be employed—as discussed in point #5 below and explained earlier [22].
2. Before use, the Sep-Pak cartridge should be washed by flushing with 5 mL of acidified acetonitrile, followed by 5 mL of acidified water using a syringe. The sample is then pushed through the syringe to capture the peptides onto C18 resin. The cartridge is washed with 5 mL of acidified water to remove the buffer and salts. Captured peptides are eluted from the cartridge into fresh Eppendorf tubes using 5 mL of acidified acetonitrile.
3. The peptides are solvent-exchanged from acidified acetonitrile into acidified water using a vacuum centrifuge to dry the initial solution and resuspend in aqueous conditions.
4. The resulting proteolytic peptides are finally suspended in 500 μL of acidified water and filtered through Durapore PVDF filters by centrifugation at 21,000 × *g* for 1 min to remove any debris.
5. For samples with low protein amounts, it may be preferable to forgo the Sep-Pak cleanup and solvent exchange step using the vacuum centrifuge. Thus, filter the samples using Durapore PVDF filter and load the samples directly onto biphasic strong cation exchange-Reverse phase C18 (SCX-C18) column. This should be followed by offline washes for 45 min. This 45 min step incorporates a gradual gradient from 95% acidified water to 75% acidified acetonitrile, three times (*see* **Note 13**).
6. The peptide sample (with or without Sep-Pak) can be stored in −80 °C until mass spectrometry analysis.

4 Notes

1. Use of a fume hood is recommended while weighing out the detergent, since the fine particles tend to disperse quickly in air and once inhaled can cause discomfort.
2. The amount of soil required for protein extraction/metaproteomics analysis is obviously influenced by the microbial biomass per gram soil. We have had success with protein extraction from 5 g soil containing at least 10^6 microbes/g soil. For soils having biomass amounts lower than that, it may be necessary to scale up thç soil amounts to 10 g or in some cases even to 20 g soil. For higher cell counts per gram soil (10^9 or higher), the amount of soil required can be scaled down to 2–3 g, which is beneficial if the soil sample is limited in quantity.
3. It is essential to break up soil clumps and disperse them uniformly in lysis buffer. If vortexing proves inadequate, sonication or a mortar and pestle (before adding buffer) can be used to break up soil clumps. Care should be taken while direct sonication of sample, since heat generated by the tip of sonicator may char the portion of sample in contact with the sonicator tip.
4. It is of utmost importance that the heat-assisted lysis step be carried out in heat-safe (i.e., pyrex) glass bottles. In our experience, the use of falcon tubes or other disposable containers tend to leach out polymeric type substances that interfere with the mass spectrometry measurements.
5. It is essential to use a water bath rather than direct heating of sample on hot plate or microwaving the sample. The latter steps can scorch/carbonize the sample at point of contact, thus destroying the proteins in the sample. The water bath for heat lysis should be maintained at gentle simmering rather than vigorous boiling. Use of heat protective gloves for sample handling is strongly recommended. Samples may be heated for longer times (up to 30 min), such as for more clay-type soils, to enhance solubilization. Heating time beyond 30 min seemed to have no positive impact on cellular lysis.
6. In case of samples consisting of finer clay-like soil (which frequently have high humic acid compositions), it may be difficult to spin down all soil debris. Under such circumstances, we recommend dispersing the supernatant to fresh tubes (or aliquot it in several Eppendorf tubes), followed by centrifugation at 21,000 × g for 5 min to pellet as much of the soil debris as possible. It is best to use a clear supernatant for the TCA precipitation step.

7. The use of a proper high speed centrifuge to attain the required speed of at least 21,000 × *g* is of utmost importance to precipitate proteins after TCA amendment. For centrifuging larger volumes of samples, we recommend using ultra-high speed centrifuges with appropriate type of centrifuges tubes. A preferred method is to divide the samples uniformly across several 2 mL Eppendorf tubes, and then centrifuge all of them at 21,000 × *g* using an Eppendorf benchtop centrifuge. While this step increases amount of tubes to be handled, it enhances protein precipitation and effective acetone washes for detergent removal. The centrifugation of proteins can be carried out at 4 °C using a refrigerated centrifuge (to continue keeping the proteins in chilled conditions) or at room temperature when a refrigerated centrifuge is not available.
8. Washing the protein pellet with acetone is a very important step. We recommend at least three rounds of chilled acetone washes of the protein pellet to remove lingering detergent. Any residual SDS in the sample will interfere with downstream analysis and will complicate (and possibly compromise) the mass spectrometric measurement. Usually by the third wash, the discarded acetone portion looks clear even in case of soil with high humics. Sometimes the protein pellet is not visible after first wash, but can be seen as translucent pellet after subsequent acetone washes. Utmost care should be taken while discarding acetone after each wash step, as the pellet tends be very mobile and can get easily discarded along with acetone.
9. The protein pellet can be air-dried by keeping the Eppendorf tubes open on bench top or 5 min incubation at 37 °C.
10. If the sample is distributed in several Eppendorf tubes, sample pooling should be carried out after solubilization in guanidine buffer. In case further processing of sample is to be temporarily halted, the protein sample can be stored in dried pellet form (preferred) or in guanidine solution by keeping it in refrigerator until further processing. For shipping of samples for MS analysis, transporting dry protein pellets at 4 °C or lower is recommended.
11. For total protein estimation, we find RC/DC protein assay kit to be reliable. Protein estimation can also be carried out after six fold dilution of the sample, since it dilutes out the guanidine and DTT, both considered interferences by most protein assay kits. If centrifugation or acetone washes are not carried out properly, contaminants from soil or SDS continue to linger in sample and interfere during protein estimation. This can be easily recognized by the instantaneous colorimetric change in

solution upon addition of Lowry's solution in final step of protein estimation.

12. Any sample cleanup step always involves some degree of unavoidable sample loss. The use of C18 Sep-Pak cartridges does desalt and clean up the sample reasonably well (resulting in a clear eluent); however there is usually noticeable sample loss with this step. When the total protein available for downstream analysis is not the limiting factor, the use of C18 Sep-Pak cartridges is highly recommended.
13. The loading of the peptide sample onto analysis column, accompanied by offline washing of the column removes the salts from the column, while retaining the peptides (and any humic acids that have made it through the sample preparation process). This minimizes sample loss associated with Sep-Pak cleanup. Peptides then are eluted sequentially off the column via gradient-elution chromatography steps during MS analysis. The downside to this approach is binding of undesirable soil constituents to the sample column that does not get removed by an offline wash. These unidentified components often elute with peptides during mass spectrometry analysis and interfere with downstream measurements.

Acknowledgments

Karuna Chourey and Robert L. Hettich acknowledge funding support from US Department of Energy, Genome Sciences Program. Oak Ridge National Laboratory is managed by UT-Battelle, LLC, for the US Department of Energy.

References

1. Abram F (2015) Systems-based approaches to unravel multi-species microbial community functioning. Comput Struct Biotechnol J 13:24–32
2. von Bergen M, Jehmlich N, Taubert M, Vogt C, Bastida F, Herbst FA, Schmidt F, Richnow HH, Seifert J (2013) Insights from quantitative metaproteomics and protein-stable isotope probing into microbial ecology. ISME J 7(10):1877–1885
3. Hettich RL, Pan C, Chourey K, Giannone RJ (2013) Metaproteomics: harnessing the power of high performance mass spectrometry to identify the suite of proteins that control metabolic activities in microbial communities. Anal Chem 85(9):4203–4214
4. Elsas JD, Jansson JK, Trevors JT (2006) Modern soil microbiology books in soils, plants, and the environment, 2nd edn. CRC Press, Taylor and Francis Group, Boca Raton
5. Nannipieri P (ed) (2006) Role of stabilised enzymes in microbial ecology and enzyme extraction from soil with potential applications in soil proteomics, vol 8. Nucleic acids and proteins in soil. Springer-Verlag, Berlin, Heidelberg
6. Solaiman Z, Kashem MA, Matsumoto I (2007) Environmental proteomics: extraction and identification of protein in soil, vol 2. Soil biology, advanced techniques in soil microbiology. Springer-Verlag, Berlin, Heidelberg
7. Bastida F, Moreno JL, Nicolas C, Hernandez T, Garcia C (2009) Soil metaproteomics: a review of an emerging environmental science. Significance, methodology and perspectives. Eur J Soil Sci 60:845–859

8. Becher D, Bernhardt J, Fuchs S, Riedel K (2013) Metaproteomics to unravel major microbial players in leaf litter and soil environments: challenges and perspectives. Proteomics 13(18–19):2895–2909
9. Ogunseitan OA (ed) (2006) Soil proteomics: extraction and analysis of proteins from soils, vol 8. Nucleic acid and proteins in soil. Berlin Heidelberg: Springer-Verlag.
10. Chourey K, Jansson J, VerBerkmoes N, Shah M, Chavarria KL, Tom LM, Brodie EL, Hettich RL (2010) Direct cellular lysis/protein extraction protocol for soil metaproteomics. J Proteome Res 9(12):6615–6622
11. Benndorf D, Balcke GU, Harms H, Bergen MV (2007) Functional metaproteome analysis of protein extracts from contaminated soil and groundwater. ISME J 1:224–234
12. Chen S, Rillig MC, Wang W (2009) Improving soil protein extraction for metaproteome analysis and glomalin-related soil protein detection. Proteomics 9(21):4970–4973
13. Schulze WX, Gleixner G, Kaiser K, Guggenberger G, Mann M, Schulze ED (2005) A proteomic fingerprint of dissolved organic carbon and of soil particles. Oecologia 142:335–343
14. Taylor EB, Williams MA (2010) Microbial protein in soil: influence of extraction method and C amendment on extraction and recovery. Microb Ecol 59:390–399
15. Benndorf D, Vogt C, Jehmlich N, Schmidt Y, Thomas H, Woffendin G, Shevchenko A, Richnow HH, Bergen MV (2009) Improving protein extraction and separation methods for investigating the metaproteome of anaerobic benzene communities within sediments. Biodegradation 20:737–750
16. Waldrop MP, Firestone MK (2006) Seasonal dynamics of microbial community composition and function in oak canopy and open grassland soils. Microb Ecol 52:470–479
17. Lau MC, Stackhouse BT, Layton AC, Chauhan A, Vishnivetskaya TA, Chourey K, Ronholm J, Mykytczuk NC, Bennett PC, Lamarche-Gagnon G, Burton N, Pollard WH, Omelon CR, Medvigy DM, Hettich RL, Pfiffner SM, Whyte LG, Onstott TC (2015) An active atmospheric methane sink in high Arctic mineral cryosols. ISME J 9(8):1880–1891
18. Glass JB, Yu H, Steele JA, Dawson KS, Sun S, Chourey K, Pan C, Hettich RL, Orphan VJ (2014) Geochemical, metagenomic and metaproteomic insights into trace metal utilization by methane-oxidizing microbial consortia in sulphidic marine sediments. Environ Microbiol 16(6):1592–1611
19. Handley KM, VerBerkmoes NC, Steefel CI, Williams KH, Sharon I, Miller CS, Frischkorn KR, Chourey K, Thomas BC, Shah MB, Long PE, Hettich RL, Banfield JF (2013) Biostimulation induces syntrophic interactions that impact C, S and N cycling in a sediment microbial community. ISME J 7(4):800–816
20. Bastida F, Hernandez T, Garcia C (2014) Metaproteomics of soils from semiarid environment: functional and phylogenetic information obtained with different protein extraction methods. J Proteome 101:31–42
21. Nissen S, Liu X, Chourey K, Hettich RL, Wagner DD, Pfiffner SM, Loffler FE (2012) Comparative c-type cytochrome expression analysis in *Shewanella oneidensis* strain MR-1 and *Anaeromyxobacter dehalogenans* strain 2CP-C grown with soluble and insoluble oxidized metal electron acceptors. Biochem Soc Trans 40(6):1204–1210
22. Sharma R, Dill BD, Chourey K, Shah M, VerBerkmoes NC, Hettich RL (2012) Coupling a detergent lysis/cleanup methodology with intact protein fractionation for enhanced proteome characterization. J Proteome Res 11 (12):6008–6018
23. Chourey K, Nissen S, Vishnivetskaya T, Shah M, Pfiffner S, Hettich RL, Loffler FE (2013) Environmental proteomics reveals early microbial community responses to biostimulation at a uranium- and nitrate-contaminated site. Proteomics 13 (18–19):2921–2930

Chapter 21

Sample Preparation for Metaproteome Analyses of Soil and Leaf Litter

Katharina M. Keiblinger and Katharina Riedel

Abstract

Soil and litter metaproteomics, assigning soil and litter proteins to specific phylogenetic and functional groups, has a great potential to shed light on the impact of microbial diversity on soil ecosystem functioning. However, metaproteomic analysis of soil and litter is often hampered by the enormous heterogeneity of the soil matrix and high concentrations of humic acids. To circumvent these challenges, sophisticated protocols for sample preparation have to be applied. This chapter provides the reader with detailed information on well-established protocols for protein extraction from soil and litter samples together with protocols for further sample preparation for subsequent MS analyses.

Key words Soil sampling, Homogenization, Matrix effects, Interfering substances, Protein separation, SDS-PAGE, In-gel digestion, Reducing sample complexity

1 Introduction

Protein extraction from soil and litter is one of the crucial steps in metaproteomic analyses and an essential prerequisite to obtain reliable and high-quality datasets. However, soil and litter are characterized by an enormous heterogeneity with respect to texture, pH, electrical conductivity, carbon–nitrogen–phosphorus stoichiometry and, most importantly, humic acid content. These compounds are known to adsorb proteins, and thus interfere with the extraction, purification, and quantification of proteins by colorimetric assays. Moreover, they may disturb the separation of proteins by SDS-PAGE and are known to hamper the chromatographic separation of tryptic peptides prior to mass spectrometry. An effective extraction protocol for soil and litter proteins should meet the following requirements: (1) proteins should be extracted quantitatively, (2) proteins should be efficiently separated from cellular debris, sample matrix, and (3) interfering chemical substances and the protein structure should not be altered by the extraction.

Dörte Becher (ed.), *Microbial Proteomics: Methods and Protocols*, Methods in Molecular Biology, vol. 1841,
https://doi.org/10.1007/978-1-4939-8695-8_21,

Protocols aiming on an enrichment of proteins from soil and litter are often based on aqueous buffers combined with mechanophysical treatments, organic solvents, strong acids or bases, and detergents (reviewed in [1, 2]). More specifically, leaf litter proteins have been successfully extracted employing SDS-containing buffer and subsequent boiling [3–5]. A detailed description of this protocol is provided to the reader in this chapter.

For soil proteins various extraction protocols have been reported, using either NaOH [6], phosphate buffer [7, 8], citric buffer [9, 10], Tris–ethylenediaminetetraaceticacid (Tris–EDTA) buffer [11, 12] or SDS-containing buffers [9, 10, 13–15]. A recent study [14] aimed on a comprehensive evaluation of well-established soil protein extraction protocols. The authors concluded that a combination of different extraction protocols results in significantly higher metaproteome coverage than applying only one specific protocol [14]. In the current book chapter we thus suggest to use a combination of two well-established protocols for soil protein, employing either SDS–phenol or NaOH. Whilst SDS-based protocols have been proven to be very efficient in terms of the total number of identified proteins from soils [14, 16], the use of NaOH leads to an effective desorption of specific proteins covalently bound to clay particles. We thus suggest combining these two complementary strategies aiming on an efficient and comprehensive metaproteomics profiling of soil environments.

The book chapter ends with a detailed description of subsequent sample preparation including the separation of proteins using SDS-PAGE, in-gel digestion of proteins, and purification of the resulting peptide mixture prior to mass spectrometry analysis.

2 Materials

Prepare all solutions using ultrapure water (prepared by purifying deionized water to attain a sensitivity of 18 $MScm^{-1}$ at 25 °C) further referred to as "MilliQ" and analytical grade reagents. Prepare and store reagents at room temperature (unless indicated otherwise). Wear gloves during all procedures.

2.1 Sample Homogenization and Storage

1. Freezer −80 °C.
2. Liquid nitrogen (*see* **Note 1**).
3. Cryo-gloves.
4. N_2 Dewar.
5. Mortar and pestle.
6. Retsch Mill Grindomax GM200 (or a kitchen knife).
7. Sieve (2 mm mesh size).

2.2 Protein Extraction

1. Schott bottles for storage of the buffers.
2. Sterile plastic ware for extractions (50 mL centrifuge tubes).
3. Ultrasonic device with a conic probe.
4. Piston pipettes with the sizes 200 μL, 1000 μL, 10 mL.
5. Pipette tips.
6. Pipette helpers.
7. Disposable graduated pipettes, 10 mL and 25 mL.
8. Autoclave.
9. Vortex.
10. Water bath.
11. Tabletop centrifuge (e.g., 5430R, Eppendorf) with a fixed angle rotor for 50 mL centrifuge tubes.
12. Freezer (−20 °C).
13. Polyvinylpolypyrrolidone PVPP, particle size 110 μm, (*see* **Note 2**).
14. Phenol solution, equilibrated with 10 mM Tris–HCl, pH 8.0, 1 mM EDTA, (e.g., commercially available as BioReagent for molecular biology, e.g., CAS Number 108-95-2, Sigma-Aldrich,), store at 4 °C, caution this reagent is toxic.
15. Tris–SDS buffer: 50 mM Tris, 1% (w/v) SDS, pH 7.0 (sterile), *see* **Note 3.**
16. Extraction buffer 1 (1:1 (v:v) SDS-phenol buffer): Mix 50 mM Tris, 1% (w/v) SDS pH 7 with the same volume of phenol solution (pH 8.0), (*see* **Notes 4** and **5**).
17. Extraction solution 2: 0.1 M NaOH in MilliQ, caustic reagent.
18. Ammonium acetate solution: 0.1 M ammonium acetate in methanol (MeOH).
19. Prechilled acetone (store at −20 °C).
20. Tris–EDTA (TE) buffer: 0.5 mL 1 M Tris–HCl, pH 8.0, 1.0 mL 0.5 M EDTA, 48.5 mL MilliQ (18.2 MΩ; TOC $\leq$ 5 ppb; sterile).
21. Speed-vacuum centrifuge (e.g., Eppendorf Vacuum Concentrator plus, Hamburg, Germany).

2.3 1D SDS PAGE

1. Piston pipettes with the sizes 20 μL, 200 μL, 1000 μL, 10 mL.
2. Pipette tips.
3. Ammoniumperoxodisulfate (APS): 10% (v/v) solution in MilliQ, prepare freshly prior to use.
4. N,N,N,N'-Tetramethylethylendiamine (TEMED): store at 4 °C.

5. Resolving gel buffer (RGB): 1.5 M Tris–HCl, pH 8.8, 1% (w/v) SDS in MilliQ.
6. Resolving gel mixture (RGM): Pipette successively 4 mL H_2O, 3 mL RGB, 5 mL acrylamide solution, add 80 μL APS and 8 μL TEMED before use.
7. Stacking gel (SGB) buffer: 0.5 M Tris–HCl, pH 6.8, 1% (w/v) SDS in MilliQ.
8. Stacking gel mixture (SGM): Pipette successively 4.75 mL MilliQ, 2 mL SGB, and 1.25 mL acrylamide solution. Add 80 μL APS and 8 μL TEMED before use.
9. Acrylamide–bisacrylamide: 30% (v/v) solution (electrophoresis grade), store at 4 °C, caution this reagent is toxic.
10. Glass plates (e.g., Bio-Rad Laboratories Inc., Hercules, USA), 0.75 mm spacer, *see* **Note 6**).
11. Rubber seals (e.g., Bio-Rad Laboratories Inc., Hercules, USA, *see* **Note 6**).
12. Metal clamps (e.g., Bio-Rad Laboratories Inc., Hercules, USA, *see* **Note 6**).
13. Spacers (e.g., Bio-Rad Laboratories Inc., Hercules, USA, *see* **Note 6**).
14. Gel Casting Chamber (e.g., PROTEAN Bio-Rad Laboratories Inc., Hercules, USA, *see* **Note 6**).
15. Pure (96%) ethanol (EtOH).
16. Comb(s) (*see* **Note 6**).
17. Loading buffer 4 ×, pH 6.8: 7.5 mL glycerol, 2.5 mL β-mercaptoethanol, 1.2 g SDS, 200 μL 1% (w/v) bromophenol blue, 0.4 g Tris, 50 mL MilliQ (*see* **Note 7**).
18. Electrophoresis buffer: 10×: Tris 30.3 g, glycine 144.1 g, 10 g SDS, 1000 mL MilliQ (*see* **Note 8**).
19. Protein molecular weight standard Precision Plus (Molecular Weight range 10–250 kDa, Bio-Rad Laboratories Inc., Hercules, USA).
20. Gel cassette Mini consisting of a buffer tank, electrode assembly, lid and power cables (e.g., PROTEAN Tetra System, Bio-Rad Laboratories Inc., Hercules, USA) to run the gels, *see* also **Note 6.**
21. Power supply, e.g., PowerPac Universal (Bio-Rad Laboratories Inc., Hercules, USA).
22. Fixing solution: 40% (v/v) ethanol, 10% (v/v) acetic acid, 50% (v/v) MilliQ.
23. Staining solution: 100 g $(NH_4)_2SO_4$, 100 mL orthophosphoric acid, 1.2 g Coomassie brilliant blue G-250 fill up to 1000 mL with MilliQ, stir overnight.

24. Sodium azide (NaN_3) solution for gel storage: 3 mM NaN_3 solution in MilliQ; caution: this reagent is toxic.
25. Aluminum heating block or water bath.
26. Orbital shaker.
27. Gel scanner for accurate quantification of gels stained with Coomassie Blue, over a large dynamic range (0–3.0 OD), and enables background correction for quality control (e.g., GS-800 Calibrated Densitometer, Bio-Rad Laboratories Inc., Hercules, USA).

2.4 In-Gel Digestion

1. Plastic tray.
2. Speed-vacuum centrifuge (Eppendorf Vacuum Concentrator plus).
3. Glass plates.
4. Scalpel.
5. Reaction tubes 1.5 mL (e.g., Eppendorf).
6. Thermomixer/thermoshaker.
7. Incubator.
8. Table top centrifuge (e.g., 5430/5430R, Eppendorf) with rotor for 1.5 mL tubes.
9. Ultrasonic bath.
10. Piston pipettes, 10 μL and 200 μL.
11. Pipette tips.
12. Washing solution: 200 mM $(NH_4)HCO_3$, 30% (v/v) acetonitrile (ACN) (HPLC-grade).
13. Trypsin solution 1:10: dissolve 20 μg trypsin (trypsin lyophilized, MS-grade) in 1 mL MilliQ, aliquot 50 μL trypsin solution and add 450 μL MilliQ, store aliquots at −20 °C.

2.5 Zip-Tip Desalting

1. Wetting solution: 70% (v/v) ACN in MilliQ.
2. Equilibration solution: 3% (v/v) ACN, 0.1% (v/v) acetic acid in MilliQ.
3. Washing solution: 0.1% (v/v) acetic acid in MilliQ.
4. Elution solution: 60% (v/v) ACN, 0.1% (v/v) acetic acid in MilliQ.
5. Zip-tip μC18 pipette tips (10 μL, e.g., Merck Millipore).
6. LoBind Eppendorf tubes 1.5 mL (coated to prevent protein absorption).
7. 2 mL glass vials with 100 μL conical glass inserts with polymer feet and screw caps (e.g., Agilent Technologies).

3 Methods

3.1 Sample Homogenization and Storage

1. Process fresh samples immediately after sampling and transportation to minimize the storage time at 4 °C [17]. For sampling strategy *see* **Notes 9** and **10**.
2. Sieve soil samples through a sieve with 2 mm mesh size (*see* **Notes 11** and **12**).
3. Shred leaf litter samples into small pieces using a Retsch knife-press (GRINDOMAX GM200) with the following settings: 7000 rpm for 1 min, repeated three times, alternatively use a knife to cut the whole leaves, but this is very laborious.
4. Store samples at −80 °C, if you do not continue with protein extraction immediately (*see* **Note 13**).

3.2 Extraction of Proteins from Soil

Protein extraction protocols differ for soil and leaf litter (*see* Figs. 1 and 2).

1. Mix 10 g of sieved (< 2 mm) soil (either processed freshly or deep frozen at −80 °C) with 10% (w/w) polyvinylpolypyrrolidone (PVPP) and ground it in a mortar with a pestle using liquid nitrogen to remove interfering substances, such as humic acids [14] (*see* **Notes 1** and **2**).
2. Split the soil sample and follow the two extraction protocols described below and targeting different soil proteins [1, 14] *see* Fig. 1a, b.
3. For **extraction of soil proteins with SDS–phenol (SP)** [14] transfer 5 g of the thoroughly ground soil material into a 50 mL sterile centrifuge tube and add extraction buffer 1 (*see* **Note 14** and Fig. 1b).
4. Vortex and sonicate the mixture with an ultrasonic homogenizer on ice, at 90% pulsing and minimum energy of 10% for 1 min twice to break soil aggregates and to lyse the cells (*see* **Note 15**).
5. This suspension is shaken vigorously at 150 rpm for 1 h at 20 °C.
6. Centrifuge the suspension at 3220 × *g* and 4 °C for 20 min.
7. Remove the aqueous phase and wash the lower phenol phase with the same amount of MilliQ by gentle vortexing for 5 min.
8. Centrifuge the suspension at 3220 × *g* and 4 °C for 20 min.
9. Transfer the purified lower phenol phase into a new 50 mL sterile centrifuge tube.
10. Precipitate the proteins by adding a fivefold amount of 0.1 M ammonium acetate in MeOH overnight at −20 °C.

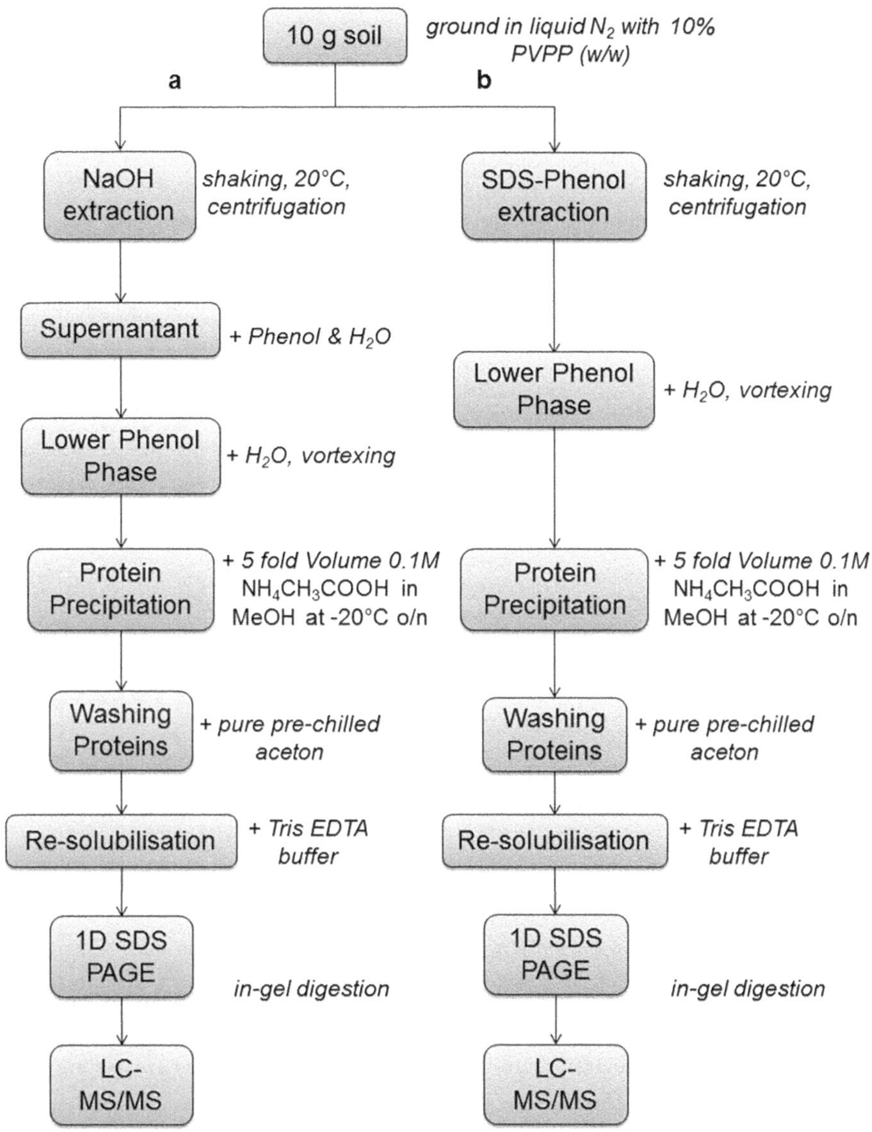

Fig. 1 Experimental design for protein extraction from soil samples using two different extraction procedures

11. For *extraction of soil protein with NaOH* [6] transfer 5 g of the thoroughly ground soil material into a 50 mL sterile centrifuge tube and add extraction solution 2 (*see* **Note 14** and Fig. 1a).
12. Vortex and sonicate the mixture with an ultrasonic probe on ice, at 90% pulsing and minimum energy of 10% twice for 1 min to break soil aggregates and to lyse the cells (*see* **Note 15**).
13. Centrifuge the suspension at 3220 × *g* and 4 °C for 20 min.
14. Transfer the supernatant into a new 50 mL sterile centrifuge tube.

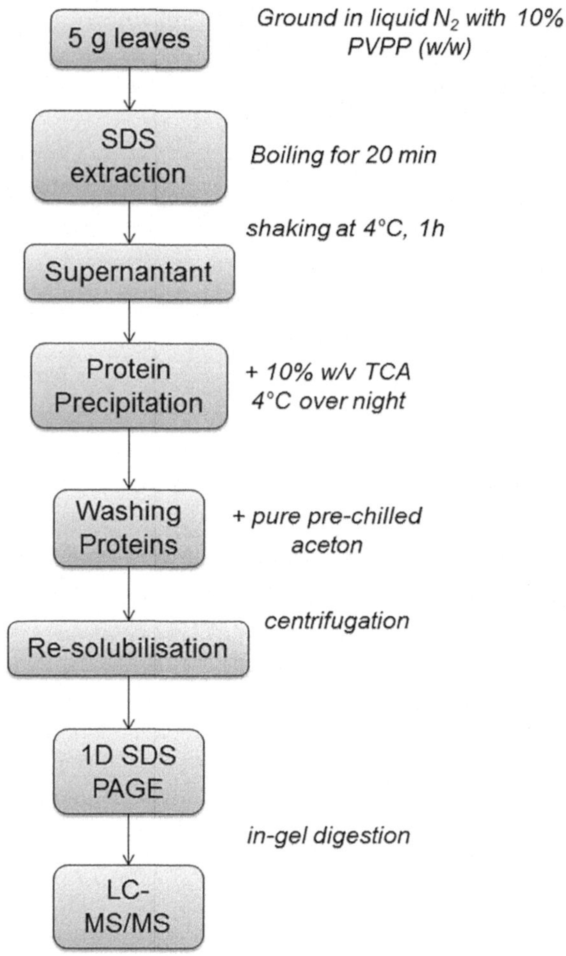

Fig. 2 Experimental design for protein extraction from leaf litter

15. Add a mixture of phenol and MilliQ (8:5) to the supernatant (*see* **Note 5**).
16. Shake the suspension vigorously at 150 rpm and 20 °C for 1 h.
17. Centrifuge the suspension at 3220 × *g* and 4 °C for 20 min.
18. Remove the aqueous phase with a pipette to wash the lower phenol phase with the same amount of MilliQ by gentle vortexing for 5 min.
19. Centrifuge the suspension at 3220 × *g* and 4 °C for 20 min.
20. Transfer the purified lower phenol phase into a new 50 mL sterile centrifuge tube.
21. Precipitate the proteins by adding fivefold amount of ammonium acetate solution overnight at −20 °C.
22. A second extraction step ("sequential extraction") can be performed to improve extraction yields. Use the initial soil pellet

and apply fresh extraction buffer as described in the protocols and continue with the procedures described above. Pool the resulting supernatants prior to precipitation.

23. For *processing of precipitates* collect the precipitated proteins by centrifugation at 10,640 × *g* for 20 min at 4 °C.
24. Discard supernatants.
25. Wash protein pellet with 100% prechilled acetone by gentle vortexing and a further centrifugation step, as described above (*see* **step 23**).
26. Discard supernatant.
27. Air-dry the remaining protein pellet.
28. Resuspend the protein pellet in 20–100 μL of Tris–EDTA (TE) buffer by gentle mixing at 4 °C overnight (*see* **Notes 16** and **17**).
29. Centrifuge for 3 min at 17,960 × *g* at 4 °C (*see* **Note 18**).
30. Keep supernatant (your protein solution) for further processing (*see* **Notes 16** and **19**).
31. For *extraction of proteins from leaf litter* mix 5 g of thoroughly homogenized leaf litter with 10% (w/w) PVPP to clean samples from interfering substances (e.g., humic substances) and ground it with a pestle in liquid nitrogen [4, 5] (*see* **Notes 1** and **2**).
32. Transfer the homogenized powder into a sterile tube and mix it with SDS buffer in a 1:5 ratio (w/v) (*see* **Note 4**).
33. Vortex and sonicate the mixture with an ultrasonic homogenizer on ice at 90% pulsing and a minimum of 10% energy twice for 1 min to break aggregates and to lyse the cells (*see* **Note 15**).
34. Continue with boiling the extract for 20 min in a water bath (*see* **Note 20**).
35. After boiling, shake the extract at 4 °C for 1 h.
36. Centrifuge extracts at 3000 × *g* at 4 °C for 10 min to remove debris.
37. Concentrate supernatant about fivefold by speed-vacuum centrifugation at 30 °C.

3.3 Protein Prefractionation via SDS PAGE

1. Preparation of 1D PAGE Minigels. Put spacers between two glass plates, fix it with metal clamps, and put it upright clamped into the gel casting chamber.
2. Fix the gel (gel in two glass plates) to the gel casting chamber using rubber seal and metal clamps.
3. Prepare resolving gel mixture (12%) and pour solution between the glass plates (*see* **Notes 21** and **22**).

4. Stop pouring when resolving gel solution is ~2 cm below the top of the plates.
5. Overlay with EtOH, let polymerize for 1 h (*see* **Notes 22** and **23**).
6. Remove EtOH with a paper strip.
7. Prepare stacking gel mixture (5%) and fill on top of the resolving gel. Add a comb. Avoid formation of bubbles.
8. Packed in wet paper towels and plastic foil, gels can be stored at 4 °C for up to 4 weeks, (*see* **Note 24**).
9. Prior to *protein separation on the gel*, determine the protein concentration by a ninhydrin assay [18, 19] (*see* **Notes 17** and **19**).
11. Clamp the gels into the gel electrophoresis chamber.
12. Fill gel chambers at anode and cathode with 1× electrophoresis buffer (*see* **Note 8**).
13. Remove comb carefully.
14. Mix protein solution with loading buffer in a 4:1 ratio (15 μL + 5 μL) containing a total of ~5 μg protein (*see* **Note 25**).
15. Load 5 μL of protein standard.
16. Denaturate protein samples in loading buffer for 5 min at 100 °C (on an aluminum heating block or water bath).
17. Pipet samples into the slots (*see* **Note 25**).
18. Start electrophoresis and run it until the bromophenol blue front reaches the boarder of the gel (constant current: 30 mA) (*see* **Notes 26** and **27**).

3.4 Gel Staining and Destaining

1. Shake the gel for at least 30 min in ~25 mL fixing solution.
2. Afterward wash the gel several times with MilliQ until no acetic acid can be smelled any more (at least twice).
3. To stain one gel, mix 20 mL of staining solution with 5 mL EtOH. Shake the gel overnight in the mixture.
4. To destain the gel, transfer it into MilliQ and shake. Exchange MilliQ several times until the background of the gel is clear.
5. Scan gel and save picture to document the obtained results (*see* **Note 28**).
6. Gels can be stored at 4 °C by adding some drops of sodium azide solution covered with a plastic foil until further proceeding (*see* **Note 29**).

3.5 Excision of Proteins and in-gel Digestion

1. Transfer the gel in a plastic tray filled with MilliQ.
2. To remove NaN_3 shake gently for 30 min at room temperature.
3. Transfer gel to solid/smooth and clean surface (e.g., a glass plate).
4. Cut gel slices containing protein smear (*see* Fig. 3) with a scalpel (*see* **Notes 28** and **30**).
5. Transfer gel slices into reaction tubes.
6. Store gel slices if necessary in reaction tubes at −20 °C.
7. Destain the gel pieces with 600 μL washing solution.
8. Incubate at 37 °C for 15 min with shaking using a Thermomixer.
9. Remove and discard washing solution from the reaction tube.
10. Repeat the washing steps until the gel pieces are completely destained.
11. Shrink the gel pieces to dryness by using a Speed-vacuum centrifuge at 30 °C for 30–40 min.
12. Cover the dried gel pieces with 50–100 μL of 1:10 trypsin solution and incubate for 15 min at room temperature.
13. Remove any excess of trypsin solution with pipette tips.
14. Incubate the swollen gel pieces at 37 °C overnight by placing the reaction tube upside down.
15. Spin down gel pieces at 10,000 × *g* for 30 s to make sure that all of them are on the bottom of the reaction tube.
16. Cover the gel pieces completely with MilliQ and incubate them in an ultrasonic bath for 15 min.

Fig. 3 Representative SDS PAGE of proteins extracted from soil or litter samples

17. Spin down the liquid at 10,000 × g for 30 s.
18. Transfer the supernatant quantitatively into a LoBind tube.
19. Concentrate the supernatants as needed (to 10 μL) in a speed-vacuum centrifuge at 30 °C.

3.6 Zip-Tip Desalting

1. Wet zip tips with 10 μL of wetting solution and discard the solution, repeat one time (*see* **Note 31**).
2. Equilibrate Zip tips with 10 μL of equilibration solution and discard the solution, repeat one time.
3. Load 10–15 μL of sample to the Zip tip (containing less than 2 μg of peptide). Aspirate 30 times (peptides bind to Zip tip matrix). Transfer the solution back into the LoBind tubes.
4. Wash Zip tip with 10 μL of washing solution and discard the solution.
5. Elute samples with 10 μL of elution solution. Aspirate 3 times and transfer the solution into a glass vial with a 100 μL conical glass insert.
6. Concentrate samples to dryness with a Speed-vacuum to remove ACN.
7. Store at −20 °C until mass spectrometric (MS) analysis.
8. For further processing fill up with MilliQ-water to 10 μL. Samples are now ready for separation via liquid chromatography and electrospray ionization mass spectrometry detection (LC-ESI MS).

4 Notes

1. Contact of liquid nitrogen with the skin or eyes may cause serious freezing (frostbite) injury. Constantly protect hands when working with liquid nitrogen with Cryo-gloves.
2. PVPP is a white granulated material, which does not dissolve in liquids and has been demonstrated to purify soil from humics [8].
3. For low protein yields a weakly alkaline buffer (pH 7.5–8) can be advantageous [2].
4. Pipetting SDS can cause foaming.
5. When pipetting volatile (organic) solvents use reverse pipetting (prewetting the tip) to ensure pipetting correct volumes and to avoid losing droplets.
6. The components ought to be compatible with each other, and the resulting gel should fit into the gel cassette for conducting the electrophoresis run.

7. Loading buffer: Prepare aliquots in 1.5 mL plastic tubes (Eppendorf) and store them at −20 °C until use.
8. Electrophoresis buffer 10×: Needs to be diluted tenfold prior to usage, store at 4 °C.
9. Sampling I: Thoroughly design your sampling strategy depending on the scientifc question you want to answer [20, 21]. Remember, that an adequate number of independent and spatially separated samples spread over the area of interest are needed to create a representative composite sample. The spatial variability of microbial population depends on land use and vegetation, which in turn determines the number of subsamples needed to generate such a sample. Moreover, sampling depth is crucial as microbial activity is highest in the upper soil layers.
10. Sampling II: In order to generate reliable quantitative information a sufficient number of biological and technical replicates (at least three of each) needs to be analyzed. Alternatively, biological samples can be pooled, which reduces the "between sample complexity" and lowers the costs due to shorter MS measuring times. However, there might be a loss of information regarding the spatial variation when pooling individual samples from a larger area.
11. Sieving I: The use of a specific mesh size (2 mm) for sieving is crucial and should not be changed, as different particle fractions vary in their microbial activity [22]. Sieving is also needed to remove larger animals and roots.
12. Sieving II: The soil texture and moisture may have a strong effect on sieving. Very wet soils and especially soils high in clay content are hard or impossible to sieve. In this case, we recommend homogenizing these samples with a knife to get rid of small stones, etc. Organic soil horizons can usually not be sieved and should be homogenized like leaf litter samples. Stones and roots have to be removed first by handpicking!
13. Individual biological replicates should be stored frozen at −80 °C if immediate processing it not applicable, for more information on storage conditions see also [2].
14. The volume of the extraction buffer needs to be adjusted depending on soil properties (soil moisture, soil texture, and humus content). For drier soils extraction buffer will be added in a 3:1 (v buffer/w soil) ratio, while for wetter soils the volume of extraction buffer can be reduced to a 2:1 ratio (v/w). Clayey soils adsorb buffer in contrast to sandy soils do not, the larger the humus content of the soil the larger the buffer volume needed to extract proteins.

15. In our hands sonication at lower energy resulted in better extraction efficiency [23].
16. Even when using PVPP and phenol for protein extraction protein pellets can remain brown after precipitation.
17. The use of Tris–EDTA buffer enables the application of a Ninhydrin assay [18, 19] for protein quantification to calculate the volume of protein solution to be loaded on the SDS gel.
18. In the case the proteins are not full precipitated, extend centrifugation time to 10 min
19. Humic substances tend to interference with protein quantification, especially with Bradford-based assays. Alternatively, proteins can be quantified with a ninhydrin assay [18, 19], or by acid hydrolysis and total amino acid assay [16].
20. SDS boiling: make sure the tubes are large enough to avoid splashing or boiling retardation, keep the lids open!
21. TEMED stabilizes free radicals and improves polymerization. When preparing the SDS gel, be quick with handling after you add APS solution.
22. Polymerization of acrylamide is faster in the presence of sunlight. Acrylamide solution is hazardous until final polymerization into polyacrylamide. Let the surplus of PAGE reagents polymerize before discarding into the waste disposal.
23. Put EtOH on the resolving gel after it is poured as EtOH removes bubbles and makes the surface smooth.
24. Alternatively, precast gels are commercially available for example 1D SDS Mini-PROTEAN® TGX™ precast gels (12%) (Bio-Rad Laboratories Inc., Hercules, USA)
25. Adding glycerol to the loading buffer simplifies protein loading as the protein solutions are quickly moving to the bottom of the slots. Bromophenol blue has a higher electrophoretic mobility than the samples and thereby allows tracking the progress of protein separation during electrophoresis.
26. The time for the electrophoretic run depends on the number of gels that run parallel, the more gels the longer it will take.
27. To get rid of humic substances, PAGE is the method of choice, as humic substances are charged and move through the gel when electric current is applied. Humic substance move faster than the proteins and thus leave the gel before the first proteins arrives the end of the gel.
28. When working with litter or soil samples often no distinct protein bands but a smear occurs on the gel (*see* Fig. 3); however, this does not hamper the further analyses.
29. Gel can be kept for several weeks at 4 °C prior to in-gel protein digestion.

30. The number and size of slices cut from the gel depend on the intensity and the expected sample complexity. In our hand cutting 6–20 slices per gel lane works fine for litter and soil samples.
31. Carefully avoid bubbles while pipetting. Tips must never run dry after wetting.

References

1. Becher D, Bernhardt J, Fuchs S, Riedel K (2013) Metaproteomics to unravel major microbial players in leaf litter and soil environments: challenges and perspectives. Proteomics 13(18–19):2895–2909
2. Keiblinger KM, Fuchs S, Zechmeister-Boltenstern S, Riedel K (2016) Soil and leaf litter metaproteomics—a brief guideline from sampling to understanding. FEMS Microbiol Ecol
3. Schneider T, Gerrits B, Gassmann R, Schmid E, Gessner MO, Richter A, Battin T, Eberl L, Riedel K (2010) Proteome analysis of fungal and bacterial involvement in leaf litter decomposition. Proteomics 10(9):1819–1830
4. Keiblinger KM, Schneider T, Roschitzki B, Schmid E, Eberl L, Hammerle I, Leitner S, Richter A, Wanek W, Riedel K, Zechmeister-Boltenstern S (2012) Effects of stoichiometry and temperature perturbations on beech leaf litter decomposition, enzyme activities and protein expression. Biogeosciences 9 (11):4537–4551
5. Schneider T, Keiblinger KM, Schmid E, Sterflinger-Gleixner K, Ellersdorfer G, Roschitzki B, Richter A, Eberl L, Zechmeister-Boltenstern S, Riedel K (2012) Who is who in litter decomposition? Metaproteomics reveals major microbial players and their biogeochemical functions. ISME J 6 (9):1749–1762
6. Benndorf D, Balcke GU, Harms H, von Bergen M (2007) Functional metaproteome analysis of protein extracts from contaminated soil and groundwater. ISME J 1(3):224–234
7. Murase A, Yoneda M, Ueno R, Yonebayashi K (2003) Isolation of extracellular protein from greenhouse soil. Soil Biol Biochem 35 (5):733–736
8. Masciandaro G, Macci C, Doni S, Maserti BE, Leo ACB, Ceccanti B, Wellington E (2008) Comparison of extraction methods for recovery of extracellular beta-glucosidase in two different forest soils. Soil Biol Biochem 40 (9):2156–2161
9. Chen SN, Rillig MC, Wang W (2009) Improving soil protein extraction for metaproteome analysis and glomalin-related soil protein detection. Proteomics 9(21):4970–4973
10. Wang HB, Zhang ZX, Li H, He HB, Fang CX, Zhang AJ, Li QS, Chen RS, Guo XK, Lin HF, Wu LK, Lin S, Chen T, Lin RY, Peng XX, Lin WX (2011) Characterization of metaproteomics in crop Rhizospheric soil. J Proteome Res 10(3):932–940
11. Singleton I, Merrington G, Colvan S, Delahunty JS (2003) The potential of soil protein-based methods to indicate metal contamination. Appl Soil Ecol 23(1):25–32
12. Williams MA, Taylor EB (2010) Microbial protein in soil: influence of extraction method and C amendment on extraction and recovery. Microb Ecol 59(2):390–399
13. Hettich RL, Chourey K, Jansson J, VerBerkmoes N, Shah M, Chavarria KL, Tom LM, Brodie EL (2010) Direct cellular Lysis/protein extraction protocol for soil metaproteomics. J Proteome Res 9(12):6615–6622
14. Keiblinger KM, Wilhartitz IC, Schneider T, Roschitzki B, Schmid E, Eberl L, Riedel K, Zechmeister-Boltenstern S (2012) Soil metaproteomics—comparative evaluation of protein extraction protocols. Soil Biol Biochem 54 (0):14–24
15. Gunnigle E, Ramond JB, Frossard A, Seeley M, Cowan D (2014) A sequential co-extraction method for DNA, RNA and protein recovery from soil for future system-based approaches. J Microbiol Meth 103:118–123
16. Bastida F, Hernandez T, Garcia C (2014) Metaproteomics of soils from semiarid environment: functional and phylogenetic information obtained with different protein extraction methods. J Proteome 101:31–42
17. ISO-10381-6 (2009) Part 6: guidance on the collection, handling and storage of soil under aerobic conditions for the assessment of microbiological processes, biomass and diversity in the laboratory. Soil quality—sampling

18. Moore S, Stein WH (1954) A modified ninhydrin reagent for the photometric determination of amino acids and related compounds. J Biol Chem 211(2):907–913
19. Jones DL, Owen AG, Farrar JF (2002) Simple method to enable the high resolution determination of total free amino acids in soil solutions and soil extracts. Soil Biol Biochem 34 (12):1893–1902
20. Quinn GP, Keough MJ (2002) Experimental design and data analysis for biologists. Cambridge University Press
21. Boeddinghaus RS, Nunan N, Berner D, Marhan S, Kandeler E (2015) Do general spatial relationships for microbial biomass and soil enzyme activities exist in temperate grassland soils? Soil Biol Biochem 88:430–440
22. Liu D, An SS, Cheng Y, Keiblinger K, Huang YM (2014) Variability in soil microbial biomass and diversity among different aggregate-size fractions of different land use types. Soil Sci 179(5):242–249
23. Keiblinger KM, Zehetner F, Mentler A (unpublished data). Low intensity ultrasonic vibration improves soil protein extraction efficiency

Chapter 22

Centrifugation-Based Enrichment of Bacterial Cell Populations for Metaproteomic Studies on Bacteria–Invertebrate Symbioses

Tjorven Hinzke, Manuel Kleiner, and Stephanie Markert

Abstract

Owing to high sample complexity, metaproteomic investigations on bacteria–animal symbioses with two or more uncultured partners can be challenging. A selective isolation or enrichment of distinct (sub-)populations within those consortia can solve this problem. Subsequent discrete proteomic analyses benefit from increased sample purity and higher proteome coverage for each of the individual organisms. Here, we describe centrifugation-based methods that allow for a separation of the host and its bacterial symbiont population(s), or even for an enrichment of distinct symbiotic cell cycle stages in the deep-sea mussels *Bathymodiolus azoricus* and *B. thermophilus*, the gutless oligochaete *Olavius algarvensis* and the deep-sea tube worm *Riftia pachyptila*, respectively.

Key words Bacteria–invertebrate symbiosis, Metaproteome, Differential pelleting, Rate-zonal density gradient centrifugation, Isopycnic density gradient centrifugation, Percoll®, HistoDenz™

1 Introduction

The study of symbioses and especially of symbiotic bacteria, most of which have not been cultivated to date, often requires a reduction of sample complexity in order to maximize gain of information: The number of expressed proteins from all partners is comparatively high and host proteins are usually massively overrepresented. Consequently, a separation of the partners is necessary to be able to detect specific molecules or functions and to attribute them to one of the organisms involved. As symbiont-bearing animals often harbor more than one bacterial species or several morphotypes of one species (e.g., [1, 2]), it can be advantageous to not only separate host from symbionts but also different symbiont populations from each other for further studies, such as proteomic analyses.

Several culture-independent methods can in principle be exploited to separate or enrich certain types of bacterial cells from

Dörte Becher (ed.), *Microbial Proteomics: Methods and Protocols*, Methods in Molecular Biology, vol. 1841, https://doi.org/10.1007/978-1-4939-8695-8_22, © Springer Science+Business Media, LLC, part of Springer Nature 2018

their surroundings, such as centrifugation, filtration, dilution, micromanipulation, use of optical tweezers and flow cytometry [3].

Centrifugation as one of these possibilities is comparatively straightforward and easily applicable even in the field, as it does not require high-end equipment. Moreover, it is fast and allows for relatively high quantities of sample to be treated at one time. A large range of centrifugation methods is available to separate cells and even subcellular compartments (for a comprehensive overview, see for example [4–10]). The most basic and fastest centrifugation procedure is differential pelleting (Fig. 1a). Here, particles are separated predominantly based on size and, to some extent, on density. Larger/denser particles sediment faster than smaller/less dense ones, which means that the latter will mostly remain in the supernatant if *g*-force and centrifugation duration are chosen

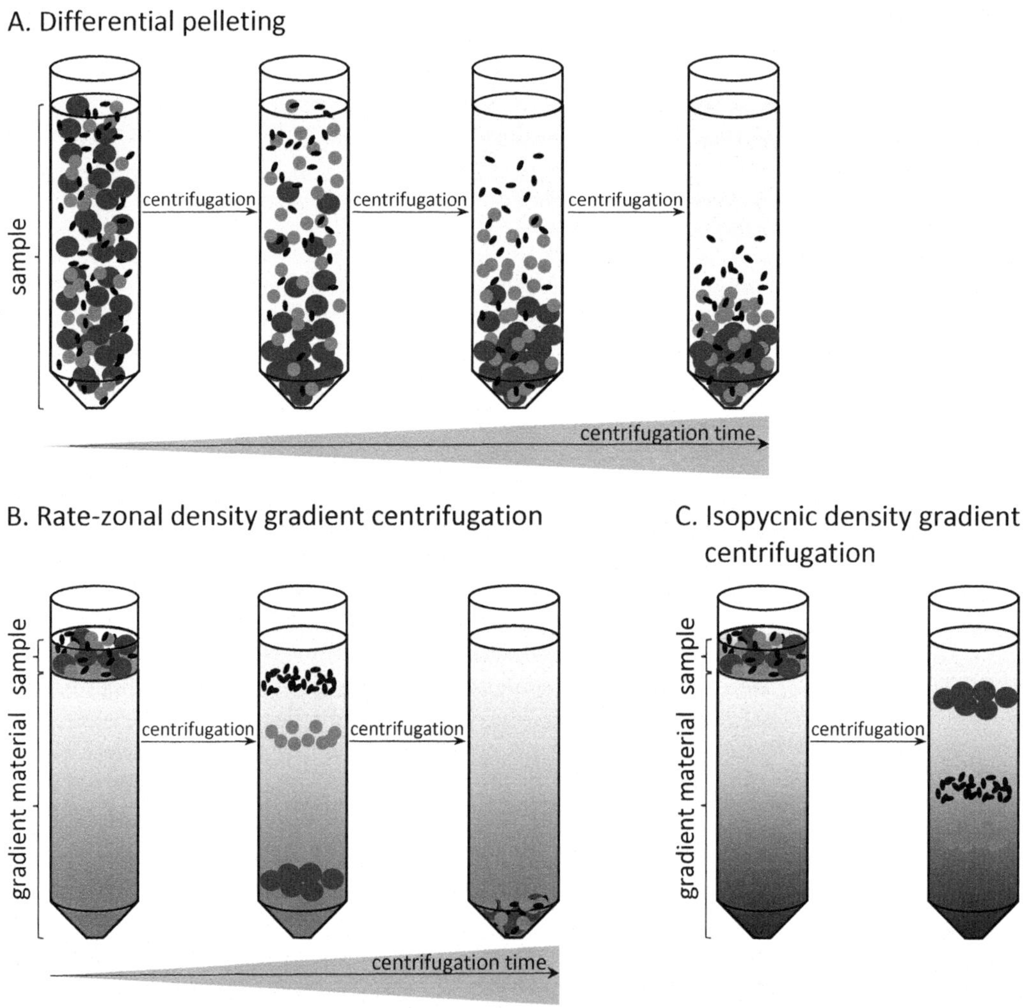

Fig. 1 Overview of centrifugation-based cell separation methods. (**a**) Differential pelleting, (**b**) Rate-zonal density gradient centrifugation, (**c**) Isopycnic density gradient centrifugation

appropriately (although some of them will sediment as well, contaminating the pellet). Thus, differential pelleting leads to a relatively crude separation of cells. To increase separation resolution, which is especially desirable for more complex separation tasks, density gradients can be employed. Two general types of density gradient-based methods can be distinguished and used according to the specific experimental question [4–12]: On the one hand, a rate-zonal centrifugation allows for separation of particles of similar density based on their size and shape, which determine their respective sedimentation speeds while moving through a shallow density gradient (Fig. 1b). During relatively short centrifugation times at low speed, smaller (i.e., slower-moving) particles accumulate in the upper part of the gradient tube, and larger (i.e., faster-moving) particles in the lower part of the tube. As for differential pelleting, all particles will pellet eventually, depending on centrifugation time and *g*-force. Particles with different densities, on the other hand, can be separated by isopycnic density gradient centrifugation. During comparatively longer centrifugation times at high speed, the particles collect in the gradient zone that matches their own density (Fig. 1c).

All centrifugation methods mentioned above have been successfully used for enrichment and separation of symbionts from marine bacteria–animal symbioses and subsequent proteomic analyses. In order to perform these experiments, whole animals (in the case of *Olavius algarvensis*) or their symbiont-containing tissues (i.e., the gills of *Bathymodiolus azoricus* and *B. thermophilus* and the trophosome of *Riftia pachyptila*) were homogenized and the homogenized tissue was subjected to one of the centrifugation methods described below (*see* overview in Table 1).

Table 1
Summary of presented centrifugation methods and previous applications

Method	Application and symbiotic system	For method description *see* Subheading
Differential pelleting	Separation of host from symbiont(s) in *Bathymodiolus azoricus* [16] and *B. thermophilus* (Ponnudurai et al., in preparation)	3.2
Isopycnic density gradient centrifugation	Separation of host and symbiont in *Riftia pachyptila* (Percoll® cushion) [14, 15]	3.3
	Enrichment of multiple symbionts in *Olavius algarvensis* [1]	3.5
Rate-zonal density gradient centrifugation	Enrichment of *Riftia pachyptila* symbiont morphotypes (Hinzke et al., in preparation)	3.4

Based on a protocol developed by Distel and Felbeck [13], a separation of host tissue and the total monospecific bacterial symbiont population of the deep sea tube worm *R. pachyptila* was obtained via isopycnic density gradient centrifugation using the density gradient medium Percoll® [14, 15].

Beyond that, an enrichment of individual life-stage dependent morphotypes of the *R. pachyptila* symbiont was recently achieved by rate-zonal density gradient centrifugation with HistoDenz™ as a density gradient medium (Fig. 2).

In recent studies with the deep-sea mussels *Bathymodiolus azoricus* and *B. thermophilus*, host and symbionts were successfully separated by differential pelleting ([16], Ponnudurai et al., in preparation). Furthermore, an effective enrichment of different symbiont species residing in the marine oligochaete *Olavius algarvensis* was achieved by means of isopycnic density gradient centrifugation [1].

If necessary symbiont enrichment resulting from the respective centrifugation approaches was analyzed and verified via fluorescence in situ hybridization (FISH), allowing for the selection of samples for subsequent proteomic analyses [1, 16, Hinzke et al., in preparation, Ponnudurai et al., in preparation].

Here, we present a compendium of these centrifugation methods to be used for bacterial symbiont enrichment and separation (*see* Table 1).

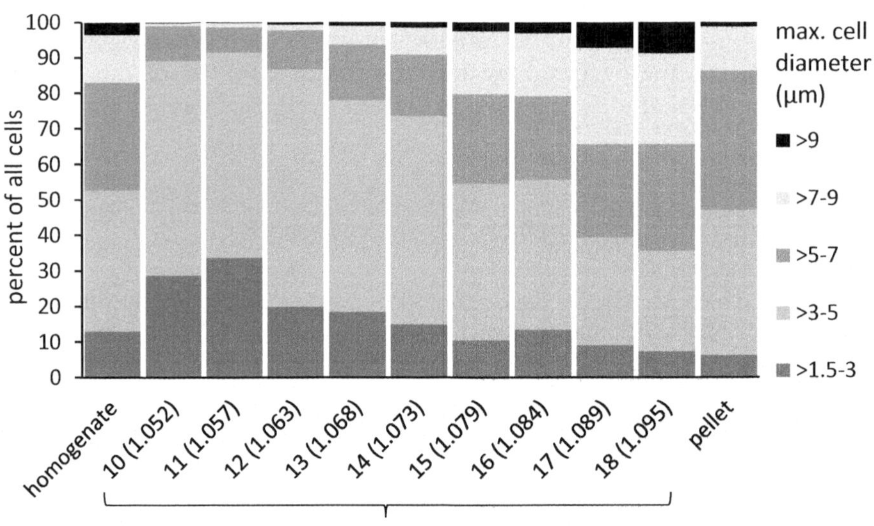

Fig. 2 Distribution of *Riftia pachyptila* symbiont cell sizes in trophosome homogenate (leftmost column) and in HistoDenz fractions of increasing concentrations after rate-zonal density gradient centrifugation (Hinzke et al., in preparation). Smaller symbiont cells (with a diameter of up to 5 m) are enriched in the less dense fractions (left columns), while larger cells of up to >9 m accumulate in the fractions with higher densities (right columns)

2 Materials

All solutions are prepared using deionized water and analytical grade reagents. Reagents are prepared at room temperature and stored at 4 °C.

2.1 Homogenization of Tissue Samples

1. 2.5× Imidazole-buffered saline (IBS): 1.225 M NaCl, 0.075 M $MgSO_4$, 0.0275 M $CaCl_2$, 0.0075 M KCl, 0.125 M imidazole, ph 7.1 [13]. Add about 300 mL water in a beaker. Weigh 28.64 g NaCl, 6.85 g $MgSO_4 \cdot 6\ H_2O$, 1.62 g $CaCl_2 \cdot 2\ H_2O$, 0.224 g KCl, 3.4 g imidazole and add reagents subsequently to the beaker. Allow each reagent to dissolve before adding the next. Adjust pH with HCl to 7.1. Fill up to 400 mL with water (*see* **Note 1**).
2. IBS: 0.49 M NaCl, 0.03 M $MgSO_4$, 0.011 M $CaCl_2$, 0.003 M KCl, 0.05 M imidazole, ph 7.1 [13]. Mix 400 mL of 2.5× IBS with 600 mL water.
3. Homogenizer: Use a 15 mL Dounce glass tissue grinder with a small and a large pestle, or a 15 mL Duall glass tissue grind tube and pestle (*see* **Note 2**).
4. Optional: rough nylon mesh (*see* **Note 3**).
5. Material for FISH analyses (*see* Subheading 2.6).

2.2 Differential Pelleting

1. IBS: *see* **items 1** and **2** in Subheading 2.1.
2. Refrigerated centrifuge with swing-out rotor for 15 mL tubes and fixed-angle rotor for 15 mL tubes (optional: fixed-angle rotor for 2 mL tubes).
3. Material for FISH analyses (*see* Subheading 2.6).

2.3 Isopycnic Gradient Centrifugation with PERCOLL® for Simple Separation of Symbionts from Host Tissue (Percoll® Cushion)

1. IBS: *see* **items 1** and **2** in Subheading 2.1.
2. 60% Percoll®: Mix 60 mL Percoll® with 40 mL of 2.5× IBS. Prepare 10 mL aliquots in 50 mL tubes. Store at 4 °C.
3. Tris–EDTA buffer (TE buffer): 10 mM Tris–HCl, 1 mM EDTA; pH 7.5. Prepare a 1 M stock of Tris–HCl by adding 60.6 g Tris to 400 mL water, adjusting pH with HCl to 7.5 and filling up to 500 mL with water. Additionally, prepare a 0.5 M EDTA stock by adding 18.6 g EDTA-$Na_2 \cdot 2H_2O$ to 80 mL water, adjusting pH to 8.0 with NaOH and filling up to 100 mL with water (*see* **Note 4**). Make TE buffer from stock solutions by adding 10 mL 1 M Tris–HCl (pH 7.5) and 2 mL 0.5 M EDTA (pH 8.0) to 950 mL water and filling up to 1 L with water.
4. 10 mL syringe with syringe needles (approx. 0.9 mm diameter, 18–20 G)

5. Refrigerated centrifuge with fixed-angle rotor for 50 mL tubes and fixed-angle rotor for 2 mL tubes (optional: fixed-angle rotor for 15 mL tubes, depending on sample volume).

2.4 HistoDenz™ Gradients for Rate-Zonal Density Gradient Centrifugation

1. IBS: *see* **items 1** and **2** in Subheading 2.1.
2. 60% (w/v) HistoDenz™ stock solution: Weigh 12 g HistoDenz™ (synonyms: Nycodenz®, iohexol) and transfer it into a 50 mL tube. Fill up to a total volume of 20 mL with IBS. Mix thoroughly by vortexing and allow for the HistoDenz™ to dissolve. Add more IBS if necessary to keep the total volume at 20 mL. The stock solution can be kept at 4 °C for several days.
3. HistoDenz™ gradients for rate zonal centrifugation (8–18%, Δ 1%): Use the 60% HistoDenz™ stock solution and IBS to prepare 7 mL of each HistoDenz™ dilution in 15 mL tubes (sufficient for six gradients; *see* Table 2 for pipetting scheme). Vortex each dilution thoroughly. HistoDenz™ dilutions can be stored at 4 °C for several days. Assemble the gradient from bottom (18%) to top (8%), starting with the highest concentration. Use 1 mL of each dilution step. Freeze gradient after assembly (*see* **Note 5**).
4. TE buffer: *see* **item 3** in Subheading 2.3.

Table 2
Pipetting scheme for dilutions to prepare 8–18% HistoDenz™ gradients for rate-zonal density gradient centrifugation

	For six gradients	
Percent HistoDenz™ in IBS (w/v)	60% HistoDenz™ stock solution (mL)	IBS (mL)
18	2.10	4.90
17	1.98	5.02
16	1.87	5.13
15	1.75	5.25
14	1.63	5.37
13	1.52	5.48
12	1.40	5.60
11	1.28	5.72
10	1.17	5.83
9	1.05	5.95
8	0.93	6.07

5. Refrigerated centrifuge with swing-out rotor for 15 mL tubes and fixed-angle rotor for 2 mL tubes.
6. Material for FISH analyses (*see* Subheading 2.6).

2.5 HistoDenz™ Gradients for Isopycnic Density Gradient Centrifugation

1. IBS: *see* **items 1** and **2** in Subheading 2.1.
2. 60% (w/v) HistoDenz™ stock solution: *see* **item 2** in Subheading 2.4.
3. 70% (w/v) HistoDenz™ stock solution: Weigh 14 g HistoDenz™ and transfer it into a 50 mL tube. Fill up to a total volume of 20 mL with IBS. Mix thoroughly by vortexing and allow for the HistoDenz™ to dissolve. Add more IBS if necessary to keep the total volume at 20 mL. The stock solution can be kept at 4 °C for several days.
4. HistoDenz™ gradients for isopycnic centrifugation (10–70%): Use the 60% HistoDenz™ stock solution and IBS to prepare 3.5 mL of each of the 60–20% HistoDenz™ dilutions and 2 mL each of the 15% and 10% HistoDenz™ dilutions in 15 mL tubes (sufficient for five gradients; *see* Table 3 for pipetting scheme). Vortex thoroughly. HistoDenz™ dilutions can be stored at 4 °C for several days. Assemble the gradient by adding 1.5 mL of 70% HistoDenz™ first, then add 0.6 mL of

Table 3
Pipetting scheme for dilutions to prepare 10–70% HistoDenz™ gradients for isopycnic density gradient centrifugation

	For five gradients		
Percent HistoDenz™ in IBS (w/v)	70% HistoDenz™ stock solution (mL)	60% HistoDenz™ stock solution (mL)	IBS (mL)
70	8.0	0	0
60	0	3.50	0
50	0	2.92	0.58
40	0	2.33	1.17
30	0	1.75	1.75
28	0	1.63	1.87
26	0	1.52	1.98
24	0	1.40	2.10
22	0	1.28	2.22
20	0	1.17	2.33
15	0	0.50	1.50
10	0	0.33	1.67

each of the subsequent 60–20% dilution steps (*see* **Note 6**). Use 0.3 mL of the 15% and the 10% dilution step (*see* **Note** 7).

5. TE buffer: *see* **item 3** in Subheading 2.3.
6. Refrigerated centrifuge with swing-out rotor for 15 mL tubes and fixed-angle rotor for 2 mL tubes.
7. Material for FISH analyses (*see* Subheading 2.6).

2.6 Preparation of Subsamples for FISH Analyses (See Note 8)

1. IBS, containing 1% paraformaldehyde (PFA): Prepare 2.5× IBS (*see* **item 1** in Subheading 2.1). Add 240 mL of 2.5× IBS, 75 mL of 8% PFA and fill up with water to 600 mL. Prepare 1 mL aliquots in tubes. The solution can be stored at 4 °C for several weeks.
2. IBS, containing 2% PFA: Prepare 2.5× IBS (*see* **item 1** in Subheading 2.1). Add 4 mL of 2.5× IBS, 2.5 mL of 8% PFA and fill up with water to 10 mL. Prepare 0.5 mL aliquots in tubes. The solution can be stored at 4 °C for several weeks.
3. Optional: Isopore™ polycarbonate, track-etched hydrophilic screen filter membranes (25 mm) with 0.2 μm pore size and filter apparatus.

3 Methods

All procedures are carried out at 4 °C or on ice.

3.1 Homogenization of Tissue Samples

1. Fill a Dounce homogenizer or a Duall homogenizer with 6 mL of prechilled (4 °C) IBS (or PBS, *see* **Note 1**). Add approximately 3 g of tissue. Alternatively, immerse sample in IBS so that it is just completely covered (*see* **Note 9**).
2. In the case of a Dounce homogenizer, use the smaller pestle first to grind the material with 5–6 strokes. Afterward, repeat with the second, larger pestle. For a Duall homogenizer, homogenize for approximately 1 min or about 20 strokes (*see* **Note 10**).
3. If necessary, filter the homogenate through a rough nylon mesh (*see* **Note 11**) into a 15 or 50 mL tube (depending on sample volume).
4. Take a 15 μL subsample of (filtered) homogenate for FISH analysis (*see* Subheading 3.6), to evaluate the enrichment procedure.

3.2 Differential Pelleting

1. Precool a refrigerated centrifuge with a swing-out rotor for 15 mL tubes to 4 °C. Additionally, precool a fixed-angle rotor for 15 mL tubes. (Optional: precool a fixed-angle rotor for 2 mL tubes).
2. Fill up tissue homogenate with IBS to a total volume of 15 mL.

3. Centrifuge tissue homogenate in swing-out rotor for 5 min at 500 × *g* and 4 °C (*see* **Notes 12** and **13**).
4. Transfer supernatant into a new 15 mL tube.
5. Take a FISH subsample (*see* Subheading 3.6) from the remaining pellet by picking approximately 5 μL out of the pellet with a pipette tip. Freeze pellet at −80 °C. (Optional: resuspend pellet in 0.5–1 mL IBS, transfer to a 2 mL tube and centrifuge at 14,000 × *g* and 4 °C for 5 min, discard supernatant before freezing.)
6. Repeat **steps 3–5** with the supernatant from **step 3** (*see* **Note 13**).
7. Use the chilled fixed-angle rotor for 15 mL tubes to centrifuge the supernatant for 10 min at 15,000 × *g* and 4 °C (*see* **Note 14**).
8. Transfer the supernatant to a new 15 mL tube. Take a 0.5 mL subsample of the supernatant for FISH analysis (*see* Subheading 3.6). Freeze supernatant at −80 °C.
9. Pick about 0.5 μL of the remaining pellet with a pipette tip as FISH subsample (*see* Subheading 3.6) without removing too much of the pellet.
10. Freeze pellet at −80 °C (*see* **Note 15**).

3.3 Isopycnic Gradient Centrifugation with Percoll® for Simple Separation of Symbionts from Host Tissue (Percoll® Cushion)

1. Precool a refrigerated centrifuge with a fixed-angle rotor for 50 mL tubes to 4 °C. Additionally, precool a fixed-angle rotor for 2 mL tubes. (Optional: also precool a fixed-angle rotor for 15 mL tubes.)
2. Centrifuge a 50 mL tube with 10 mL of chilled (4 °C) 60% Percoll® for 30 min at 5,400 × *g* (or higher) and 4 °C to allow the Percoll to self-assemble into a gradient (*see* **Note 16**).
3. Carefully layer about 5 mL homogenate on top of the Percoll® gradient.
4. Centrifuge for up to 30 min at 4 °C and 5,400 × *g* (*see* **Note 17**).
5. Remove the pelleted symbiont fraction carefully from the bottom of the tube using a 10 mL syringe with needle. Take care not to disturb the host layer on top.
6. Place the symbiont fraction in a 15 mL tube or 50 mL tube (depending on volume). To remove residual Percoll®, dilute the sample 1:2 with IBS and mix. Centrifuge for 10 min at 4 °C and 5,400 × *g*. Discard supernatant containing host tissue debris.
7. Resuspend the pellet in 5 mL TE buffer. Aliquot symbiont suspension in 1 mL portions in 2 mL tubes. Pellet cells for

5 min at 4 °C and 14,000 × *g*. Discard supernatant. (Optional: repeat washing step with 1 mL TE per tube.)

8. Freeze pellets at −80 °C.

3.4 Rate-Zonal Gradient Centrifugation with HistoDenz™ for Separation of Multiple Symbionts or Cell Types

1. Let HistoDenz™ gradients for rate-zonal centrifugation thaw at room temperature, then place them at 4 °C or on ice.
2. Precool a refrigerated centrifuge with a swing-out rotor for 15 mL tubes to 4 °C. Also precool a fixed-angle rotor for 2 mL tubes.
3. Layer 0.5 mL of homogenate on top of the gradient (*see* **Note 18**).
4. Centrifuge the gradients using the swing-out rotor. 1,000 × *g* and 5 min at 4 °C can already be sufficient (*see* **Note 19**).
5. Disassemble the gradient in 0.3–1 mL fractions by holding it upright and placing the pipette tip in the middle of the meniscus. Pipette slowly while following the meniscus downward (*see* **Note 20**). Resuspend the pellet—should there be any—in 0.5 mL IBS or in remaining gradient fluid, respectively. Transfer each zone (fraction) into an individual 2 mL tube. Place the tubes on ice.
6. Take a 20 μL subsample of each fraction for FISH analysis (*see* Subheading 3.6).
7. Add 1 mL TE buffer to each fraction to remove remaining HistoDenz™. Vortex thoroughly. Centrifuge for 5 min at 14,000 × *g* and 4 °C. Discard supernatant. (Optional: repeat washing step.) Freeze cell pellets at −80 °C.

3.5 Isopycnic Centrifugation with HistoDenz™ for Separation of Multiple Symbionts or Cell Types

1. Let HistoDenz™ gradients for isopycnic density gradient centrifugation thaw at room temperature (if frozen before use), then place them at 4 °C or on ice.
2. Precool a refrigerated centrifuge with a swing-out rotor for 15 mL tubes to 4 °C. Additionally, precool a fixed-angle rotor for 2 mL tubes.
3. Layer 1 to 2.5 mL of homogenate on top of the gradient (*see* **Note 21**).
4. Centrifuge the gradients for at least 30 min at maximum speed (5,400 × *g* or higher) and 4 °C (*see* **Note 22**).
5. Disassemble the gradient as described in Subheading 3.4, **steps 5–7**.

3.6 FISH Subsampling and Further Analyses

1. Transfer subsamples for FISH immediately into tubes with 1 mL IBS, containing 1% PFA (0.5 mL of IBS containing 2% PFA when taking 0.5 mL subsamples).
2. Mix thoroughly, fix at 4 °C overnight.

3. After fixation, dilute sample with 5 mL IBS and filter onto a 0.22 μm polycarbonate filter, wash filter once by adding additional 5 mL IBS, dry and label filter and store at −20 °C. Alternatively, freeze sample in fixative at −80 °C and process as soon as possible (*see* **Note 23**).
4. Select cell pellets for the specific research question and analysis method (e.g., proteomics) according to the results of the FISH analysis.

4 Notes

1. Alternatively, it is also possible to use phosphate-buffered saline (PBS) instead of IBS for all methods described here, depending on the organism's salinity preferences (for *Riftia*, IBS or another buffer with similar osmolarity is required). PBS: 137 mM NaCl, 2.7 mM KCl, 10 mM Na_2HPO_4, 1.8 mM KH_2PO_4. Fill approximately 800 mL water in a beaker. Weigh 8.0 g NaCl, 0.2 g KCl, 1.42 g Na_2HPO_4, 0.24 g KH_2PO_4. Allow each salt to dissolve before adding the next. Adjust pH with HCl to 7.4. Fill up to a final volume of 1 L.
2. The ball-shaped ends of the pestles used with Dounce homogenizers have a clearance that is designed to leave the symbiont cells intact, but destroy the host cells. Duall homogenizers with a ground inner surface are better suited for tougher material (such as *Bathymodiolus* gill tissue), whereas Dounce homogenizers are a good choice for softer tissue, e.g., the *Riftia* trophosome. *See* [10, 17] for further information regarding tissue homogenization.
3. Use for example nylon screening material with approximately 2 mm pore width. This is only necessary if the homogenate still contains chunks of host material, which cannot be crushed by homogenization (e.g., blood vessels), and is not recommended for low homogenate quantities, as some loss of material is inevitable during filtering.
4. The EDTA will not completely dissolve as long as pH is below approximately 8. Furthermore, solving will need time and proceeds better with vigorous stirring and moderate heat.
5. Hold the tube at an angle of 30–45° and slowly pipette each dilution by holding the pipette tip against the tube wall and as near to the meniscus of the previous layer as possible without touching it. It is strongly recommended to practice the gradient assembly and centrifugation process without adding a sample at first. Staining the density steps with different food colors helps to judge the quality of the gradient produced—the zones should mix as little as possible. Also try out different piston-

operated pipettes and serological pipettes with rubber two- or three-valve pipette fillers (*see* also Chap. 4 in [6]). Avoid rotating (vertically and, of course, horizontally) the gradient, as zones can get mixed otherwise. Tilt the gradient very slowly so as not to disturb the zonation due to the friction of the zones at the tube wall. To avoid mixing, it might help to "fix" each zone by freezing the tube standing upright at −80 °C or in dry ice before adding the next dilution step. The gradients can be prepared in advance and stored frozen at −80 °C for at least several months. Freezing and thawing of the gradient will cause the discontinuous gradient to become continuous (*see* Chap. 4 in [6], [12]). The ions in the gradient can change their distribution during freezing and thawing. Thus, the resulting gradient will not be isotonic [5] and will not necessarily be isosmotic (*see* Chap. 4 in [6]), but this did not seem to impair separation of the *Riftia* symbiont morphotypes (Hinzke et al., in preparation). A continuous gradient is necessary for rate-zonal centrifugation to allow for smooth particle movement (*see* Chap. 1 in [6]), while in a discontinuous gradient particles can get stuck at the interfaces between the density steps [5, 7, 12]. Nevertheless, as the shape of the gradient depends both on the tube volume and on the freeze and thaw rates, these factors should be as similar as possible throughout all experiments. Moreover, it is not recommended to repeatedly freeze and thaw the gradient, as this might alter its shape and reduce reproducibility (*see* [5], Chap. 4 in [6]). If the cells (or some of them) have the same density as one of the density steps in the gradient, they will not move further and will thus be separated by density, not by size. Therefore, the maximum density of the gradient has to be lower than the density of the cells (*see* Chap. 1 in [6], [7]). The density steps (HistoDenz™ dilutions) described here have been successfully used for enrichment of individual *Riftia* symbiont morphotypes (Hinzke et al., in preparation). However, other organisms may require other densities. Individual adaptations of the protocol are therefore indispensable for new applications.

6. Assemble the gradient as described in **Note 5**. Note that, as the originally discontinuous gradient becomes continuous during freezing and thawing and also during diffusion when it stands for longer time (*see* [5], Chap. 4 in [6]), it might be better to assemble the gradient immediately before use. Discontinuous gradients are well suited for preparative purposes, as they allow for a good separation of a restricted number of particle types with varying, known densities if the gradient densities are adapted accordingly. However, discontinuous gradients give less precise results, if a large number of different particle types are present. If a higher resolution for sample separation is

needed, a continuous gradient may work better (*see* Chap. 1 in [6]). An additional potential limitation of discontinuous gradients is that a barrier can form if many particles collect at a density border, causing denser, slower sedimenting material to get stuck [5, 7, 12].

7. It is also possible to use larger volumes of each dilution step and thus generate longer gradients. Collecting precise fractions might be easier, if density steps are larger. However, longer gradients come with higher cost for HistoDenz™ and the centrifugation time increases as it takes longer for the cells to reach the density that corresponds to their own buoyant density.
8. It is advisable to take FISH samples of the obtained centrifugation fractions in order to verify whether and to which degree the respective enrichment procedure was successful. Depending on the research question, it might be desirable to also take other subsamples, such as for DNA or protein analyses, in addition to or instead of FISH subsamples.
9. As a rough estimate of 3 g tissue: The total volume of IBS in the homogenizer should increase by 3 mL when adding the tissue sample. The amount of buffer should be adjusted for the respective type of tissue to ensure that the homogenate is not too viscous but contains enough material for subsequent analyses.
10. Use rotating movements while moving the pestle. The first few strokes with each pestle may be very difficult, depending on the texture of the tissue used. Also, it might be necessary to use more strokes for sufficient homogenization of the respective sample. But do not "overhomogenize" so as not to damage the bacterial cells.
11. Fold the nylon mesh into a cone and place it tip down in the opening of the falcon tube. It might be necessary to use more than one piece of mesh for filtering the whole material, depending on the viscosity of the homogenate.
12. These conditions were used for enrichment of symbionts from *B. thermophilus* (Ponnudurai et al., in preparation). In the case of *B. azoricus*, host and symbiont fractions were separated by centrifugation for 10 min at 4 °C and 700 × *g* [16]. The exact conditions have to be adapted for the sample used. It is always important to centrifuge relatively slowly in this step so as not to remove too many symbiont cells from the supernatant. For the *B. azoricus* symbiosis, which contains two bacterial symbionts, the larger and denser methanotrophic symbiont is enriched in the first pellet (R. Ponnudurai, personal communication). Differential pelleting can also be used to separate symbionts

containing storage granules (e.g., sulfur or polyhydroxyalkanoates) from those without granules.

13. This step serves to remove (remaining) large host tissue fragments and host nuclei [16].
14. This centrifugation step serves to pellet small tissue fragments, symbionts and mitochondria, whereas the final supernatant is enriched in cytosolic host proteins [16].
15. Alternatively, it is possible to resuspend the pellet in a low buffer volume (to obtain a concentrated suspension; approximately 0.5–1.5 mL) and use the material for subsequent density gradient centrifugation to achieve further enrichment. However, a rate-zonal centrifugation that was performed for *B. azoricus* after differential pelleting did not increase symbiont purity significantly (R. Ponnudurai, personal communication).
16. Shorter centrifugation times are possible, if the used centrifuge allows for higher speed—the higher the *g*-force, the lower the centrifugation time needed.
17. **Note 16** applies here, too—but a maximum of 25,000 × *g* is recommended for gradients loaded with sample material so as not to damage the symbionts. If the cell pellet is too soft, increase centrifugation time and/or speed. Note that this simplified version of an isopycnic density gradient centrifugation only works for symbionts with dense storage inclusions, such as sulfur granules. Less dense cells may not be sufficiently enriched by this method.
18. Do not use too much homogenate for one gradient as overloading the gradient will lead to poor separation—the starting position of the cells should be as uniform as possible, i.e., the total volume of homogenate added should be small for good results in rate-zonal centrifugation. With regard to pipetting the homogenate on top of the gradient, the same remarks as for assembling the gradient do apply here (*see* **Note 5**). Upon loading, the homogenate may sink down approximately 0.5 cm below the surface of the gradient, depending on concentration of the homogenate and density of the uppermost density gradient step. Diluting the homogenate might help circumvent this, though this sinking seems not to significantly hamper cell separation (Hinzke et al., in preparation).
19. This step might take some adjustments in order to adapt centrifugation conditions for the type of sample used. It is important that, after centrifugation, the cell material is distributed over the whole gradient and thus maximally separated. If centrifugation time and/or speed is too low, the material is still mostly in the upper part of the gradient, while a too long centrifugation time and/or too high speed will cause the cell material to pellet at the bottom of the tube.

The appearance of the gradient gives some hints whether the centrifugation conditions where appropriate: a slight turbidity also in the lower part of the gradient and a nonvisible or (very) small pellet at the bottom indicate an equal distribution of cells and thus a good separation (but there might be no turbidity at all). Decrease the centrifuge's acceleration and deceleration to approximately 2/3 of the highest possible value in order to reduce perturbations of the gradient.

20. The fraction size depends on the desired sample resolution. A fraction size of 0.5 mL was found to allow for good resolution while keeping expenditure of time and material acceptable (Hinzke et al., in preparation). First remove a volume equivalent to the amount of sample that was layered on top of the gradient. Then proceed with disassembling the gradient. The same pipette tip can be used for the whole gradient as the HistoDenz™ density increases downward. Alternatively, piercing a hole in the tube wall at the bottom of the tube and letting the gradient flow out stepwise (for example through a syringe needle) into subsequent tubes might work, too, but is not recommended for higher HistoDenz™ densities of 50% and more, as the high viscosity will hamper the flow. Practice gradient disassembly before working with samples, e.g., with colored density steps (*see* also **Note 5**).

21. Unlike in rate-zonal centrifugations (*see* **Note 18**), in isopycnic centrifugation the amount of homogenate is mainly limited by the remaining free space in the tube, although loading too much or too dense homogenate might impair separation here, too. The homogenate may sink down a bit under the surface.

22. Decrease the centrifuge's acceleration and deceleration to approximately 2/3 of the highest possible value in order to reduce perturbations of the gradient. Once the cells have reached the gradient density that matches their own buoyant density they will not move further downward. Therefore, it has to be made sure that centrifugation time is long and speed is high enough for the cells to reach the respective zones. Longer centrifugation times will then not alter the particles' positions any further. If a higher speed is used, centrifugation time can be reduced.

23. It is recommended to perform FISH analyses as soon as possible after taking the samples to avoid overfixation. Nevertheless, we obtained good results with samples stored in fixative at −80 °C for 6 months and longer.

Acknowledgments

We thank Horst Felbeck for valuable help with sample preparation, Ruby Ponnudurai for data collection and analysis, and Thomas Schweder for helpful input. This work was supported by the Deutsche Forschungsgemeinschaft (DFG grant MA 6346/2-1).

References

1. Kleiner M, Wentrup C, Lott C et al (2012) Metaproteomics of a gutless marine worm and its symbiotic microbial community reveal unusual pathways for carbon and energy use. Proc Natl Acad Sci U S A 109:E1173–E1182
2. Bright M, Sorgo A (2003) Ultrastructural reinvestigation of the trophosome in adults of *Riftia pachyptila* (Annelida, Siboglinidae). Invertebr Biol 122:345–366
3. Hugenholtz P (2002) Exploring prokaryotic diversity in the genomic era. Genome Biol 3: reviews0003.1–reviews0003.8
4. Hinton R, Dobrota M (1976) Density gradient centrifugation. Elsevier/North-Holland Biomedical Press, Amsterdam
5. Sharpe PT (1988) Chapter 3: Centrifugation. In: Methods of cell separation. Elsevier Science Publishers B.V. (Biomedical Division), Amsterdam
6. Graham J (2001) Biological centrifugation. BIOS Scientific Publishers Limited, Oxford
7. Hinton RH, Mullock BM (1997) Isolation of subcellular fractions. In: Graham JM, Rickwood D (eds) Subcellular fractionation: a practical approach. The practical approach series. Oxford University Press, Oxford, pp 31–69
8. Spragg SP, Steensgaard J (1992) Theoretical aspects of practical centrifugation. In: Rickwood D (ed) Preparative centrifugation: a practical approach. The practical approach series. Oxford University Press, Oxford, pp 1–42
9. Dobrota M, Hinton R (1992) Conditions for density gradient separations. In: Rickwood D (ed) Preparative centrifugation: a practical approach. The practical approach series. Oxford University Press, Oxford, pp 77–142
10. Evans WH (1992) Isolation and characterization of membranes and cell organelles. In: Rickwood D (ed) Preparative centrifugation: a practical approach. The practical approach series. Oxford University Press, Oxford, pp 233–270
11. Brakke MK (1951) Density gradient centrifugation: a new separation technique. J Am Chem Soc 73:1847–1848
12. Pertoft H (2000) Fractionation of cells and subcellular particles with Percoll. J Biochem Biophys Methods 44:1–30
13. Distel DL, Felbeck H (1988) Pathways of inorganic carbon fixation in the endosymbiont-bearing lucinid clam *Lucinoma aequizonata*. Part 1. Purification and characterization of the endosymbiotic bacteria. J Exp Zool 247:1–10
14. Markert S, Arndt C, Felbeck H et al (2007) Physiological proteomics of the uncultured endosymbiont of *Riftia pachyptila*. Science 315:247–250. https://doi.org/10.1126/science.1132913
15. Markert S, Gardebrecht A, Felbeck H et al (2011) Status quo in physiological proteomics of the uncultured *Riftia pachyptila* endosymbiont. Proteomics 11:3106–3117
16. Ponnudurai R, Kleiner M, Sayavedra L et al (2017) Metabolic and physiological interdependencies in the *Bathymodiolus azoricus* symbiosis. ISME J 11:463–477
17. Graham JM (1997) Homogenization of tissues and cells. In: Graham JM, Rickwood D (eds) Subcellular fractionation: a practical approach. The practical approach series. Oxford University Press, Oxford, pp 1–29

Correction to: Thiol-Redox Proteomics to Study Reversible Protein Thiol Oxidations in Bacteria

Martina Rossius, Falko Hochgräfe, and Haike Antelmann

Correction to:
Chapter 18 in: Dörte Becher (ed.), *Microbial Proteomics: Methods and Protocols*, Methods in Molecular Biology, vol. 1841, https://doi.org/10.1007/978-1-4939-8695-8_18

This protocol was originally published © Springer Science+Business Media, LLC, part of Springer Nature 2018, but has now been made available © The Author(s) under a CC BY 4.0 license.

The updated online version of this chapter can be found at
https://doi.org/10.1007/978-1-4939-8695-8_18

Dörte Becher (ed.), *Microbial Proteomics: Methods and Protocols*, Methods in Molecular Biology, vol. 1841,
https://doi.org/10.1007/978-1-4939-8695-8_23, © Springer Science+Business Media, LLC, part of Springer Nature 2018

INDEX

A

AAA$^+$ proteases ... 189, 190
Absolute protein quantification (AQUA), *see* Quantification
Acrylamide–bisacrylamide ... 137, 138, 268, 270, 306
Alignment ... 106, 171, 221
Alkylation ... 4, 23, 30, 31, 110, 144, 148, 151, 234, 239, 269
Amino acids
- arginine ... 22, 29, 61, 109, 116, 123, 204
- cysteine ... 61, 116, 151, 165
- heavy ... 164, 180
- labeled ... 143, 151, 163, 202
- light ... 164, 185
- lysine ... 22, 29, 90, 96, 109, 116, 123, 204
- ^{15}N labeling ... 164
- phosphorylated ... 231
- ^{34}S and ^{36}S labeling ... 163–173
- [^{35}S]-L-methionine ... 114, 117–119, 134, 140
- stable-isotope labeled ... 143

Amino groups ... 41, 96, 108, 109, 155, 231
Amoebae ... 61–63, 65, 69, 72
Ampholytes ... 89, 98, 108, 110
Analysis of variance (ANOVA) ... 108, 124, 125
Andromeda ... 201
ANOVA, *see* Analysis of variance (ANOVA)
Antibacterial ... 113
Arginine ... 29, 61, 87, 90, 109, 116, 123, 204
- *See also* Amino acids

Artificial
- carbamylation ... 89
- colocalization ... 125
- degradation ... 133
- oxidation ... 273
- spots ... 89
- stabilization ... 133
- transfer ... 90

ATP-dependent proteolysis ... 138
Autoradiography ... 137, 138
Auxotrophic ... 61

B

Bacillus subtilis ... 41, 131, 132, 137–139, 176–178, 180–182, 244, 249, 250, 262
Bacteria-invertebrate symbiosis ... viii
Bacterial
- adaptation ... 207, 209, 217, 222, 223
- cell sorting ... 210, 215, 226
- contamination ... 72
- culture ... 176, 213, 214, 225, 269, 272
- effector ... 63
- growth ... 118, 119, 168, 178
- infection ... 190
- kinases ... 232
- metabolism ... 62, 63
- pathogen ... 207–209
- protease ... 22, 189, 190, 232
- samples ... 226, 231
- signaling ... 231
- symbiont ... 322, 331

Binding
- capacity ... 135
- low-binding plastics ... 82
- low-binding tubes ... 37, 41, 146, 147
- Zn^{2+}-binding ... 190–192

Biological
- availability ... 142
- effects ... 124
- follow up ... 231
- function ... 249
- mass spectrometry ... 175
- replicates ... 101, 109, 111, 123, 155, 156, 164, 195, 203, 204, 254, 257, 315
- role ... 249
- samples ... 4, 156, 175, 315
- variances ... 204
- variation ... 107, 108

Biomolecular
- extract ... 282, 284, 286–288
- fraction ... 281, 284–286, 288
- isolation ... 284–286

Biotechnological application ... 47, 95
Biotin
- biotin switch assay ... 265
- biotinylation ... 35–43

Dörte Becher (ed.), *Microbial Proteomics: Methods and Protocols*, Methods in Molecular Biology, vol. 1841, https://doi.org/10.1007/978-1-4939-8695-8, © Springer Science+Business Media, LLC, part of Springer Nature 2018

Bottom-up
approach 22
proteomics 263
Bradford Protein Assay 234, 239
Buffer
carbonate 25, 27, 31, 50, 52
digestion 6, 25, 29, 42, 143
extraction 49, 286, 296, 297, 308, 311, 315
high salt 23, 25, 27, 49, 50, 52
homogenization 66, 232
IBS 323–329, 331
lysis 6–8, 24, 25, 30, 97, 102, 103, 108, 109, 133, 135, 179, 183, 194, 196, 197, 234, 238, 242, 252, 254, 258, 281, 284, 285, 294, 299
MOPS 116, 237
PBS 65, 155, 196, 237, 240
phosphate 37, 65, 166, 168, 193, 211, 213, 237, 304, 329
redox 262
resolving 98, 99, 134, 306
running 25, 27, 99, 105, 106, 117, 120, 121, 134, 167, 194, 198, 267, 268, 271, 272
sample 25, 27, 30, 56, 98, 104, 134, 136, 147, 167–169, 267, 272
separation 170
TBS 24, 25
TE 12, 116, 119, 127, 193, 196, 265, 267, 269, 272, 305, 311, 323, 324, 326–328
TEAB 13, 25, 29, 38, 50, 155, 156, 179, 183, 234
washing 97, 166, 178, 179, 182, 183, 196, 232

C

Calcium phosphate precipitation (CPP) 250, 252, 254–256, 258
Carbamidomethylation 27, 171, 257
Carbamylation 89, 109
Cell
culture 42, 65–67, 72, 135, 140, 164, 176, 195, 196, 208, 210, 211, 214, 224, 250, 269
disruption 42, 47, 49, 51, 55, 156, 179, 183, 191, 193, 196, 263
homogenate 69, 73, 322
host vii, 60–63, 207, 208, 210, 211, 213, 218, 226, 329
lysis 24, 30, 41, 42, 55, 66, 97, 102, 103, 108, 109, 117, 135, 136, 182, 183, 232, 233, 238, 242, 282, 284, 285, 287
membrane vii, 37, 39, 191, 223
sorting viii, 208, 210, 215, 226
surface vii, 35–43
types 281, 328
wall 7, 8, 36, 37, 232
Centrifugation
density gradient 61, 66, 69, 70, 320–322, 324, 325, 328, 332
ultra 23, 25, 27, 29–31, 47, 50, 52, 53, 55, 56, 97, 102
Charges 37, 96, 151, 169–171, 185, 200, 217, 218, 241
3-[(Cholamidopropyl)-dimethylammonio]-propanesulfonic acid (CHAPS) 25, 38–40, 81, 97, 98, 117, 265, 266
Chromatography
affinity 166, 231
liquid 3, 5, 22, 163, 176, 194, 208, 218, 233, 238, 241, 314
reversed phase 31, 184, 218, 233
SCX 233, 298
Cleavage
chymotryptic 32
enzymatic 4, 22
LysC 4, 5
missed sites 32, 43, 149, 159
proteinase K 31
rRNA 30
tryptic 31, 144
unspecific 32, 42, 172
Collision-induced dissociation (CID) 149, 200, 235, 257
COMET 180, 208, 212, 218, 219
Coomassie, *see* Dye

D

1D, *see* One-dimensional (1D)
2D, *see* Two-dimensional (2D)
Data
acquisition 180, 184, 200, 219
analysis 111, 123, 126, 164, 167, 169, 171, 173, 190, 195, 200–202, 208, 212, 218, 219, 221, 222, 225, 227, 238, 242, 257, 286
Database
decoy 204, 219, 220, 227, 235
expression 115, 126–128, 293
search 31, 123, 169, 180, 184, 185, 212, 218, 219, 221, 235, 257
2D Difference Gel Electrophoresis (2D DIGE) 95–97, 100–103, 106, 107, 109–111

Degradation....30, 60, 62, 131–133, 138, 189, 284, 287
Density gradient centrifugation....61, 66, 69, 70, 320–322, 324, 325, 328, 332
Desalting....13, 15–16, 167, 169, 194, 200, 234, 237, 243, 244, 252, 253, 256, 296, 298, 306, 314
Destaining....273, 312
Detection....22, 23, 79, 85, 87, 118, 120, 122, 123, 142, 143, 149, 164, 194, 198, 208, 262, 314
Detergents....4, 8, 12, 22, 23, 29, 31, 104, 143, 232, 242, 293–301, 304
Dialysis....242, 252, 258
Dictyostelium discoideum....61–63, 65–69, 71–73
Differential
  alkylation....261
  analysis....108, 175, 327
  pelleting....320–323, 326, 331, 332
  proteomics....95, 175
Differentiation....80, 232, 249–258
Digestion
  in-gel....14, 22, 23, 28, 37, 82, 118, 168, 172, 199, 232, 263, 264, 268, 272, 273, 304, 306, 313–314, 316
  in-solution....4, 12, 13, 15–16, 142, 144, 146–150, 179, 183, 184, 232, 255 (*see also* Cleavage)
Dimension
  first....96, 102, 121
  second....92, 96, 102, 120, 127, 270–271
1 Dimensional electrophoresis (1DE/1D PAGE/1D SDS PAGE)....22, 288, 289, 305–307, 309
2 Dimensional electrophoresis (2DE/2D PAGE/2D SDS PAGE)....vii, 87, 96–98, 102, 104, 108, 110, 111, 114, 117–122, 141, 143, 145, 153, 262, 263, 266–267, 270
  See also Two-dimensional (2D)
Disulfide bonds....15, 110, 148, 298
Dithiothreitol (DTT)....6, 12, 81, 98, 99, 104, 108, 110, 117, 167, 169, 193, 194, 203, 234, 239, 252, 255, 266, 267, 272, 296, 298, 300
Dyes
  BODIPY....262, 263, 273
  coomassie....29, 142, 263
  CyDyes....96, 98, 102, 103, 107–110
  fluorescence....81, 90, 93, 98, 102–104, 108, 120, 156, 223, 224, 262, 263, 273
  Krypton™....120
  SERVA Purple....81, 83, 84, 90, 92

E

Effector proteins....60, 63, 69
Electrophoresis
  one-dimensional....22
  two-dimensional....114, 262
Electrospray ionization (ESI)....170, 234, 253, 256, 271–272, 314
Endoproteinase Lys-C, *see* Cleavage
Epithelial cells....208, 213, 223, 224, 289
Escherichia coli (E. coli)....5, 181, 190–192, 195, 201, 203, 233, 244, 249, 250
ESI, *see* Ionization
Exposition....119, 132, 135–139

F

FA, *see* Formic acid (FA)
Filter aided sample preparation (FASP)....3–9
Flow rate....170, 200, 217, 233, 241
Fluorescence
  based redox proteomics....262–273
  labelling....98, 100, 101, 103, 104, 108, 269
  stains....81, 212, 223, 263
Formic acid (FA)....151, 167, 169, 195, 200, 237, 238, 241, 252, 253, 255, 256, 296
Fractionation
  cellular....47, 54
  protein....37, 272, 274, 309–312
Fragmentation....22, 79, 170, 184, 200, 217, 235, 257, 263

G

Gel
  electrophoresis....22, 47, 114, 153, 194, 196–198, 204, 262, 288, 312
  gel-based proteomics....113
  gel-free proteomics....12, 79
  staining....83, 111, 120, 121, 124, 129, 147, 169, 196, 198, 204, 267, 271, 312
  systems....82, 86, 96, 117, 120, 121, 129
Glutathione....261, 262

Gradient 61, 66, 69–71, 81, 83, 96, 98, 104, 109, 110, 120, 152, 170, 200, 217, 218, 241, 257, 298, 301, 320–325, 327–333
Gram-negative 261
Gram-positive 21–32, 114, 176, 208, 262
Growth
conditions 48, 115–119, 125, 189, 203
deficiencies 134
medium 48, 54, 64, 116, 178, 181–183
phase 51, 67, 72, 109, 114, 191, 195, 196, 225, 271

H

HCD, *see* Higher-energy collisional dissociation (HCD)
Heavy
nitrogen 175–188
peptide 153, 176, 184, 185, 187
stable isotopes 175–188
sulfur 165
Higher-energy collisional dissociation (HCD) 170, 217, 235, 241
Highly dilute samples 11–17
High performance electrophoresis (HPE) 79, 85, 87, 98
High-performance liquid chromatography (HPLC) 25, 133, 134, 194, 195, 199, 200, 210–213, 216, 218, 226, 227, 234, 238, 252–254, 296, 306
High-resolution 5, 163, 176, 200, 238, 280
High-throughput 93, 164, 165, 179
HistoDenz™ 322, 324, 325, 328, 330, 331, 333
Homogenization 30, 42, 66, 68, 73, 119, 232, 304, 308, 323, 326, 329, 331
Host-microbe interactions 280

I

IAA, *see* Iodoacetamide (IAA)
Identification
peptide 4, 80, 108, 152, 164, 180, 184, 200, 202, 204, 208, 212, 218–220, 227
protein .... vii, 5, 24, 79, 85, 87, 118, 122, 123, 125, 145, 165, 180, 184, 200, 203, 208, 219, 263, 265, 267–268, 271–272, 274
IEF, *see* Isoelectric focusing (IEF)
IgG 66, 70
IMAC, *see* Immobilized metal affinity chromatography (IMAC)
Image
analysis 81, 82, 84, 100, 107, 123, 125, 264
analysis software 81, 84, 87, 90, 100, 107, 108, 123, 153, 264
Immobilized metal affinity chromatography (IMAC) 233, 250
Immobilized pH gradient (IPG) 81, 83, 90–92, 96, 98, 104–106, 109, 110, 117, 119–121, 265–267, 270, 271, 273, 274
Immuno-magnetic separation 69
Immunoprecipitation (IP) 133, 135, 136, 139, 237, 240, 243
In-gel digestion, *see* Cleavage
Inhibitor 24, 25, 30, 66, 68, 73, 133, 193, 196, 234, 252
Integral membrane proteins 22, 23
Integrated omics 279–290
Interfering substances 308, 311
Internal
calibration 87, 91, 122, 123
protein standard 96, 103, 157
standard 96, 103, 109, 111, 151, 286
Internalization 41
Intracellular
bacteria 60–63, 208, 213–216, 223, 224, 226, 227
compartments 45, 60, 62
In vivo 114, 132, 133, 151, 207, 209, 217, 222, 223, 280
Iodoacetamide (IAA) 4, 6–8, 13, 15, 17, 25, 27, 30, 38, 39, 92, 99, 105, 110, 117, 147, 148, 156, 167, 169, 179, 184, 239, 252, 255, 258, 262
Ion
fragments 143, 151, 180, 218, 257
immonium 257
injection time 170, 217
mass tolerance 258
metal 30, 233
peptide 145, 149, 151, 152, 176, 218, 241, 263
precursor 143, 151, 218, 257, 258, 263
scan 180, 241, 257
source 195, 241
total ion chromatogram (TIC) 204, 286, 288, 289
trap 110, 149, 235, 241, 257
Ionization
electrospray ionization (ESI) 234, 253, 256, 271–272, 314

matrix-assisted laser desorption ionization (MALDI) ... 79, 85, 87, 108, 115, 122, 234, 258, 263, 314
Isoelectric focusing (IEF) ... 81, 83, 89, 90, 92, 96, 98, 102, 104, 110, 119, 120, 265, 266, 269–270, 273
Isoforms ... vii, 79, 80, 89, 125, 145, 158, 159
Isopycnic density gradient centrifugation, *see* Density gradient centrifugation
Isotope
dilution ... 143, 145, 147, 157
heavy ... 175–188
labeling ... 143, 151, 163–173, 175–188
pattern ... 184, 185
reference ... 143, 151
stable ... 145, 147, 151, 163–173, 175–188, 250
Isotope Distribution Calculator (IDCalc) ... 180, 185

K

Kinase ... 62, 63, 232, 250

L

Label
biotin ... 36, 37, 39, 41, 42
^{13}C ... 61, 151, 164
^{14}C ... 61
CyDyes ... 96, 98, 102, 103, 108, 110
^{15}N ... 151, 164, 175–188
^{34}S ... 163–173
^{35}S-methionine ... 114, 117, 119, 120, 133–135
^{36}S ... 163–173
Label-free
analysis ... 190, 195, 200, 202
approach ... 191
proteomics ... 12, 189–205, 250
quantification (LFQ) ... 12, 200–202, 204
Labeling
conditions ... 118, 119, 164
efficiency ... 41, 164
fluorescent ... 61, 90, 262, 263, 265–273
heavy ... 164, 175–188
incorporation rate ... 181, 184, 185
isotopic ... 140, 164, 176, 185
media ... 134, 164, 176–183, 185
metabolic ... viii, 42, 114, 163–165, 175–188
proteins ... viii, 114, 118, 119, 133, 140, 142, 164
pulse ... 114, 115, 118, 119, 133–135, 139
pulse-chase ... 133–135, 139
radioactive ... 114, 118, 119, 133, 136, 140, 142
reaction ... 118, 119
spot ... 114
stable isotope ... 143, 151, 163–173, 175, 176, 250
switching ... 164, 185
thiol ... 108, 265–273
Legionella pneumophila ... 60, 64
Limited
amounts ... 5, 29, 30, 79, 281, 299, 333
sample material ... 281
stability ... 90
Lipid
anchor ... 36, 61
bilayer ... 22
Lipopolysaccharide (LPS) ... 190
Liquid chromatography (LC) ... 16, 152, 159, 163, 167, 169, 170, 216–218, 227, 233, 314
Liquid chromatography-mass spectrometry (LC-MS) ... vii, 12–16, 71, 143, 146, 175, 176, 180, 198, 200, 211, 233, 235, 236
Liquid chromatography tandem mass spectrometry (LC-MS/MS) ... 5, 23, 25, 29, 31, 143, 169, 175, 180, 184, 187, 216, 241, 253, 256, 257, 263, 271–274
Listeria monocytogenes ... 207, 249
Low-abundant proteins ... 79, 85, 87, 190
LTQ-Orbitrap ... 235, 241, 253, 263
Lysis
buffer ... 6–8, 24, 25, 30, 41, 42, 97, 102, 103, 108, 109, 135, 179, 183, 194, 196, 197, 232, 234, 238, 242, 252, 254, 258, 281, 284, 285, 294, 299
Lysyl endopeptidase (LysC) ... 4–8, 237, 239

M

Macrophages ... 60–63, 65–68, 71, 73
Magnetic separation ... 52, 69
Magnetite nanoparticles ... 47
Magnetosomes ... 45–56
MALDI, *see* Ionization
MASCOT ... 82, 89, 123, 180, 218, 227, 235, 253, 257
Mass
accuracy ... 219, 234, 235
analyzer ... 82, 118, 122, 219, 235
charge ... 96, 151, 185, 200, 257
differences ... 142, 151, 164, 257

Mass (*cont.*)
monoisotopic.. 122, 171, 185
precursor.. 89, 123, 171, 184, 217, 219, 257, 258
range.. 87, 89, 122, 123, 149, 170, 200, 217, 234, 257
resolution.. 163, 170, 176, 184, 200, 256, 257
shifts.. 96, 143, 165, 173
spectrometer.. 82, 143, 163, 170, 175, 184, 195, 208, 216, 217, 233–235, 253, 254, 256, 263, 293
spectrometry.. 3, 11, 22, 35–43, 62, 71, 113, 118, 122, 123, 142, 153, 165, 170, 175, 176, 179, 180, 184, 194, 195, 200, 211, 212, 216, 217, 232, 238, 241, 250, 262–265, 267–268, 273, 274, 295, 298, 299, 301, 303, 304, 314
spectrum.. 87, 122, 123, 176, 185, 263
tolerance.. 173, 201, 202, 219, 257, 258
trace.. 176
Mass spectrometry (MS).. 3, 11, 22, 35–43, 62, 71, 113, 118, 122, 123, 142, 153, 165, 170, 175, 176, 179, 180, 184, 194, 195, 200, 211, 212, 216, 217, 232, 238, 241, 250, 262–265, 267–268, 273, 274, 295, 298, 299, 301, 303, 304, 314
Matrix effects.. 303
MaxQuant.. 195, 200–202, 204, 238, 242, 245
Membrane
fraction.. vii, 22, 23, 27, 30, 31, 47, 54, 56
protein loops.. 35
shaving.. 22–25, 27, 29
spanning domains.. 22
Metabolic labeling, *see* Labeling
Metabolism.. viii, 60, 62, 69, 163
Metaproteome
analysis.. 303–317
coverage.. 304
profiling.. 304
studies.. 304
Microorganisms.. vii, viii, 64, 80, 95, 100, 101, 107, 163, 166, 175–188
Minimal medium.. 133, 177, 181, 187, 268, 271
Missed cleavage.. 32, 43, 143, 149, 159, 201, 219, 221, 257
Modification
covalent.. 37
fixed.. 180, 201, 202, 257
post-translational.. viii, 80, 151, 249
variable.. 169, 201, 219, 257
Mouse models.. 280
MS, *see* Mass spectrometry (MS)
MS-based.. 22, 63, 96, 108, 142, 143, 175, 184, 190
MsConvert.. 167, 171
MS/MS
analysis.. 4, 25, 31, 87, 169, 184, 216, 253, 256, 263, 271–274
spectrum.. 87, 263
Multiplexing.. vii, 96
Mycobacterium tuberculosis.. 207, 249
m/z range.. 257

N

NanoDrop.. 289
Nano-electrospray ionization source (Nano ESI).. 170
Nano high performance liquid chromatography (NanoLC).. 169, 241, 256
Nanoparticles.. 47
Neutral loss.. 263
Ni-Nitrilotriacetic acid (NTA) agarose.. 193
^{15}N labeling.. 164
Non-covalent.. 36
Nonessential.. 210
Non-growing.. 226
Non-ionic.. 96, 104
Non-linear (NL).. 98, 109, 138, 266, 270
Non-linearity.. 187
Nonphosphorylated.. 245, 255
Non-polar.. 284, 287
Non-proteinaceous.. 107
Non-radioactive.. 133, 135
Non-reducing.. 263, 267, 272, 274
Normalization.. 96, 124, 159, 172, 173, 221, 222

O

OMSSA.. 235
One-dimensional (1D).. 22, 27, 47, 190, 191, 197, 265, 268, 272, 288, 289, 305–307, 309, 316
gel (*see* 1 Dimensional electrophoresis)
PAGE.. 309
SDS PAGE (*see* 1 Dimensional electrophoresis)
Orbitrap, *see* LTQ-Orbitrap
OxICAT.. 265
Oxidation.. viii, 151, 171, 201, 202, 219, 257, 262–264
Oxidative stress.. viii, 132, 133, 268–273

P

PAGE, *see* Polyacrylamide gel electrophoresis (PAGE)
Pathogen vacuole 60, 62, 69
Peak
 capacity 234
 intensity 200, 288
 isotope 143, 185, 219
 lists 89, 122, 123, 235, 257
 monoisotopic 185, 186
PeakExtractor 167, 171
PeakPicking 171
Peptide
 concentration 7–9, 16, 42, 204
 data 218
 desalting 167, 169, 200, 234, 237, 239, 296, 298
 extraction 26, 122, 199, 234
 fractions 5, 22
 fragments 123
 heavy 153, 175, 176, 184–187
 identification 4, 152, 169, 171, 173, 204, 208, 212, 218–220, 227
 intensity 176, 227
 ions 218, 241
 labeled 143, 152, 164, 171
 level 221
 light 176, 184, 185, 187
 mass fingerprinting 108
 mixtures 29, 82, 88, 123, 200, 208, 232, 256, 298, 304
 modified 80, 151, 204, 221
 peak 9, 91
 phospho- 234, 235, 250, 252–258
 phosphorylated 233, 234, 243
 protein-specific 79
 proteolytic 298
 purification 31, 144, 147, 149, 216, 304
 quantification 80, 149–152, 166, 171, 200, 202, 204, 219
 recovery 31
 sequence 22, 143, 151–153, 176, 185, 212, 218, 221, 236
 settings 219
 solutions 122, 216, 218, 240, 255
 tryptic 144, 151, 218, 263, 273, 303
 unique 5, 143, 149, 202, 203
 values 221
PeptideProphet 219, 220
Percoll® 321–323, 327
Phosphopeptide enrichment 231, 234, 235, 250, 255, 256
Phosphoproteomes 232, 235, 249, 250, 258
Phosphoproteomics 5, 232, 233, 249–258
Phosphorylation .... viii, 63, 151, 231–233, 235, 236, 243, 249, 250, 253, 257, 258
Poly acryl amide gel electrophoresis (PAGE) 96, 99, 102, 105, 106, 166, 167, 175, 262, 263, 265, 267–268, 270–274, 288, 289, 295, 305–307, 309–313, 316
Posttranslational modification, *see* Modification
Prokaryotic organelles 45
Proteases .... viii, 4, 5, 22, 24, 25, 30, 37, 46, 61, 68, 73, 132, 143, 144, 149, 189–193, 195, 196, 199, 204, 232, 234, 243, 252, 258
Protease substrates 189–205
Protein
 abundance .... vii, viii, 21, 90, 95, 96, 102, 107, 111, 141, 145, 163, 171, 222, 233
 activity 131, 134, 144, 261
 amount .... vii, 5, 8, 22, 30, 41, 79, 85, 91, 103, 127, 136, 141–143, 145–148, 153–159, 184, 197, 198, 250, 263, 264, 271, 298–300, 310
 assay 25, 29, 97, 102, 103, 296, 298, 300, 303, 312, 316
 class 35, 36
 cleavage 4, 5, 22, 30, 96, 144, 154, 243
 concentration 3, 4, 8, 9, 25, 27, 31, 56, 60, 92, 97, 102, 103, 119, 124, 127, 141, 142, 146–148, 157, 172, 179, 183, 184, 188, 193, 197, 234, 238, 239, 242, 252, 255, 265, 269, 271, 272, 312
 degradation .... viii, 30, 131, 132, 189
 denaturation 23, 30, 31, 234, 239, 242, 265, 267
 desalting 234, 237, 239
 digestion 3–5, 31, 37, 84, 111, 142–144, 146, 157, 232, 234, 237, 239, 263, 272, 297, 298, 304, 313–314, 316
 dyes 29, 90, 142, 147, 156, 263
 effector 60, 63, 69
 enrichment .... vii, 11–17, 22, 35–43, 203, 243, 304
 expression 114, 115, 125–128
 extracellular 294
 extraction .... vii, viii, 31, 42, 156, 232, 252, 254, 255, 294–299, 303, 305, 308–311, 316
 extracts 102, 104, 118, 119, 122, 125, 127, 184, 262, 263, 269–270, 273, 274, 289

Protein (*cont.*)
fraction 22, 23, 30, 47, 54, 141, 197, 198, 264, 272, 279–290
gel 114, 115, 288
identification .... vii, 5, 24, 79, 85, 87, 96, 111, 118, 122, 123, 125, 145, 165, 180, 200, 267–268, 271–272
isoforms 80, 125, 145
labeling 142
lifetime 164
lysates 4, 136
membrane .... vii, 21–24, 27–31, 46, 47, 50, 52, 56, 72, 190, 232
precipitation 40, 104, 238, 239, 242, 269, 294, 297, 300, 308, 310, 311
purification 144, 193, 196, 198, 304
quantification .... vii, viii, 12, 141, 143, 145, 153, 164, 202, 208, 219, 316
ratio 7, 8, 28, 29, 153, 172, 199
regulatory 131, 132, 137–139, 142
sample 17, 27, 29, 83, 100, 102, 103, 110, 119, 136, 142, 143, 146–148, 157, 184, 188, 198, 199, 239, 265–267, 274, 297, 300, 312
secreted vii
secretion 60
separation 28, 56, 84, 96, 102, 198, 238, 263, 272, 303, 304, 312, 316
solubilization 3, 27, 31, 50, 52, 54, 56, 89, 299, 300
solution 91, 119, 127, 239, 297, 311, 312, 316
species 21–32, 153, 154, 158, 159, 218, 242, 319
spot 79, 80, 82, 85, 89, 93, 96, 107–110, 114, 115, 122–125, 127, 129, 145, 147, 154, 158, 159, 264
stability .... vii, 90, 131, 132, 137–139
S-thiolations 261–265, 267, 272, 274
structure 142, 293, 303
surface .... vii, 5, 35–43, 47, 110, 172
synthesis .... vii, viii, 113–115, 119, 124, 125, 127, 129, 132, 133, 135, 142
target 37, 41, 47, 134, 142, 143, 153, 158, 262, 308
turnover 131
Proteinase K cleavage 32
Proteolysis 133, 138, 189, 298
Proteolytic 61, 132, 144, 190, 297, 298
Proteome .... vii, 3–9, 12, 22–24, 35, 36, 45–56, 61–63, 71, 96, 118, 122, 149, 163–173, 189, 207, 209, 212, 219, 222, 225, 232, 257, 263, 264, 293–301
ProteomeDiscoverer 253
Proteomics 3, 11, 21, 47, 60, 79, 95, 113, 142, 163, 175, 200, 207, 232, 262, 280, 295, 319
analysis 3–9, 11, 21, 47, 53, 60, 235, 245, 319, 321, 322
approach 21–32, 176, 207, 209, 217, 222, 223, 263
changes .... vii, 163, 175, 207
coverage 36, 164
data 149, 164, 200, 222
profiling 95, 304
quantification .... vii, 142, 163–173, 176
reactor 3
sample preparation .... vii, 3–9, 176, 264
signatures 113–115, 126, 127
Proteowizard 167, 212
p-value 108, 203

Q

Quantification
absolute .... vii, viii, 141, 143, 145, 153
label-free 12, 200–202, 204, 250
relative .... vii, 80, 141, 154, 175
Quantitation, *see* Quantification
Quantitative
analysis 163, 173, 219, 221, 245, 257, 264
data 154, 172, 173, 219, 221, 222
proteome profiling 95, 163
proteomics 21, 163, 175, 176, 245, 264

R

Rate-zonal density gradient centrifugation 320–322, 324
Ratio
H/L (heavy/light) 171, 172
S/N (signal to noise) 87, 89, 122, 123, 153
Raw files 171, 201, 221
Reducing sample complexity 315, 319
Reduction 23, 30, 31, 37, 89, 144, 148, 234, 239, 265, 269, 319
Reference
gel 96, 102, 103, 107, 111, 115, 288
map 115, 125, 149
pooled 102, 185, 187
state 96, 102, 103, 107, 111
Regulated proteolysis 189, 190
Regulatory proteins 131, 132, 137–139
Rehydration 81, 83, 91, 92, 98, 104, 110, 117, 119, 124, 265, 266, 270
Relative vii

abundance 141, 257, 258
quantification (*see* Quantification)
Replicates 60, 102, 109, 111, 123, 140, 147, 148, 153–157, 164, 172, 185, 195, 203, 204, 213, 221–223, 226, 254, 257, 258, 286, 315
Reproducibility 5, 72, 96, 121, 164, 200, 258, 330
Resolution 54, 79, 84, 96, 106, 107, 109, 110, 138, 163, 170, 176, 180, 184, 200, 217, 219, 238, 241, 256, 257, 321, 330, 333
Retention times 152, 159, 169, 171, 173, 184, 218, 221, 258, 288
Reversed phase liquid chromatography 31, 218, 233
Reversible thiol-oxidations .... viii, 262

S

Salmonella enterica 207
Sample
complexity 166, 315, 317, 319
heterogeneity 143, 281
material 281, 282, 287, 332
preparation .... vii, 3, 4, 12, 98, 100, 156, 158, 168, 176, 180, 199, 211, 212, 216, 232, 252, 254, 255, 264, 295, 301, 303–317
Scanning conditions 79
Score
localization 236
Mowes 123
PeptideTeller 171
probability 236
z-score 124
SCX, *see* Strong cation exchange (SCX) chromatography
SDS, *see* Sodium dodecyl sulfate (SDS)
SDS-PAGE, *see* 1 Dimensional electrophoresis
Search
criteria 126
database 31, 123, 169, 180, 184, 212, 218, 219, 221, 235, 257
engine 89, 123, 164, 169, 212, 218, 219, 235, 236
parameter 32, 219, 257
pattern 87, 122, 123
tolerance 122, 258
Selected reaction monitoring (SRM) 143, 145, 147, 151, 152, 159, 219
Selectivity 233
SEQUEST 218
SEQUEST/Sorcerer 171
Shaving
approach 23, 24, 37, 38, 40, 41
membrane 23–25, 27, 29
proteinase K 25
trypsin 35–43
Shotgun 143, 219, 227, 263, 264
Signal
intensity 122, 143, 175, 176, 200
transduction 232, 233
Signal to noise ratio (S/N) 87, 89, 122, 123, 152, 153
Significance 111, 126, 172, 173
SILAC, *see* Stable isotope labelling of amino acids in cell culture (SILAC)
[^{35}S]-L-methionine pulse labeling 115
S/N, *see* Signal to noise ratio (S/N)
Sodium dodecyl sulfate (SDS) 4, 6–9, 12, 22, 25, 30, 38, 40, 41, 96, 98, 99, 102, 105, 106, 117, 118, 120, 127, 134, 135, 137, 138, 166–168, 175, 194, 196–198, 204, 211, 214, 215, 226, 232, 234, 252, 254, 258, 266–268, 272, 294, 296, 300, 305, 306, 309–312, 314, 316
Soil
protein extraction 294–297, 299, 303, 304, 308, 309
proteomics 293–301
sampling 293–301, 303–317
Solid phase enrichment (SPE) 11, 234
Source 235
Spectrum
fragment ion (MS/MS) 143, 152
mass 176
sub 87, 122, 123
theoretical 185, 236
Sporulation 253
Spot
detection 84, 86, 107–109, 123, 124
excision 111
intensities 124, 125, 145
quantification 80, 125, 145, 147, 154, 158, 274
volume 84, 124, 154, 159
SRM, *see* Selected reaction monitoring (SRM)
Stable isotope dilution (SID) 143, 145, 147
Stable isotope labelling of amino acids in cell culture (SILAC) 163–165, 167, 176, 250
Staining, *see* Gel staining
Standard
deviation 87, 122, 138, 154, 155, 290
internal 96, 102, 103, 107, 109, 111, 149, 151, 157, 286
peptides 150–152
protein 193, 234, 312
Standardization 124, 281
Staphylococcus aureus 23–25, 30, 37, 41, 113–115, 126, 127, 164, 176, 178–180, 182, 183, 190, 208, 213–217, 222–225, 249
StrataClean 11–17

Streptomyces coelicolor 249–258
Strong cation exchange (SCX) chromatography 233
SULAQ, *see* Sulfur stable isotope labeling of amino acids for quantification (SULAQ)
Sulfur-^{34}S 163–173
Sulfur-^{36}S 163–173
Sulfur stable isotope labeling of amino acids for quantification (SULAQ) 163–173

T

Tandem mass spectrometry, *see* MS/MS
Target
- peptide 149
- protein 37, 134, 142, 143, 158

Theoretical
- protein 24
- spectrum 185

Thiol
- alkylation 269
- modifications 274
- trapping 265

TIC, *see* Total ion chromatogram (TIC)
Time of flight (TOF) 82, 93, 123
TiO_2 chromatography 231
Total ion chromatograms (TICs) 286, 288, 289
TPP, *see* Trans Proteomics Pipeline (TPP)
Transformation 172, 173, 203, 218
Trans Proteomic Pipeline (TPP) 208, 212, 218, 219, 225
Trapping 189–205
Trypsin
- immobilized 37
- shaving 35–43

Tryptic 28
- digestion .... 28, 31, 82, 85, 87, 88, 111, 199, 208, 263, 268, 272, 273 (*see also* Cleavage)
- in-gel 28, 263, 264, 268, 272, 273
- in solution 26, 28, 272, 273

Two dimensional (2D) .... vii, 12, 47, 79–81, 95–97, 100–103, 106–111, 114, 115, 123, 125, 127, 129, 175, 190, 262–267, 270–271, 274
- difference in gel electrophoresis (*see* 2D difference gel electrophoresis (2D DIGE))
- electrophoresis (*see* 2DE)
- gel-based proteomics 189
- gel electrophoresis (*see* Two-dimensional gel electrophoresis (2DE)))
- gel(s) 12, 79, 80, 115, 123, 127, 129, 145, 147, 149, 153–156, 158, 190, 264, 265, 270, 271, 274
- PAGE .... vii, 114, 117–122, 141, 143, 145, 153, 262, 263, 266–267 (*see also* Polyacrylamide gel electrophoresis (PAGE))

Type IV secretion 60, 62

U

Ultracentrifugation (UC) 23, 25, 27, 30, 31, 47, 52, 56, 102
UniProt 124, 201, 235

V

Variability
- technical 142, 164

Variance
- biological 204
- technical 109, 204

Variation
- biological 111
- gel-to-gel 96, 123, 125
- statistically significant 126
- technical 111, 124

Virulence factors 35

W

Western blots 71, 132

Y

Yeast
- extract 48, 64, 65, 72, 134, 193, 211, 250

Z

Z-score 124

Printed by Printforce, the Netherlands